CELL AND MOLECULAR BIOLOGY

D1797857

CELL AND MOLECULAR BIOLOGY

Pragya Khanna

Ph.D., FSLSc.
Sr. Lecturer
Govt. Degree College
Jammu & Kashmir

I.K. International Publishing House Pvt. Ltd.

NEW DELHI • BANGALORE

Published by

I.K. International Publishing House Pvt. Ltd.
S-25, Green Park Extension
Uphaar Cinema Market
New Delhi – 110 016 (India)
E-mail: info@ikinternational.com
Website: www.ikbooks.com

ISBN 978-81-89866-59-4

Reprint 2013

Published by Krishan Makhijani for I.K. International Publishing House Pvt. Ltd., S-25, Green Park Extension, Uphaar Cinema Market, New Delhi – 110 016 and Printed by Rekha Printers Pvt. Ltd., Okhla Industrial Area, Phase II, New Delhi – 110 020.

To
MY PARENTS

Preface

Studies on *Cell and Molecular Biology* have grown rapidly in recent years, generating huge quantity of new and important information. This book is an attempt to provide a balanced coverage on different areas of cell biology ranging from classical cytology to modern molecular biology. It gives an overview of the cell and its main structural components, a general introduction to the biochemistry of the cell, introduction of tools and techniques employed in the study of cell, the molecular structure of genes and their modes of replication, DNA repair and recombination (Genetic engineering), genetic disorders, the structure and function of oncogenes and proto-oncogenes provide an insight into the molecular genetics of cancer. Besides, the book is well rounded up with a separate chapter that attempts to convey the excitement of genetics in the present times and its future prospects.

The book is intended to be a comprehensive guide on cell and molecular biology with easy-to-read text and focuses on the syllabi and curriculum of most Indian universities with appropriate tools of self-assessment for various competitive examinations like NET/SLET. An effort has been made to provide appropriate illustrations/diagrams/photographs throughout the text for easy understanding of the concepts.

At the end of the book, an extensive reference list has been given to provide a good starting point for the students for more advanced research on any topic. A glossary of important terms and an index complete the book. This book has been written for a wide range of students and it should prove useful for B.Sc., B.Sc. (Hons.) and M.Sc. students.

My thanks are due to my parents for inspiring me and for granting me the time it took to work on this project.

I gratefully acknowledge the cooperation and support rendered by I.K. International Publishing House Pvt. Ltd. in completing this task.

PRAGYA KHANNA

Contents

The Cell

1

INTRODUCTION

Cells are the fundamental building blocks of life. Their numbers vary to form individual 'single-cell' organisms (bacteria) to 'multi-cellular' structures (tissue, organs) and organisms (animals and plants).

Cells are 90% fluid (called cytoplasm) which consists of free amino acids, proteins, carbohydrates, fats, and numerous other molecules delimited by a semipermeable membrane. Different cell biologists have proposed different definitions of cell from time to time. Loewy and Siekevitz in 1963 defined cell as a unit of biological activity delimited by a semipermeable membrane and capable of self-reproduction in a medium free of other living systems. John Paul in 1970 defined the cell as the simplest integrated organization in living systems, capable of independent survival. Therefore, in a more simplified way we can define cell as "the smallest but complete expression of the fundamental structure and function of all organisms delimited by a semipermeable membrane and capable of self-reproduction in a suitable nonliving medium".

Discovery of Cell

Most cells are so small that they cannot be observed with the naked eye. For this reason, even the existence of cells escaped notice until scientists first learned to harness the magnifying power of lenses in the second half of the seventeenth century. At that time a Dutch clothing dealer named Antony van

Leeuwenhoek (1632–1723) (Fig. 1.1) designed extraordinarily accurate single-lens microscopes. Looking into the lens of these microscopes, he discovered single-celled organisms, which he called 'animalcules' and today, we call bacteria and protists.

FIG. 1.1 Antonie van Leeuwenhoek (1632-1723)

Englishman Robert Hooke (1635–1703) expanded on Leeuwenhoek's observations with the newly developed compound microscope, which uses two or more aligned lenses to increase magnification. When Hooke turned the microscope on a piece of cork, the term 'cell' was born. Within a decade, researchers had determined that cells were not empty but, instead, filled with a watery substance called cytoplasm. Over the next 175 years, research led to the formation of the cell theory, first proposed by the German botanist Matthias Jacob Schleiden and the German zoologist Theodore Schwann in 1838 (Fig. 1.2).

(a)   (b)

FIG. 1.2 (a) Matthis Jacob Schleiden (b) Theodore Schwann

Cell Theory

As microscope technology developed, scientists were able to study cells in even more detail. By 1839, enough evidence had accumulated for German biologists Matthias Jacob Schleiden and Theodore Schwann to proclaim that "cells are the elementary particles of organisms." But many researchers still did not believe that cells arose from other cells, until 1855, when famous German pathologist Rudolph Virchow pronounced, "All cells come from cells." Nearly two centuries after the discovery of cells, the observations of Virchow, Schleiden, and Schwann were formulated in the form of cell theory which states that:

1. All living things are made of cells and their products.
2. The cell is the structural and functional unit of life.
3. All cells arise from pre-existing cells.

In its modern form, this theory has four main postulates:

1. The cell is the basic structural and functional unit of life; all organisms are composed of cells.
2. All cells are produced by the division of pre-existing cells (in other words, through reproduction). Each cell contains genetic material that is passed down from cell to cell during this process.
3. All basic chemical and physiological functions, for example, repair, growth, movement, immunity, communication, and digestion are carried out inside the cells.
4. The activities of cells depend on the activities of subcellular structures within the cell (these subcellular structures include the organelles, the plasma membrane, and, if present, the nucleus).

The cell theory leads to two very important generalizations about cells and life in general:

1. Cells are alive. This means cells can take energy (in the form of light, sugar, or other compounds depending on the cell type) for building materials (proteins, carbohydrates and fats), and use these to repair themselves and make new generations of cells (reproduction).
2. The characteristics and needs of an organism are in reality the characteristics and needs of the cells that make up the organism. For example, you need water because your cells need water.

A Brief History of Cell and Molecular Biology

1655 Robert Hooke	coined the word 'cell' from microscopic observation of slices of cork bark
1674 Antonie van Leeuwenhoek	studied blood cells, spermatozoa, protozoans and microorganisms
1745 Charles Bonnet	discovered natural parthenogenesis
1781 Fontana	described nucleolus

1831 Robert Brown	described the presence of nucleus
1838 Matthias Jacob Schleiden	proposed the cell theory based upon plants followed by Schwann's study in animals
1839 Theodor Schwann	all organisms consist of one or more cells and the cell is the basic unit of structure for all organisms
1840 J.E. Purkinje	proposed the term protoplasm for cell contents
1855 Rudolph Virchow	a German physician, stated that all living cells come only from other living cells
1861 Schultz	proposed the 'protoplasm theory'
1865 Gregor Mendel	phenotypic basis of inheritance
1868 Friedrich Miescher	isolated DNA from pus and fish sperm cells, which he called 'nuclein'
1875 Van Benden	observed the centriole
1882 F. Strasburger	introduced the terms 'cytoplasm' and 'nucleoplasm'
1887 August Weismann	described reduction division of chromosomes
1888 Waldeyer	introduced the term chromosome
1892 A. Weisman	proposed 'germplasm theory'
1897 Graier	named and described ergastoplasm
1898 Camillo Golgi	developed staining technique (silver nitrate) that allows identification of the cellular organelle 'Golgi apparatus'
1898 B. Benda	coined the term 'Mitochondrion'
1901 Strasburger	coined the term 'Plasmodesmata'
1903 Walter Sutton	described meiosis and spermatogenesis in insects and also established "Chromosomal Theory of Inheritance"
1905 Farmer and Moore	coined the term meiosis
1906 M. Tswett	discovered chromatography
1909 Thomas Hunt Morgan	studied chromosomal linkage and crossing over in *Drosophila*
1911 Alfred Stuartevant	mapping of genes on chromosomes
1924 Robert Fuelgen	cytochemical staining and identification of DNA
1926 T. Svedberg	Nobel Prize for ultracentifugation technique
1928 Frederick Griffith	described gene transformation using *Pneumococcus*

1933	Ted Painter	identified polytene chromosomes
1930's	George Beadle and Edward Tatum	"one gene - one enzyme" hypothesis
1940's	Linus Pauling	identified alpha helix structure of proteins
1940's	Maurice Wilkins and Rosy Franklin	discovered structure of DNA by using X-diffraction technique
1944	Oswald Avery, Maclyn MacLeod and Colin McCarty	chemically suggested that "DNA is the genetic material"
1950	Erwin Chargaff	"Chargaff's Rule" - DNA base complimentarity A:T G:C
1950's	Frederick Sanger	sequenced 1st protein – insulin
1952	de Duve	identified lysosome
1952	Alfred Hershey and Martha Chase	^{32}P viral DNA replication = genes are DNA
1953	James Watson and Francis Crick	identified model structure of DNA
1956	Arthur Kornberg	described action of DNA polymerase
1958	Matthew Meselson and Frank Stahl	semiconservative replication of DNA
1960	Paul Doty and Jay Marmur	Tm-hyperchromicity DNA-DNA hybridization
1961	Holley, Khorana, Nirenberg and Matthaei	identified the Genetic Code
1963	Jerome Vinograd	identified super-coiled structure of DNA
1968	Stanley Cohen	discovered plasmids and antibiotic resistance
1969	Edmonds and Caramelai	identified poly-A polymerase
1970	Herbert Boyer	discovered restriction endonuclease
1971	Gunter Blobel	proposed signal hypothesis
1972	Paul Berg	constructed first recombinant DNA's - splices SV40 and *E coli*
1975	Walter Gilbert, Allan Maxam and Frederick Sanger	developed DNA sequencing technique
1975	Cesar Milstein, Georges Kohler and Niles Jeme	developed monoclonal antibodies
1976	Robert Swanson and Herbert Boyer	created GENENTECH Biopharmaceutical Co.

1977	Richard Roberts and Philip Sharp	identified split genes (introns and exons)
1978	Genentech, Inc.	produced Humulin 1st recombinant DNA drug
1985	Kary Mullis	created PCR - Polymerase Chain Reaction
1989	HGP Project	project to sequence entire human genome
1990	French Anderson	1st use of Recombinant DNA drug (ADA)
1992	Harry Noller	peptidyl transferase is a ribozyme
1992	Edmund Fischer and Edwin Krebs	identified protein phosphorylation
1993	Kerry Mullis	created PCR reaction
1996	Ian Wilmut	mammalian cloning and Dolly
1997	Paul Boyer and John Walker	described ATP synthase mechanism
1998	James Thompson and John Gearhart	pluripotent (stem) cells cultured
1999	Craig Venter	identified complete gene sequence of *Drosophila*
2000	Craig Venter	announced human genome sequence

The cells are all composed chiefly of molecules containing carbon, hydrogen, oxygen, nitrogen, phosphorus and sulfur in the following approximations:

  (i) 59% Hydrogen (H)
  (ii) 24% Oxygen (O)
  (iii) 11% Carbon (C)
  (iv) 4% Nitrogen (N)
  (v) 2% Others - Phosphorus (P), Sulphur (S), etc.

The other molecules present in the cell are:

  (i) 50% Protein
  (ii) 15% Nucleic acid
  (iii) 15% Carbohydrates
  (iv) 10% Lipids
  (v) 10% Other

Different Types of Cells

Living organisms are subdivided into 5 major kingdoms, including the Monera, the Protista (Protoctista), the Fungi, the Plantae, and the Animalia. Each kingdom is further subdivided into separate phyla or divisions. Generally 'animals' are subdivided into phyla, while 'plants' are subdivided into divisions.

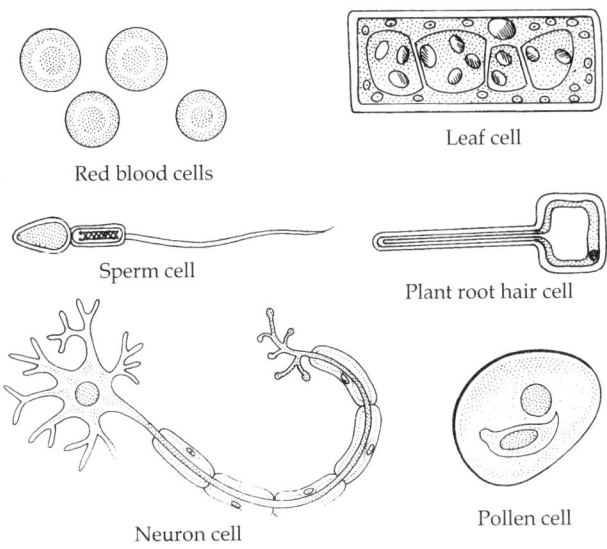

Red blood cells

Leaf cell

Sperm cell

Plant root hair cell

Neuron cell

Pollen cell

FIG. 1.3 Different types of cells

There are two main groups of cells, prokaryotic and eukaryotic cells. They differ not only in their appearance but also in their structure, reproduction, and metabolism (Fig. 1.3). The terms Prokaryotic and Eukaryotic were given by Hans Ris in 1960's.

Prokaryotic Cells lack **Nuclei and Membrane-Bound Organelles**

1. **Kingdom Monera [10,000 species]:** Unicellular and colonial— including the true bacteria (eubacteria), cyanobacteria (blue-green algae) and the archaebacteria.

Eukaryotic Cells possess **Nuclei and Membrane-Bound Organelles**

2. **Kingdom Protista [250,000 species]:** Unicellular protozoans and unicellular and multicellular (macroscopic) algae with 9 + 2 cilia and flagella (called undulipodia).

3. **Kingdom Fungi [100,000 species]:** Haploid and dikaryotic (binucleate) cells, multicellular, generally heterotrophic, without cilia and eukaryotic (9 + 2) flagella (undulipodia).

4. **Kingdom Plantae [250,000 species]:** Haplo-diploid life cycles, mostly autotrophic, retaining embryo within female sex organ on parent plant.

5. **Kingdom Animalia [1,000,000 species]:** Multicellular animals, without cell walls and without photosynthetic pigments, forming diploid blastula.

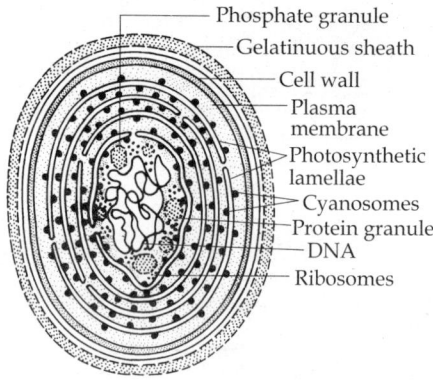

FIG. 1.4 Prokaryotic cell of Cyanobacteria

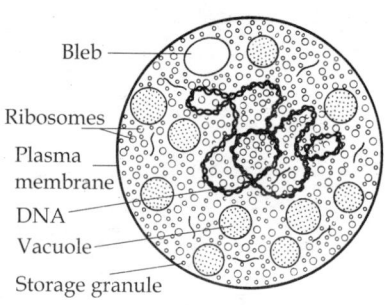

FIG. 1.5 Diagram of a Typical PPLO cell

THE PROKARYOTIC CELLS (Pro = before; Karyon = nucleus)

Prokaryotes are the cells that lack a membrane-bound nucleus but the chromosomal DNA that carries the genetic information is present in the form of a circular loop (from the Greek meaning before nuclei). These cells have few internal structures that are distinguishable under the microscope. Cells in the Monera kingdom such as bacteria and cyanobacteria (also known as blue-green algae) (Fig. 1.4) are prokaryotes. The other examples of prokaryotes include rickettsiae, mycoplasma or pleuropneumonia like organisms (PPLO) (Fig. 1.5). Bacterial cells are very small, roughly the size of an animal mitochondrion (about 1-2µm in diameter and 10 µm long). Prokaryotic cells feature three major shapes: rod shaped, spherical, and spiral. Instead of going through elaborate replication processes like eukaryotes, bacterial cells divide by binary fission.

BACTERIA

Size of Bacteria

Bacteria are small organisms which are not visible with a naked eye. They are measured in units of length called micrometers, or microns; one millimeter is equal to 1,000 microns. Generally, they have an average diameter of 1.25m. The smallest bacterium is *Dialister pneumosintes* (0.15 to 0.3 m) and largest bacterium is *Spirillum volutans* (13 to 15m) in length.

Shapes of Bacteria

Based on their shape, bacteria are classified into the following shapes (Fig. 1.6):

(a) **Spherical or *cocci* (sing. *coccus*):** These bacteria are round in shape. The cells may occur in pairs (*diplococi*), in groups of four (*tetracocci*), in bunches (*staphylococci*), in a bead-like chain (*streptococci*) or in a cubical arrangement of eight or more (*sarcinae*). For example, *Diplococcus pneumoniae*, *Staphylococcus aureus*, *Streptococcus pyogenes* etc.

(b) Rod-like or *bacilli* (sing. *bacillus*): They generally occur singly, but may occasionally be found in pairs (*diplobacilli*) or in chains (*streptobacilli*). For example, *Bacillus tuberculosis, Clostridium tetani, Corynebacterium diptheriae, Mycobacterium leprae* etc.

(c) Spiral or *spirilla* (sing. *spirillum*): These are also called spirochaetes. For example, *Treponema pallidum* etc.

(d) Comma shaped or *Vibrios* (sing. vibrio): These have a characteristic comma shape or bent rod like shape. For example, *Vibrio cholerae* etc.

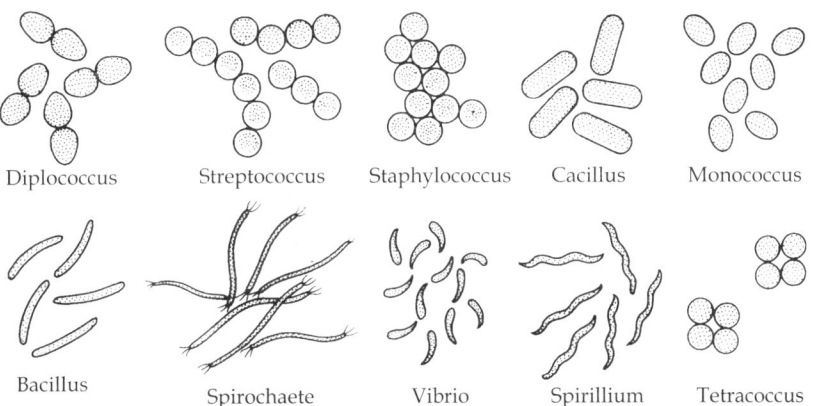

Diplococcus Streptococcus Staphylococcus Cacillus Monococcus

Bacillus Spirochaete Vibrio Spirillium Tetracoccus

FIG. 1.6 Different forms of Bacteria

Structure of Bacteria

Bacteria have a very simple internal structure with no membrane bound organelles. A typical bacterial cell has the following structural components (Fig. 1.7).

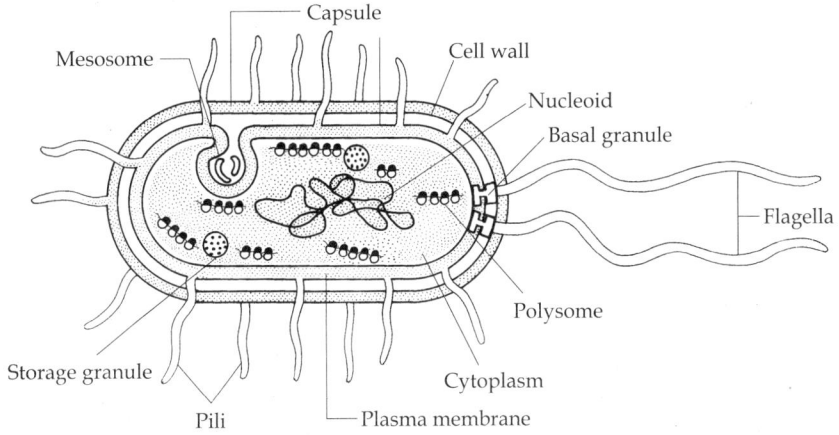

FIG. 1.7 Digram of a Bacterial Cell

Nucleoid

The nuclear material in the bacterial cell is generally confined to a central region called nucleoid. Though it is not bounded by a membrane, it is visibly distinct (by transmission microscopy). It consists of a single, circular and double stranded DNA molecule which is permanently attached to the plasma membrane at one point. It lacks the proteins associated with the DNA of eukaryotic cells. But like a eukaryotic chromosome, a prokaryotic chromosome bears the genes that control the metabolism of the cell as well as its hereditary traits. Prokaryotic DNA directs protein synthesis via mRNA in ribosomes; however, prokaryotic ribosomes are much smaller than eukaryotic ribosomes and have a different structure.

Plasmids

In addition to the circular DNA molecule, a small amount of extrachromosomal DNA is also present in bacteria in the form of plasmids or episomes. These are defined as "autonomously replicating, extrachromosomal circular DNA molecules, distinct from the normal bacterial genome and non-essential for cell survival under non-selective conditions". A plasmid contains genes normally not essential for cell growth or survival. Some plasmids can be integrated into the host genome and artificially constructed in the laboratory and serve as vectors (carriers) in cloning.

Ribosomes

Ribosomes give a granular appearance to the cytoplasm of bacteria in electron micrographs. Though smaller than the ribosomes in eukaryotic cells, these inclusions have a similar function in translating the genetic message in messenger RNA into the production of peptide sequences (proteins). Ribosomes of bacteria are 70S type and consist of two subunits (a larger 50S and a small 30S subunit) and are composed of ribonucleic acid (RNA) and proteins.

Storage Granules

Nutrients and reserves may be stored in the cytoplasm in the form of glycogen, lipids, polyphosphate, or in some cases, sulfur or nitrogen.

Endospore

Some bacteria, like *Clostridium botulinum*, form spores that are highly resistant to drought, high temperature and other environmental hazards. In the formation of endospore, a part of protoplasmic material is used to form an impermeable coat or cyst wall around the chromosome along with some cytoplasm. The rest of the cell degenerates once the hazard is removed and the spore germinates to form a new population.

Surface Structure: Beginning from the outermost structure and moving inward, bacteria have some or all of the following structures:

Capsule

This layer of polysaccharide (sometimes proteins) protects the bacterial cell and is often associated with pathogenic bacteria because it serves as a barrier against phagocytosis by white blood cells.

Outer Membrane

This lipid bilayer is found in Gram-negative bacteria and is the source of lipopolysaccharide (LPS) in these bacteria.

Cell Wall

Composed of peptidoglycan (polysaccharides + protein), the cell wall maintains the overall shape of a bacterial cell. The three primary shapes in bacteria are coccus (spherical), bacillus (rod-shaped) and spirillum (spiral). Mycoplasma are bacteria that have no cell wall and therefore have no definite shape. The cell walls of all bacteria contain a unique type of peptidoglycan called murein. Peptidoglycan is a polymer of disaccharides (a glycan) cross-linked by short chains of amino acids (peptides).

In the Gram-positive bacteria (those that retain the purple crystal violet dye when subjected to the Gram-staining procedure) the cell wall is thick (15-80 nanometers), consisting of several layers of peptidoglycan. In the Gram-negative bacteria (which do not retain the crystal violet) the cell wall is relatively thin (10 nanometers) and is composed of a single layer of peptidoglycan surrounded by a membranous structure called the outer membrane (Fig. 1.8). The outer membrane of Gram-negative bacteria invariably contains a unique component, lipopolysaccharide (LPS or endotoxin), which is toxic to animals. In Gram-negative bacteria the outer membrane is usually thought of as part of the cell wall. Gram stains will determine if antibiotics will work (Gram-positive) or if they will not (Gram-negative).

Periplasmic Space

This cellular compartment is found only in those bacteria that have both an outer membrane and plasma membrane (i.e., Gram-negative bacteria). In this space enzymes and other proteins are present that help digestion and movement of nutrients into the cell.

Plasma Membrane

This is a lipid bilayer like the cytoplasmic or plasma membrane of other cells. There are numerous proteins moving within or upon this layer that are

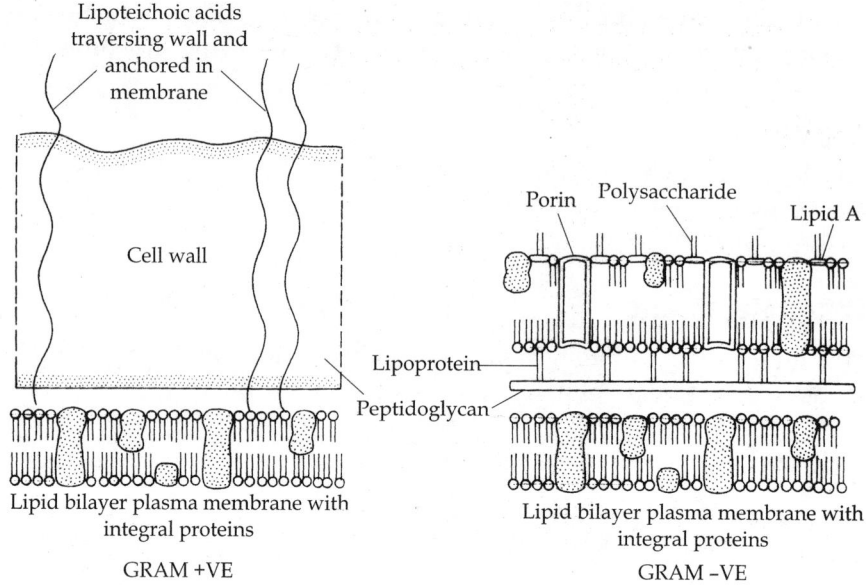

FIG. 1.8 Bacterial cell walls

primarily responsible for transport of ions, nutrients and waste across the membrane.

The plasma membrane is the most dynamic structure of a prokaryotic cell. Its main function is to act as a selective permeability barrier that regulates the passage of substances into and out of the cell. The bacterial membrane allows passage of water and uncharged molecules up to molecular weight of about 100 daltons, but does not allow the passage of larger molecules or any charged substances except by means of special membrane transport processes and transport systems.

Since prokaryotes lack any intracellular organelles for processes such as respiration or photosynthesis or secretion, the plasma membrane performs these processes for the cell and consequently has a variety of functions in energy generation, and biosynthesis. For example, the electron transport system that couples aerobic respiration and ATP synthesis is found in the prokaryotic membrane. The photosynthetic chromophores that harvest light energy for conversion into chemical energy are located in the membrane. Hence, the plasma membrane is the site of oxidative phosphorylation and photophosphorylation in prokaryotes, analogous to the functions of mitochondria and chloroplasts in eukaryotic cells. Besides the transport proteins that selectively mediate the passage of substances into and out of the cell, prokaryotic membranes may contain sensing proteins that measure concentrations of molecules in the environment or binding proteins that translocate signals to genetic and metabolic machinery in the cytoplasm. Membranes also contain enzymes involved in many metabolic processes such as cell wall synthesis, septum formation, membrane synthesis, DNA replication, CO_2 fixation and ammonia oxidation. The predominant functions of bacterial membranes (prokaryotic plasma membrane) are listed below:

1. It acts as an osmotic or permeability barrier.
2. It provides a site for location of transport systems for specific solutes (nutrients and ions).
3. It performs energy generating functions, involving respiratory and photosynthetic electron transport systems, establishment of proton motive force, and transmembranous ATP- synthesizing ATPase.
4. It helps in the synthesis of membrane lipids (including lipopoly-saccharide in Gram-negative bacteria) and in synthesis of murein (cell wall peptidoglycan).
5. It helps in the assembly and secretion of extracytoplasmic proteins.
6. It helps in the coordination of DNA replication and segregation with septum formation and cell division.
7. It functions in the chemotaxis.
8. It is the location of specialized enzyme system.

Bacterial membranes are composed of 40% phospholipid and 60% protein. The phospholipids are amphoteric molecules with a polar hydrophilic glycerol head attached, via an ester bond, to the two non-polar hydrophobic fatty acid tails, which naturally form a bilayer in aqueous environments. Dispersed within the bilayer are various structural and enzymatic proteins which carry out most membrane functions. The arrangement of proteins and lipids to form a membrane is called the fluid mosaic model.

Appendages: Bacteria may have the following appendages:

Fimbriae (Pili)

Fimbriae and pili are interchangeable terms used to designate short, hair-like structures on the surfaces of bacterial cells. Like flagella, they are composed of protein. Fimbriae are shorter and stiffer than flagella, and slightly smaller in diameter. Generally, fimbriae have nothing to do with bacterial movement (there are exceptions, e.g., twitching movement on *Pseudomonas*). Fimbriae are very common in Gram-negative bacteria, but occur in some archaea and Gram-positive bacteria as well. Fimbriae are most often involved in adherence of bacteria to surfaces, substrates and other cells or tissues in nature. In *E. coli*, a specialized type of pilus, the F or sex pilus, mediates the transfer of DNA between mating bacteria during the process of conjugation, but the function of the smaller, more numerous common pili is quite different.

Common pili (almost always called fimbriae) are usually involved in specific adherence (attachment) of bacteria to surfaces in nature. In medical situations, they are the major determinants of bacterial virulence because they allow pathogenic bacteria to attach to (colonize) tissues and/or to resist attack by phagocytic white blood cells. For example, pathogenic *Neisseria gonorrhoeae* adheres specifically to the human cervical or urethral epithelium by means of its fimbriae; enterotoxigenic strains of *E. coli* adhere to the mucosal epithelium of the intestine by means of specific fimbriae; the M-protein and associated fimbriae of *Streptococcus pyogenes* are involved in adherence and to resistance to engulfment by phagocytes.

Flagella

The purpose of flagella (*sing.*, flagellum) is motility. Flagella are long appendages which rotate by means of a 'motor' located just under the cytoplasmic membrane. Bacteria may have one or many flagella in different positions on the cell. They are filamentous protein structures attached to the cell surface that provide the swimming movement for most motile prokaryotes. The diameter of a prokaryotic flagellum is about 20 nm, well-below the resolving power of the light microscope. The flagellar filament is rotated by a motor apparatus in the plasma membrane allowing the cell to swim in fluid environments. Bacterial flagella are powered by proton motive force (chemiosmotic potential) established on the bacterial membrane. About half of the bacilli and all of the spiral and curved bacteria are motile by means of flagella. Very few cocci are motile, which reflects their adaptation to dry environments and their lack of hydrodynamic design (Fig. 1.9).

Bacteria perform many important functions on earth. They serve as decomposers, agents of fermentation, and play an important role in our digestive system. Also, bacteria are involved in many nutrient cycles such as the nitrogen cycle, which restores nitrate into the soil for plants. Unlike eukaryotic cells that depend on oxygen for their metabolism, prokaryotic cells enjoy a diverse array of metabolic functions. For example, some bacteria use sulfur instead of oxygen in their metabolism.

Inorganic ions present in a growing bacterial cell	
Ion	**Function**
K^+	Maintenance of ionic strength; cofactor for certain enzymes
NH_4^+	Principal form of inorganic N for assimilation
Ca^{++}	Cofactor for certain enzymes
Fe^{++}	Present in cytochromes and other metalloenzymes
Mg^{++}	Cofactor for many enzymes; stabilization of outer membrane of Gram-negative bacteria
Mn^{++}	Present in certain metalloenzymes
Co^{++}	Trace element constituent of vitamin B12 and its coenzyme derivatives and found in certain metalloenzymes
Cu^{++}	Trace element present in certain metalloenzymes
Mo^{++}	Trace element present in certain metalloenzymes
Ni^{++}	Trace element present in certain metalloenzymes

Contd.

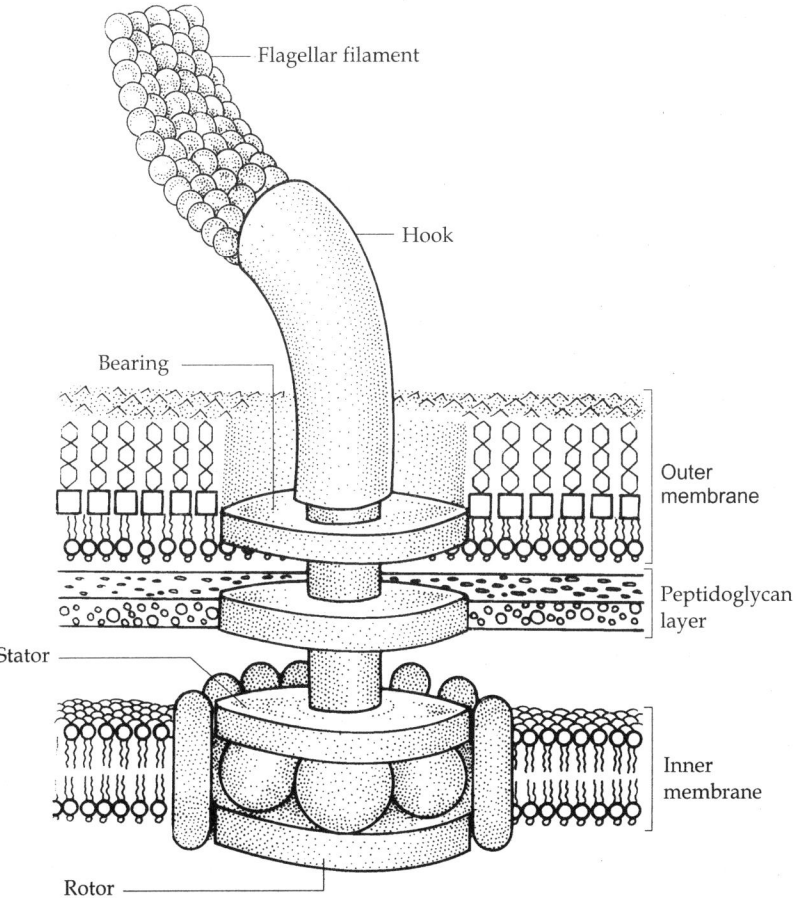

FIG. 1.9 Schematic representation of the motor mechanism of *E. Coli*

Contd.

Zn^{++}	Trace element present in certain metalloenzymes
SO_4^{--}	Principal form of inorganic S for assimilation
PO_4^{---}	Principal form of P for assimilation and a participant in many metabolic reactions

Nutrition in Bacteria

Most bacteria are heterotrophic (saprophytic or parasitic). The bacteria that cause disease are heterotrophic parasites. There are also many non-disease causing bacterial parasites, many of which are helpful to their hosts. These include the 'normal flora' of the human body. For example, *Bacteroides*, *Peptococcus*, *Escherichia*, *Eubacterium*, etc.

Autotrophic bacteria manufacture their own food by the processes of photosynthesis and chemosynthesis. The photosynthetic bacteria include the green and purple bacteria and the cyanobacteria. Many of the thermophilic archaebacteria are chemosynthetic autotrophs.

Respiration and Fermentation

Respiration is a process of oxidation in which energy is released. Respiration in bacteria is both aerobic as well as anaerobic. Some end products of bacterial anaerobic respiration are useful to man and hence, they are used in the manufacture of various foods such as butter, cheese and vinegar.

Reproduction in Bacteria

In bacteria the genetic material is organized in a continuous strand of DNA. This circle of DNA is localized in an area called the nucleoid, but there is no membrane surrounding a defined nucleus as present in the eukaryotic cells of protists, fungi, plants, and animals. In addition to the nucleoid, the bacterial cell may include one or more plasmids, separate circular strands of DNA that can replicate independently, and that are not responsible for the reproduction of the organism. However, drug resistance is often conveyed via plasmid genes.

Reproduction in bacteria is chiefly by binary fission — a process of cell division resulting in identical daughter cells. Some bacteria reproduce by budding or fragmentation. Despite the fact that these processes produce identical generations, the rapid rate of mutation possible in bacteria makes them very adaptable. Some bacteria are capable of specialized types of genetic recombination, which involves the transfer of nucleic acid by individual contact (conjugation). It is defined as the ability of bacterial cells to transfer DNA between cells that are in physical contact (Fig. 1.10), by exposure to nucleic acid remnants of dead bacteria (transformation) (Fig. 1.11), by

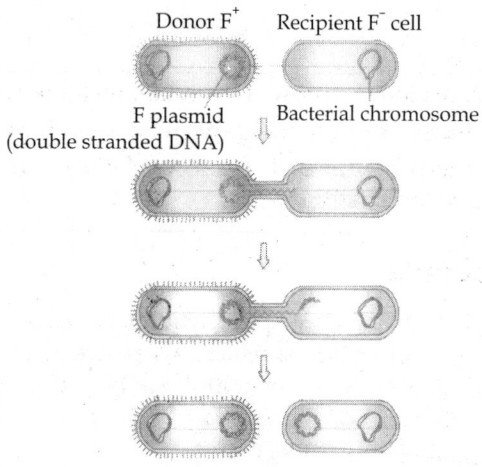

FIG. 1.10 Conjugation

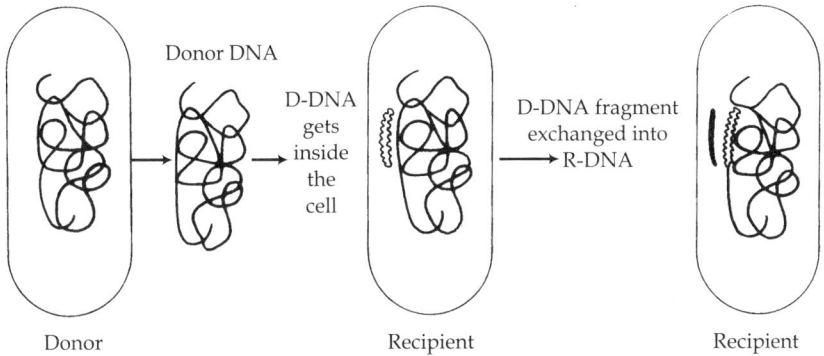

FIG. 1.11   Bacterial Transformation

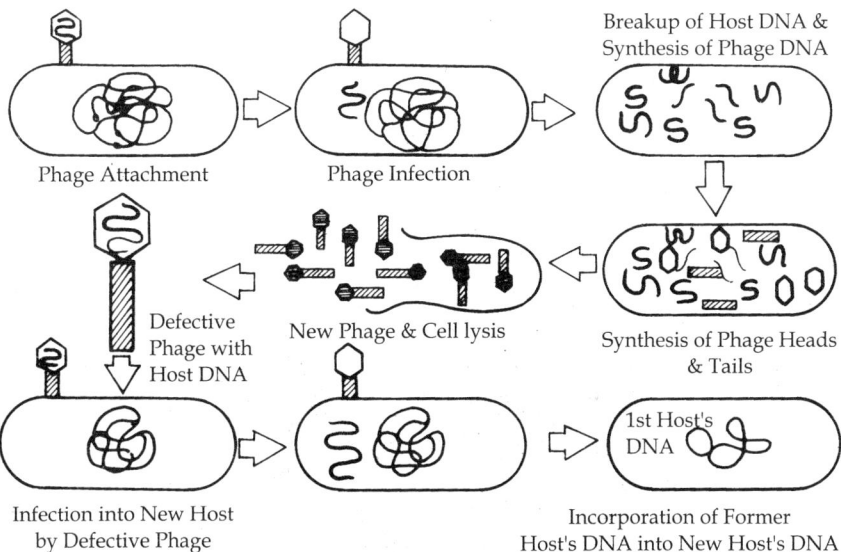

FIG. 1.12   Bacterial Transduction

exchange of plasmid genes, or by a viral agent, the bacteriophage (transduction), defined as the way of transporting DNA between organisms involving the mediation of viruses (Fig. 1.12). Under unfavorable conditions some bacteria form highly resistant spores with thickened coverings, within which the living material remains dormant in altered form until conditions improve. Others, such as the radioactivity-resistant *Deinococcus radiodurans,* can withstand serious damage by repairing their own DNA.

THE EUKARYOTIC CELLS (Eu = true; Karyon = nucleus)

Eukaryotic cells (from the Greek meaning truly nuclear) comprise of all the life kingdoms except monera (i.e., from algae to angiosperms in plants and from protozoa to mammals in animals). They can be easily distinguished through a membrane-bound nucleus. Eukaryotic cells also contain many internal membrane-bound structures called organelles. These organelles such as the mitochondrion, Golgi bodies, endoplasmic reticulum and nucleus serve to perform metabolic functions and energy conversion. Other organelles like intracellular filaments provide structural support and cellular motility. Other important features of the eukaryote family can be seen in the plant cells. The plant cells function essentially in the same manner as other eukaryotic cells, but there are three unique structures which set them apart. Plastids, cell walls, and vacuoles are present only in plant cells and not seen in animal cells (Fig. 1.13 and Fig. 1.14).

Basic Structure

The basic structure of eukaryotic cell contains the following:

Cell Wall

It is present in most of the plant cells. The presence of cell wall distinguishes them from the animal cells. It is a dead and rigid structure composed mainly of

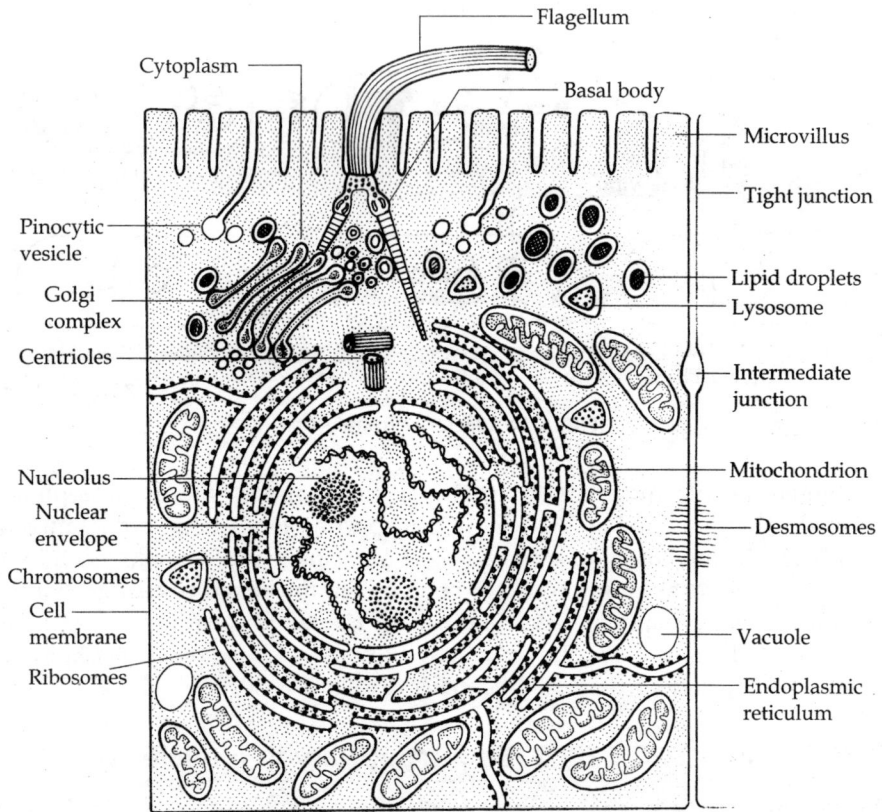

FIG. 1.13 Ultrastructure of an Animal cell

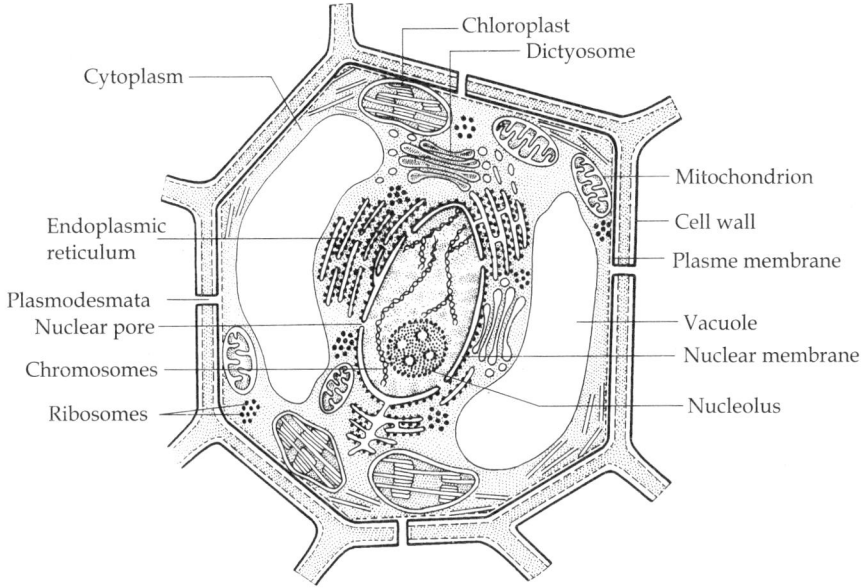

Cytoplasm

Chloroplast

Dictyosome

Mitochondrion

Endoplasmic reticulum

Cell wall

Plasme membrane

Plasmodesmata

Nuclear pore

Vacuole

Nuclear membrane

Chromosomes

Ribosomes

Nucleolus

FIG. 1.14 Ultrastructure of a Plant cell

carbohydrates such as cellulose, pectin, hemicellulose, lignin and certain fatty substances like wax. It is supportive and protective in function and can be distinguished into three layers, viz., the middle lamella, the primary cell wall and the secondary cell wall. Sometimes a tertiary cell wall can be seen. The cell wall often contains pore like structures called plasmodesmata that allow communication between the adjacent cells in a tissue.

Plasma Membrane

One universal feature of all cells is an outer limiting membrane called the plasma membrane. It is a lipid, protein and carbohydrate complex, providing a barrier and containing transport and signaling systems. In addition, all eukaryotic cells contain elaborate systems of internal membranes which set up various membrane-enclosed compartments within the cell. The plasma membrane serves as the interface between the machinery in the interior of the cell and the extracellular fluid (ECF) that bathes all cells. The lipids in the plasma membrane are chiefly phospholipids like phosphatidyl ethanolamine and cholesterol. Phospholipids are amphiphilic with the hydrocarbon tail of the molecule being hydrophobic and its polar head hydrophilic. As the plasma membrane faces watery solutions on both sides, its phospholipids accommodate this by forming a phospholipid bilayer with the hydrophobic tails facing each other. Many of the proteins associated with the plasma membrane are tightly bound to it (See Chapter 5).

Cytoplasm

Cytoplasm is a homogeneous, generally clear jelly-like material that fills the cells. The cytoplasm consists of cytosol and the cellular organelles, except the cell nucleus. The cytosol is made up of water, salts, organic molecules and many enzymes that catalyze reactions. The cytoplasm plays an important role in a cell, serving as a 'molecular soup' in which the organelles are suspended and held together by a fatty membrane. It is found within the plasma membrane of a cell and surrounds the nuclear envelope and the cytoplasmic organelles.

The cytoplasm plays a mechanical role, that is to maintain the shape and consistency of the cell, and to provide suspension to the organelles. It is also a storage place for chemical substances indispensable to life, which are involved in vital metabolic reactions, such as anaerobic glycolysis and protein synthesis.

Endoplasmic Reticulum

The endoplasmic reticulum or ER (endoplasmic means 'within the cytoplasm', reticulum means 'little net') is an organelle found in all eukaryotic cells. ER membranes may assume the form of cisternae, tubules or vesicles. The ER modifies proteins, makes macromolecules, and transfers substances throughout the cell. The basic structure and composition of the ER is similar to the plasma membrane, although it is actually an extension of the nuclear membrane. The ER is the site of the translation, folding, and transport of protein that are to become part of the cell membrane (e.g., transmembrane receptors and other integral membrane proteins) as well as proteins that are to be secreted from the cell (e.g., digestive enzymes). There are two basic morphological types of ER—the rough ER which has ribosomes attached on the outer surface and the smooth ER that does not have the attached ribosomes.

Golgi Apparatus

Golgi apparatus is composed of a stack of about half a dozen sac like structures, whose purpose in the cell is to prepare and store chemical products produced in the cell, and then to secrete these outside the cell. Golgi bodies are formed when small sac like pieces of membrane are pinched away from the cell. The number and size of Golgi bodies found in a cell depends on the quantity of chemicals produced in the cell. For example, a large number of Golgi bodies are found in cells that produce saliva and other materials for digestion.

Lysosomes

Lysosomes are tiny, spherical or irregular-shaped membrane bound vesicles that function as the garbage disposal system of the cell. They degrade the products of ingestion by their lysosomal enzymes, worn out organelles such as mitochondria are also degraded by a similar action and they can handle the products of receptor-mediated endocytosis such as the receptor, ligand and associated membrane. Lysosomes carry hydrolases that degrade nucleotides, proteins, lipids, phospholipids, and also remove carbohydrate, sulfate, or phosphate groups from molecules. Lysosomal morphology varies with the state of the cell and its degree of degradative activity. Lysosomes have pieces of membranes, vacuoles, granules and parts of mitochondria inside. Phagolysosomes may have parts of bacteria or the cell it has ingested.

Peroxisomes

This organelle is responsible for protecting the cell from its own production of toxic hydrogen peroxide. As an example, white blood cells produce hydrogen peroxide to kill bacteria. The oxidative enzymes in peroxisomes breakdown the hydrogen peroxide into water and oxygen.

Mitochondria

Mitochondria are membrane-bound organelles, and like nucleus they also have a double membrane. The outer membrane is fairly smooth. But the inner membrane is highly convoluted, forming folds called cristae. The cristae greatly increase the inner membrane's surface area. It is on these cristae that food (sugar) is combined with oxygen to produce ATP—the primary energy source for the cell. It contains its own DNA, and is believed to have originated as a captured bacterium. Mitochondria provide the energy as a cell needs to move, divide, produce secretory products and contract thus, they are the power centers of the cell. They are about the size of bacteria but may have different shapes depending on the cell type. Mitochondria are also known as the 'power house' of the cell.

Plastids

In most plant cells structures called plastids are found. They are found in the cytoplasmic matrix of plant cells only. These structures are generally spherical or ovoid in shape and they are clearly visible in living cells. Major three types of plastids found in plant cells are:

Chloroplasts: These are probably the most important among the plastids since they are directly involved in photosynthesis. They are usually situated near the surface of the cell and occur in those parts that receive sufficient light, e.g., the palisade cells of leaves. The green colour of chloroplasts is caused by the green pigment chlorophyll. The chloroplasts are the sites for photosynthesis and contain enzymes and co-enzymes necessary for the process of photosynthesis.

Chromoplasts: They are red, yellow or orange in colour and are found in petals of flowers and in fruit. Their colour is due to two pigments, carotene and xanthophyll. Their primary function in the cells of flowers is to attract agents of pollination and in fruit to attract agents of dispersal.

Leucoplasts: They are colorless plastids due to the absence of pigments and occur in plant cells not exposed to light, such as roots and seeds. They are the centers of starch grain formation and are also involved in the synthesis of oils and proteins.

Ribosomes

Ribosomes are made up of proteins and ribonucleic acid (RNA). These molecules are arranged into two subunits called the large and small subunits. The subunits are attached to each other and together forming the entire ribosome. They are often associated closely with endoplasmic reticulum (ER), forming rough ER and are the site of protein synthesis in cells.

Centrioles

Centrioles are barrel shaped microtubule structures found only in animal cells, these paired organelles are typically located together near the nucleus in the centrosome, a granular mass that serves as an organizing center for microtubules. Within the centrosome, the centrioles are positioned so that they are at right angles to each other. Each centriole is made of nine bundles of microtubules (three per bundle) arranged in a ring.

Microtubules

Microtubules are filamentous intracellular structures that are responsible for various kinds of movements in all eukaryotic cells. Microtubules are involved in the cell division, organization of intracellular structure, and intracellular transport, as well as ciliary and flagellar motility.

Cilia and Flagella

Cilia and flagella have the same basic structure. They are attached to the basal bodies, which in turn are anchored to the cytoplasmic side of the plasma membrane. To form cilia or flagella, microtubules arrange themselves in a '9 + 2' arrangement. Each of the two central microtubules consists of a single microtubule with 13 protofilaments arranged to form the wall of a circular tube. Each of the outer nine consists of a pair of microtubules that share a common wall. In cilia and flagella, tubulin forms a core structure to which other proteins contribute structures called dynein arms, radial spokes, and nexin links. The arms, spokes, and links hold microtubules together and allow interaction between microtubules that is superficially similar to the sliding of actin and myosin filaments in muscle contraction (See Chapter 14).

Nucleus

The nucleus is found in all eukaryotic cells and contains the nuclear genes which form most of the cell's genetic material. The nucleus is a highly specialized organelle that serves as the information processing and administrative center of the cell. This organelle has two major functions: it stores the cell's hereditary material, or DNA, and it coordinates the cell's activities, which include growth, intermediary metabolism, protein synthesis, and reproduction (cell division).

Nucleolus

The nucleus contains a conspicuous, darkly stained, spherical organelle called nucleolus where ribosome subunits are assembled. The nucleolus consists primarily of ribosomal precursor RNA, ribosomal RNA, their associated proteins, and enzymatic equipment like polymerase, RNA methylase, RNA cleavage enzymes which are required for synthesis, conversion and assembly of ribosomes.

Vacuoles

Plant cells have large structures containing water surrounded by a membrane in the centre of their cells. These are vacuoles and they act as a store house of water and food (in seeds) and a place to dump wastes. It acts as a structural support for the cell to maintain turgor. When the plant loses water the vacuoles also lose their water, and when plants have a lot of water the vacuoles get filled up.

Comparison of Prokaryotic and Eukaryotic cells

	Prokaryotes	Eukaryotes
Typical organisms	Bacteria, archaebacteria	Protists, fungi, plants, animals
Typical size	~ 1-10 μm	~ 10-100 μm
Type of nucleus	nucleoid region; no real nucleus	real nucleus with double membrane
DNA	circular (usually)	linear molecules (chromosomes) with histone proteins
Ribosomes	70S type	80S type
Cytoplasmatic structures	very few structures	highly structured by endomembranes and cytoskeleton
Cell movement	flagellum made of flagellin	flagella and cilia made of tubulin
Mitochondria	none	one to several dozen
Chloroplast	none	present in algae and plants
Organization	usually single cells organisms with specialized cells	single cells, colonies, higher multicellular organisms
Cell division	binary fission (simple division)	mitosis and meiosis

Structural differences between Plant and Animal Cells

Plant Cells	Animal Cells
1. Cell wall almost present	No cell wall present outside the cell membrane
2. Plastids occur in cytoplasm	No plastids are found
3. Lysosomes are usually not evident	Lysosomes occur in cytoplasm
4. Centrioles present only in cells of lower plant forms	Centrioles always present
5. Large vacuoles filled with cell sap	Vacuoles, if present, are small and contractile or temporary vesicles
6. Animal cells have centrosome and centrioles	Plant cells lack centrosome and centrioles

VIRUSES

Introduction

A virus is a submicroscopic, obligate intracellular parasite that lacks the cellular machinery for self-reproduction. Typically, viruses carry a small amount of genetic material, either in the form of DNA or RNA, but not both, surrounded by some form of protective coat consisting of proteins, lipids, glycoproteins or a combination. The viral genome codes for the proteins that form this protective coat, as well as for those proteins required for viral reproduction, which are not provided by the host cell. Viruses are non-living particles that can only replicate when an organism reproduces the virulent RNA or DNA. They are considered non-living by the majority of virologists because they do not meet all the criteria of the generally-accepted definition of life. Among other factors, viruses do not move, metabolize, or decay on their own. Viruses infect both eukaryotes and prokaryotes (such as bacteria). Viruses infecting bacteria are also known as *bacteriophages* or *phages*.

History

It was in May 1796 that Edward Jenner used cowpox-infected material obtained from the hand of a milkmaid Sarah Nemes, from his home village of Berkley in Gloucestershire to successfully vaccinate an eight year old boy James Phipps. To everybody's surprise the boy did not become infected! However, this was the first evidence of the infectious particle.

It was not until Robert Koch and Louis Pasteur jointly proposed the 'germ theory' of disease in the 1880s that the significance of these organisms became apparent.

Koch defined the four famous criteria now known as Koch's postulates which are still generally regarded as the proof that an infectious agent is responsible for a specific disease:

(a) The infectious agent must be present in every case of the disease.

(b) The agent must be isolated from the host and grown *in vitro.*

(c) The disease must be reproduced when a pure culture of the agent is inoculated into a healthy susceptible host.

(d) The same agent must be recovered once again from the experimentally infected host.

Later, Pasteur worked extensively on the disease called rabies, which he identified as being caused by a 'virus' (from the Latin word 'poison') but in spite of this, he could not discriminate between bacterial and other agents of disease.

The Russian botanist, Dmitri Iwanowski, presented a paper to the St. Petersburg Academy of Science in February 1892 which showed that the extracts from diseased tobacco plants could transmit disease to other plants after passage through ceramic filters which were fine enough to retain the smallest known bacteria. This is generally recognized as the beginning of Virology. Unfortunately, Iwanowski did not fully realize the significance of these results.

A few years later, in 1898, Martinus Beijerinick confirmed and extended Iwanowski's results on tobacco mosaic virus and was the first to develop the modern idea of the virus, which he referred to as *contagium vivum fluidum* ('soluble living germ').

Also in 1898, Freidrich Loeffler and Paul Frosch showed that a similar agent was responsible for foot-and-mouth disease in cattle. Thus, it was certain that these new agents caused disease in animals as well as plants.

Frederick Twort in 1915 and Felix d'Herelle in 1917 were the first to recognize viruses which infect bacteria, which d'Herelle called bacteriophages.

In 1933, Schelsinger determined the composition of a virus and showed that it consists of only protein and DNA.

STRUCTURE OF VIRUSES

A virus particle, known as a virion, consists of a genome contained within a protective casing of protein called a capsid. The nucleic acid genome varies among different viruses and may be either DNA or RNA, which may be either single- or double-stranded; linear or circular. The capsid is composed of proteins encoded by the viral genome and may be either spherical or helical. These proteins are associated with the nucleic acid and are hence known as nucleoproteins, thus the combination of nucleoproteins and nucleic acid is known as a nucleocapsid. It is composed of a number of subunits called capsomeres. The terms 'capsid' and 'capsomeres' were proposed by Lwoff, Anderson and Jacob (1959) to represent, respectively, the protein shell and the units comprising it. The term 'virion' was also proposed by them to denote the complete infective virus particle (i.e., a capsid enclosing the nucleic acid). This terminology was generally accepted although it later proved to be inadequate.

In an attempt to clarify the terminology for virus components, Caspar *et al.* (1962) made a number of proposals which were generally accepted. Briefly, the proposals are as follows:

(a) The capsid denotes the protein shell that encloses the nucleic acid. It is built of structure units.

(b) Structure units are the smallest functional equivalent building units of the capsid.

(c) Capsomeres are morphological units seen on the surface of particles and represent the clusters of structure units.

(d) The capsid together with its enclosed nucleic acid is called the nucleocapsid.

(e) The nucleocapsid may be invested in an envelope which may contain material of host cell as well as viral origin.

(f) The virion is the complete infective virus particle.

In addition to a capsid, some viruses are able to hijack a modified form of the plasma membrane surrounding an infected host cell, thus gaining an outer lipid bilayer known as a viral envelope. This extra membrane is studded with proteins synthesized by the host cell, which the virus may have modified at a genetic level. This gives the virion a few distinct advantages over 'naked' virions - the plasma membrane provides a degree of protection for the virus, especially from harmful agents such as enzymes and chemicals. The proteins studded upon it include glycoproteins, which serve as receptor molecules, allowing healthy cells to recognize virions as 'friendly' and resulting in the possible uptake of the virion into the cell.

Viruses occur in three main shapes, viz., spherical, helical and complex.

Spherical virus capsids completely enclose the viral genome and do not generally bind as tightly to nucleic acid as helical capsid proteins do. These structures can range in size from less than 20 nm up to 400 nm and are composed of viral proteins arranged with Icosahedral symmetry. Icosahedral architecture is the same principle employed by R. Buckminster-Fuller in his geodesic dome, and it is the most efficient way of creating an enclosed robust structure from multiple copies of a single protein. The number of proteins required to form a spherical virus capsid is denoted by the 'T-number', where 60t proteins are necessary. In the case of the Hepatitis B virus, the T-number is 4, therefore 240 proteins assemble to form the capsid. Many spherical viruses forgo a lipid envelope, leaving the capsid proteins to be directly involved in attachment and entry into the host cell. The examples of viruses having the icosahedral symmetry of Bacteriophage ϕ (phi), Polyoma virus, Polio virus, Turnip Yellow Mosaic virus, Papilloma virus, Reovirus, Herpes virus (Fig. 1.15).

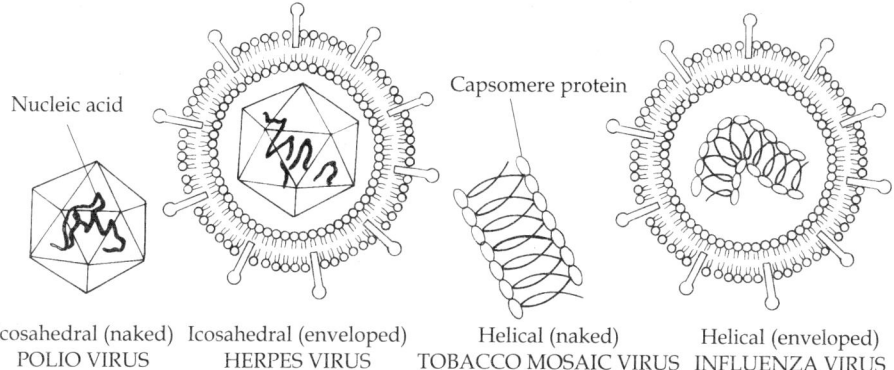

Nucleic acid

Capsomere protein

Icosahedral (naked)   Icosahedral (enveloped)   Helical (naked)   Helical (enveloped)
POLIO VIRUS   HERPES VIRUS   TOBACCO MOSAIC VIRUS   INFLUENZA VIRUS

FIG. 1.15   The common viral morphologies

Helical capsids are composed of identical proteins stacked at a constant amplitude and pitched to one another around a central circumference, much like a spiral staircase, which effectively forms an enclosed tube housing the genetic material. This arrangement results in rod-shaped virions which can either be short and rigid or long and flexible; the nature of long helical particles neccessitates flexibility, as they are prone to damage if they are too rigid. For example, Tobacco mosaic virus (TMV), Bacteriophage M13, Influenza virus, etc. (Fig. 1.15).

Complex viruses are divided into two groups: (i) those without identifiable capsids, for example, vaccinia virus, cow pox virus etc., and (ii) those with tadpole-shaped structures, for example, bacteriophages of T-even series (T_2, T_4, T_6) (Fig. 1.16).

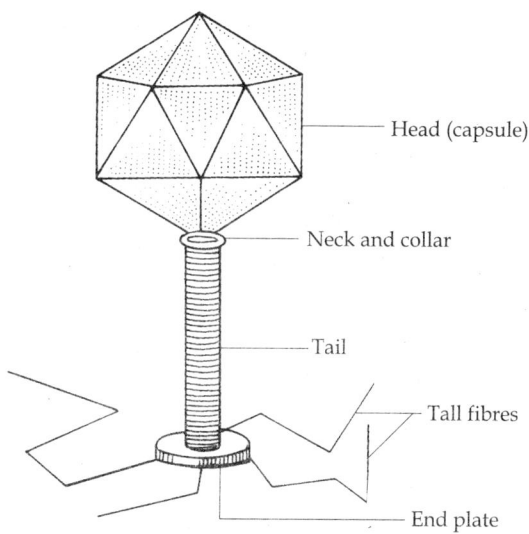

Head (capsule)

Neck and collar

Tail

Tall fibres

End plate

FIG. 1.16   A bacteriophage

At the simplest level, the function of the outer shells (capsid) of a virus particle is to protect the fragile nucleic acid genome from:

(i) Physical damage—Shearing by mechanical forces.

(ii) Chemical damage—UV irradiation (from sunlight) leading to chemical modification.

(iii) Enzymatic damage—Nucleases derived from dead or leaky cells or deliberately secreted by vertebrates as defence against infection.

The protein subunits in a virus capsid are multiply redundant, i.e., present in many copies per particle. Damage to one or more subunits may render that particular subunit non-functional but does not destroy the infectivity of the whole particle.

Furthermore, the outer surface of the virus is responsible for recognition of the host cell. Initially, this takes the form of binding of a specific virus-attachment protein to a cellular receptor molecule. However, the capsid also has a role to play in initiating infection by delivering the genome from its protective shell in a form in which it can interact with the host cell.

To form an infectious particle, a virus must overcome two fundamental problems:

(i) To assemble the particle utilizing only the information available from the components which make up the particle itself (capsid + genome).

(ii) Virus particles form regular geometric shapes, even though the proteins from which they are made are irregularly shaped.

How do these simple organisms solve these difficulties? The information to answer this problem lies in the rules of symmetry.

In 1957, Fraenkel-Conrat and Williams showed that when mixtures of purified tobacco mosaic virus (TMV) RNA and coat protein were incubated together, virus particles formed. The discovery that virus particles could form spontaneously from purified subunits without any extraneous information indicated that the particle was in the free energy minimum state and was therefore the favored structure of the components. This stability is an important feature of the virus particle.

Although some viruses are very fragile and are essentially unable to survive outside the protected host cell environment, many are able to persist for long periods, in some cases for years.

Replication

Viruses are acellular and do not have their own metabolism, they utilize the machinery and metabolism of the host for the purpose of self-replication. A virus is called a virion before it has entered a host cell— a package of viral genetic material. Virions can be passed from host to host either through direct contact or through a vector, or carrier. Inside the organism, the virus can enter a cell in various ways. Bacteriophages or bacterial viruses attach to the cell wall surface in specific places. Once attached, enzymes make a small hole in the cell wall, and the virus injects its DNA into the cell. Other viruses, such as HIV, enter the host via endocytosis, the process by which cells take in material from the external environment. After entering the cell, the genetic material of virus begins the destructive process of causing the cell to produce new viruses.

Some viruses have DNA which is replicated by the host cell along with the host's own DNA.

There are two different replication processes for viruses containing RNA. In the first process, the viral RNA is directly copied using an enzyme called RNA replicase. This enzyme then uses that RNA copy as a template to make hundreds of duplicates of the original RNA. A second group of RNA-containing viruses, called the retroviruses, uses the enzyme reverse transcriptase to synthesize a complementary strand of DNA so that the genetic information of virus is contained in a molecule of DNA rather than RNA. The viral DNA can then be further replicated using the resources of the host cell.

Bacteriophages have two cycles called the lytic and lysogenic cycle. In the lytic cycle the host cell undergoes lysis (the breaking open of the cell to release viral particles), biosynthesis, maturation and finally, release. In the lysogenic cycle viral replication does not immediately occur, but replication may take place sometime in the future. Some viruses, for example, lambda are capable of carrying out both cycles. In the lytic cycle, the cell is soon destroyed and the newly formed viruses have to find new host. However, during the lysogenic stage, the DNA of the bacteria is spliced using restriction enzymes and the viral DNA (or RNA turned to DNA by the enzyme reverse transcriptase) is integrated into the spliced section of the host DNA. The virus remains dormant but after the host cell has replicated many times the virus will become active and will enter the lytic phase. The lysogenic cycle allows the host cell to continue to survive and reproduce. As the host reproduces its genome with that of the virus, the virus itself is reproduced as a by-product. Therefore many progeny cells containing the virus get developed (Fig. 1.17).

Steps of Life Cycle

1. Attachment or adsorption: The virus attaches to receptors on the host cell wall.

2. Injection: The nucleic acid of the virus moves through the plasma membrane and into the cytoplasm of the host cell. The capsid of a phage remains outside. In contrast, many viruses that infect animal cells enter the host cell intact.

3. Transcription: Within minutes of phage entry into a host cell, a portion is transcribed into mRNA, which is then translated into proteins specific for the infecting phage.

4. Replication: The viral genome contains all the information necessary to produce new viruses. Once inside the host cell, the virus induces the host cell to synthesize the necessary components for its replication.

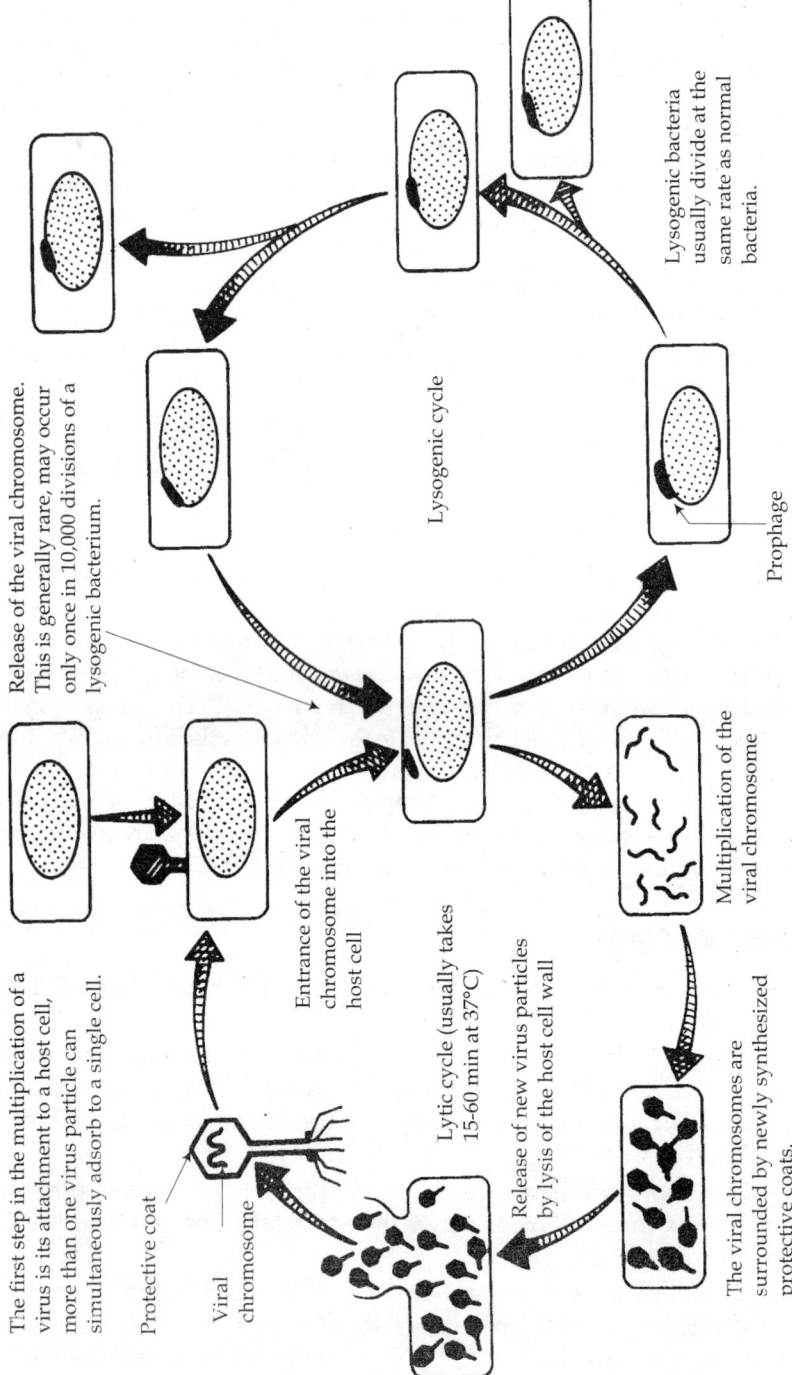

The first step in the multiplication of a virus is its attachment to a host cell, more than one virus particle can simultaneously adsorb to a single cell.

Protective coat

Viral chromosome

Entrance of the viral chromosome into the host cell

Lytic cycle (usually takes 15-60 min at 37°C)

Release of new virus particles by lysis of the host cell wall

The viral chromosomes are surrounded by newly synthesized protective coats.

Multiplication of the viral chromosome

Release of the viral chromosome. This is generally rare, may occur only once in 10,000 divisions of a lysogenic bacterium.

Lysogenic cycle

Lysogenic bacteria usually divide at the same rate as normal bacteria.

Prophage

FIG. 1.17 Lytic and Lysogenic cycles

5. Assembly: The newly synthesized viral components are assemble into new viruses.

6. Release: Assembled viruses are released from the cell and can now infect other cells, and the process begins again.

When the virus has infected the cell, it immediately causes the host to begin manufacturing the proteins necessary for virus reproduction. Some viruses, like herpes, cause the host to produce three kinds of proteins: early proteins, enzymes used in nucleic acid replication; late proteins, proteins used to construct the virus coat; and lytic proteins, enzymes used to break open the cell for viral exit. The final viral product is assembled spontaneously, that is, the parts are made separately by the host and are joined together by chance. This self assembly is often aided by proteins made by the host that help the capsid parts come together.

The new viruses then leave the cell either by exocytosis or by lysis. Envelope-bound animal viruses cause the host's endoplasmic reticulum to make certain proteins, called glycoproteins, which then collect in clumps along the cell membrane. The virus is then discharged from the cell at these exit sites, referred to as exocytosis. On the other hand, bacteriophages must break open, or lyse, the cell to exit. To do this, the phages have a gene that codes for an enzyme called lysozyme. This enzyme breaks down the cell wall, causing the cell to swell and burst. The new viruses are released into the environment, killing the host cell in the process.

OBJECTIVE TYPE QUESTIONS

1. Part of the cell between the plasma membrane and the nucleus is called
   A. Nucleoplasm
   B. Cytoplasm
   C. Vacuole
   D. Ergastoplasm

2. Which of the following is a correct difference between prokaryotic and eukaryotic cells?
   A. Prokaryotes have a cell wall, eukaryotes do not
   B. Prokaryotes are larger than eukaryotes
   C. Prokaryotes have larger ribosomes than eukaryotes
   D. Prokaryotes have no membrane-bound organelles

3. The largest organelle in a eukaryotic cell is _____
   A. Nucleus
   B. Mitochondrion
   C. Golgi body
   D. Lysosome

4. Which of the following molecules is most likely to be synthesized in the Smooth Endoplasmic Reticulum?
   A. Phospholipids
   B. Proteins
   C. Carbohydrates
   D. Nucleic acids

5. The main function of ribosomes is _____
   A. Energy generation
   B. Carbohydrate metabolism
   C. Protein synthesis
   D. Formation of nucleolus

6. The Golgi body has a major role in which of the following membrane transport processes?
   A. Exocytosis
   B. Endocytosis
   C. Pinocytosis
   D. Phagocytosis

7. The inside of the phospholipid bilayer is _____
   A. Hydrophilic
   B. Hydrophobic
   C. Lipophilic
   D. All of the above

8. Water cannot diffuse through a bilayer because:
   A. A water molecule is too large
   B. The inside of the bilayer repels charged molecules like water
   C. A water molecule is too slow-moving
   D. None of the above

9. In osmosis, water moves from a
   A. Hypotonic solution to a hypertonic solution
   B. Hypertonic solution to a hypotonic solution
   C. Isotonic solution to isotonic solution
   D. Isotonic solution to a hypotonic solution

10. The main role of the nucleolus, a region of the nucleus is _____
    A. Maintain cell shape
    B. Cell trafficking
    C. Making ribosomes
    D. Heredity and variation

Answers

1. B	2. D	3. A	4. A	5. C
6. A	7. B	8. B	9. A	10. C

SHORT ANSWER TYPE QUESTIONS

1. What are the major functions of the nucleus, and what parts of the nucleus carry out these functions?
2. How can you determine whether a unicellular organism is a prokaryote or a eukaryote?
3. Why are the mitochondria important to the functioning of eukaryotic cells?
4. Draw a labeled diagram of a typical eukaryotic cell.
5. Write a brief account on the structure of a bacterium.

LONG ANSWER TYPE QUESTIONS

1. If a cell has a high energy requirement, would you expect it to have many or few mitochondria? Explain your answer.
2. Describe differences between prokaryotic cells and eukaryotic cells.

3. Describe the structure and function of the cytoskeleton.
4. Explain how the nucleolus, ribosomes, endoplasmic reticulum, and golgi apparatus function together in protein synthesis.
5. What are the three postulates of Cell Theory?
6. Explain the role of DNA in cells.
7. Describe three differences between plant and animal cells.
8. Describe the structure, composition, and function of the cell membrane.
9. Write an essay on viruses.
10. Explain the life cycle of viruses.

Tools and Techniques in Cell Biology

2

INTRODUCTION

The limitation of the human eye has led to the concentration of much of the early biological research on the development of tools that help us see very small things. The first remarkable invention in this field was that of the light microscope: an instrument that enables the human eye, by means of a lens or combinations of lenses, to observe enlarged images of tiny objects. It made visible the fascinating details of the cell. This was followed by the introduction of many other tools which have given birth to many sophisticated instruments and techniques. A brief discussion of some of these tools and techniques will be taken up in this chapter.

MICROSCOPY

A microscope (Greek: micron = small and scopes = aim) is an instrument for viewing objects that are too small to be seen by the naked or unaided eye. The science of investigating small objects using such an instrument is called microscopy, and the term microscopic means minute or very small, not easily visible with the unaided eye.

History

Zaccharias Janssen and his son Hans, two Dutch spectacle makers, in 1590s, discovered that nearby objects appeared greatly enlarged while experimenting with several lenses in a tube. Later, Antony van Leeuwenhoek (1632-1723) taught himself new methods of grinding and polishing tiny lenses of great curvature which gave magnifications up to 270 diameters, the finest known at that time (Fig. 2.1). This led to the building of his microscopes and the biological discoveries for which he is famous. He was the first to see and describe bacteria, yeast plants, the life in a drop of water and the circulation of blood corpuscles in capillaries. During a long life he used his lenses to make pioneer studies on an extraordinary variety of things, both living and non-living, and reported his findings in over a hundred letters to the Royal Society of England and the French Academy.

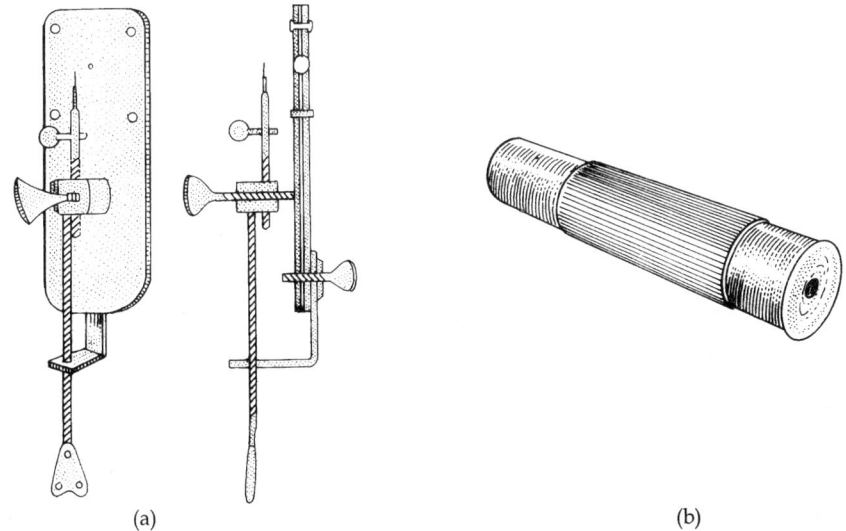

(a)   (b)

FIG. 2.1   (a) A Replica of van Leeuwenhoek's Primitive one lens Microscope (b) Microscope of Zaccharias Janssen (1588-1631)

Robert Hooke, the father of microscopy (Fig. 2.2), re-confirmed Antony von Leeuwenhoek's discoveries of the existence of tiny living organisms in a drop of water. Hooke made a copy of Leeuwenhoek's light microscope and then improved upon his design (Fig. 2.3).

Later, few major improvements were made until the middle of the 19th century. Then several European countries began to manufacture fine optical equipment but none was finer than the marvelous instruments built by the American, Charles A. Spencer, and the industry he founded. Present day instruments, changed but little, give magnifications up to 1250 diameters with ordinary light and up to 5000 with blue light.

FIG. 2.2 Robert Hooke (1635-1703)

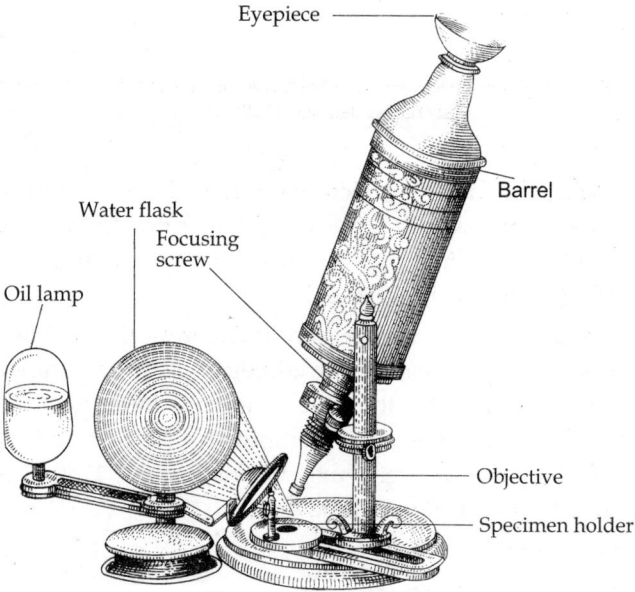

FIG. 2.3 Hooke's Microscope

Light Microscopy

This is the simplest and most widely used form of microscopy. In this the specimens are illuminated with light that is focused using glass lenses and viewed using the eye. The specimens can be living or dead, but they often need to be stained with a colored dye to make them visible. Many different stains are available that stain specific parts of the cell such as DNA, lipids, cytoskeleton, etc. All light microscopes today are compound microscopes and they use several lenses to obtain high magnification (Fig. 2.4). Light microscopy has a resolution of about 200 nm, which is good enough to see cells, but not the details of cell organelles. There has been a recent resurgence in the use of light microscopy, partly due to technical improvements, which have improved the resolution far beyond the theoretical limit. For example, fluorescence microscopy has a resolution of about 10 nm, while interference microscopy has a resolution of about 1 nm.

A light microscope, whether a simple student microscope or a complex research microscope, has the following basic systems:

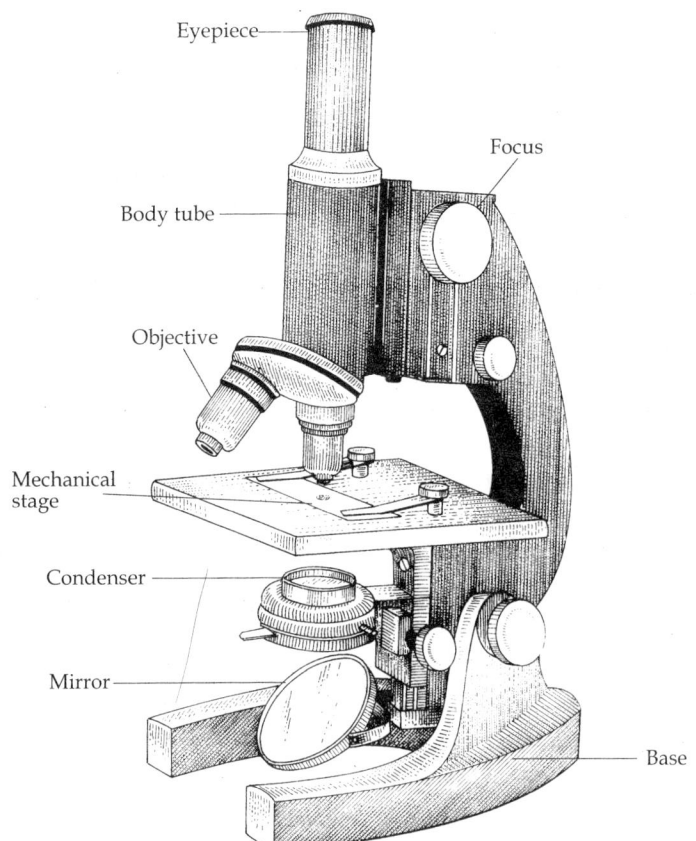

FIG. 2.4 A Compound Microscope

Specimen control: that holds the slide containing the specimen to be viewed.

Stage: where the specimen rests

Clips: used to hold the specimen still on the stage (because you are looking at a magnified image, even the smallest movements of the specimen can move parts of the image out of your field of view).

Micromanipulator: device that allows you to move the specimen in controlled, small increments along the x and y axes (useful for scanning a slide)

Illumination: to shed light on the specimen. (The simplest illumination system is a mirror that reflects room light up through the specimen.)

Lamp: produces the light. (Typically, lamps are tungsten-filament light bulbs. For specialized applications, mercury or xenon lamps may be used to produce ultraviolet light. Some microscopes even use lasers to scan the specimen.)

Rheostat: alters the current applied to the lamp to control the intensity of the light produced

Condenser: lens system that aligns and focuses the light from the lamp onto the specimen

Diaphragms or **pinhole apertures:** placed in the light path to alter the amount of light that reaches the condenser (for enhancing contrast in the image)

Body tube: holds the eyepiece at the proper distance from the objective lens and blocks out stray-light

Focus: position the objective lens at the proper distance from the specimen

Coarse-focus knob: used to bring the object into the focal plane of the objective lens

Fine-focus knob: used to make fine adjustments to focus the image

Lenses

Objective lens: gathers light from the specimen

Eyepiece: transmits and magnifies the image from the objective lens to your eye

Nosepiece: rotating mount that holds many objective lenses

Support and Alignment

Arm: curved portion that holds all of the optical parts at a fixed distance and aligns them

Base: supports the weight of all of the microscope parts

The tube is connected to the arm of the microscope by way of a rack and pinion gear. This system allows you to focus the image when changing lenses or observers and to move the lenses away from the stage when changing specimens.

Thus, light microscopes contain two lens systems, an objective and an ocular. The total magnification of an image is calculated by multiplying the magnification of the ocular by the magnification of the objective. Such microscopes generally have a 10× ocular lens and four different objective lenses listed in the table below.

Objective	Magnification	Total Magnification
Scanning	4×	40×
Low Power	10×	100×
High Power	40×	400×
Oil Immersion	100×	1000×

Light bends when it passes from glass to air or from air to glass because air and glass have different refractive indices. The bending of light as it passes through the glass slide to the air and then to the glass lens decreases the resolving power. At high magnification (1000×) it can prevent a clear image from being viewed. This decrease in resolution can be prevented by putting *immersion oil* between the slide and the lens because immersion oil has the same refractive index as glass.

The condenser also increases the resolving power of the microscope. When using the oil-immersion lens, the condenser (located beneath the stage) should be raised to a position very close to the stage for maximum resolution (Fig. 2.5).

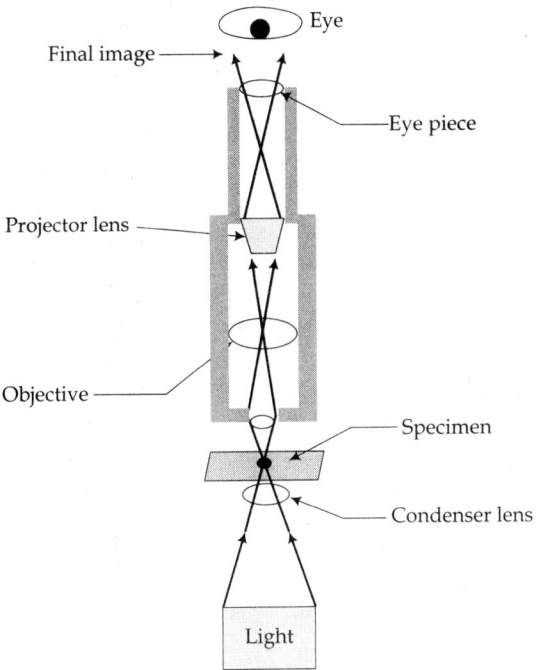

FIG. 2.5 Light traveling through the Compound Microscope

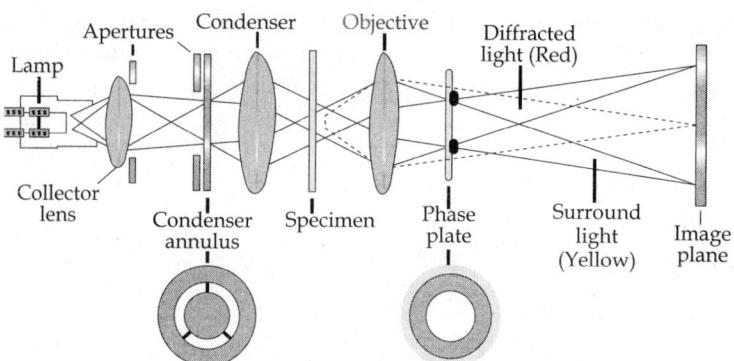

FIG. 2.6   Phase Contrast Microscope Optical Train

Phase Contrast Microscopy (Fig. 2.6)

Due to very little contrast between structures with similar transparency and insufficient natural pigmentation most of the detail of living cells is undetectable in bright field microscopy. However, the various organelles show wide variation in refractive index, i.e., the tendency of the materials to bend light, providing an opportunity to distinguish them.

Phase contrast microscopy is prefered over bright field microscopy when high magnifications (400×, 1000×) are needed and the specimen is colorless or the details so fine that color does not show up well. Cilia and flagella, for example, are nearly invisible in bright field but show up in sharp contrast under phase contrast. Amoebae look like vague outlines in bright field, but show a great deal of detail under phase contrast microscope. Most living microscopic organisms are much more obvious in phase contrast.

Highly refractive structures bend light to a much greater angle than do structures of low refractive index. The same properties that cause the light to bend also delay the passage of light by a quarter of a wavelength or so. In a light microscope with bright field mode, light from highly refractive structures bends farther away from the center of the lens than light from less refractive structures and arrives about a quarter of a wavelength out of phase.

Light from most objects passes through the center of the lens as well as to the periphery. Now, if the light from an object to the edges of the objective lens is retarded a half wavelength and the light to the center is not retarded at all, then the light rays are out of phase by a half wavelength. They cancel each other when the objective lens brings the image into focus. A reduction in brightness of the object is observed. The degree of reduction in brightness depends on the refractive index of the object.

Dark Field Microscopy (Fig. 2.7)

Dark field microscopy is especially useful for finding cells in suspension. Dark field makes it easy to obtain the correct focal plane at low magnification for

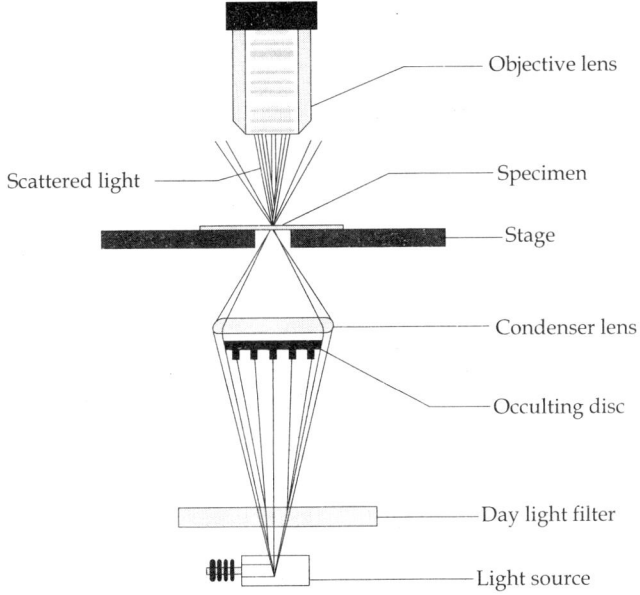

FIG. 2.7 Optical path of Dark field Microscopy

small, low contrast specimens. To view a specimen in dark field, an opaque disc is placed underneath the condenser lens, so that only light that is scattered by objects on the slide can reach the eye. Instead of coming up through the specimen, the light is reflected by particles on the slide. Everything is visible regardless of color, usually bright white against a dark background. Pigmented objects are often seen in 'false colors', that is, the color of reflected light is different than the color of the object. This type of light microscopy can give better resolution as opposed to bright field viewing.

Electron Microscopy

Electron microscopes use a beam of highly energetic electrons to examine objects on a very fine scale. The basic steps involved are:

1. A stream of electrons is formed by the electron source and accelerated toward the specimen using a positive electrical potential.
2. This stream is confined and focused using metal apertures and magnetic lenses into a thin, focused, monochromatic beam.
3. This beam is focused onto the sample using a magnetic lens.
4. Interactions occur inside the irradiated sample, affecting the electron beam. These interactions and effects are detected and transformed into an image.

Electron Microscope was developed due to the limitations of light microscope to 500x or 1000x magnification and a resolution of 0.2 μm. In the early 1930's this theoretical limit had been reached and there was a scientific desire to see the fine details of the interior structures of organic cells (nucleus, mitochondria, etc.). This required 10,000 × plus magnification which was not possible using light microscopes.

The Transmission Electron Microscope (TEM) was the first type of electron microscope developed. It is patterned exactly on the light transmission microscope except that a focused beam of electrons is used instead of light to see through the specimen. It was developed by Max Knoll and Ernst Ruska in Germany in 1931. Later, the Scanning Electron Microscope (SEM) was developed in 1942.

Transmission Electron Microscope (TEM) (Fig. 2.8)

A TEM works much like a slide projector. A projector transmits a beam of light through the slide, as the light passes through it is affected by the structures and objects on the slide. These effects result in only certain parts of the light

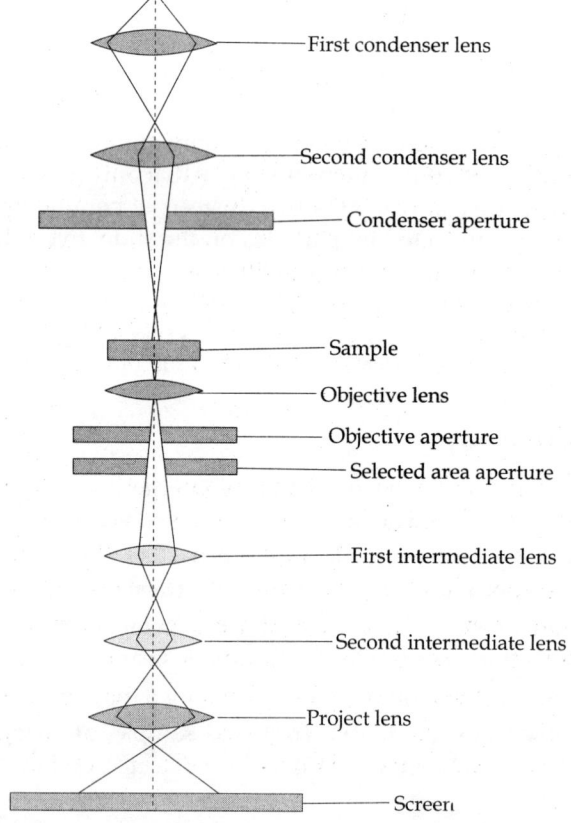

FIG. 2.8   Light traveling through the Transmission Electron Microscope

beam being transmitted through certain parts of the slide. This transmitted beam is then projected onto the viewing screen, forming an enlarged image of the slide.

TEMs work the same way except that they shine a beam of electrons (like the light) through the specimen (like the slide). Whatever part is transmitted is projected onto a phosphor screen for the user to see. A more technical explanation of a typical TEM working is as follows:

1. The virtual source at the top represents the electron gun, producing a stream of monochromatic electrons.
2. This stream is focused to a small, thin, coherent beam by the use of two condenser lenses. The first lens (usually controlled by the 'spot size knob') largely determines the 'spot size'; the general size range of the final spot that strikes the sample. The second lens (usually controlled by the 'intensity or brightness knob' actually changes the size of the spot on the sample; changing it from a wide dispersed spot to a pinpoint beam.
3. The beam is restricted by the condenser aperture (usually user selectable), knocking out high angle electrons.
4. The beam strikes the specimen and parts of it are transmitted.
5. This transmitted portion is focused by the objective lens into an image.
6. Optional objective and selected area metal apertures can restrict the beam; the objective aperture enhancing contrast by blocking out high-angle diffracted electrons, the selected area aperture enabling the user to examine the periodic diffraction of electrons by ordered arrangements of atoms in the sample.
7. The image is passed down the column through the intermediate and projector lenses, being enlarged all the way.
8. The image strikes the phosphor image screen and light is generated, allowing the user to see the image. The darker areas of the image represent those areas of the sample where fewer electrons were transmitted through (they are thicker or denser). The lighter areas of the image represent those areas of the sample where more electrons were transmitted through (they are thinner or less dense).

Scanning Electron Microscope (SEM) (Fig. 2.9)

The scanning electron microscope (SEM) scans a fine beam of electron onto a specimen and collects the electrons scattered by the surface. This has poorer resolution, but gives excellent 3-dimensional images of surfaces. A detailed explanation of working of a typical SEM is as follows:

1. The virtual source at the top represents the electron gun, producing a stream of monochromatic electrons.
2. The stream is condensed by the first condenser lens (usually controlled by the 'coarse probe current knob'). This lens is used to form the beam

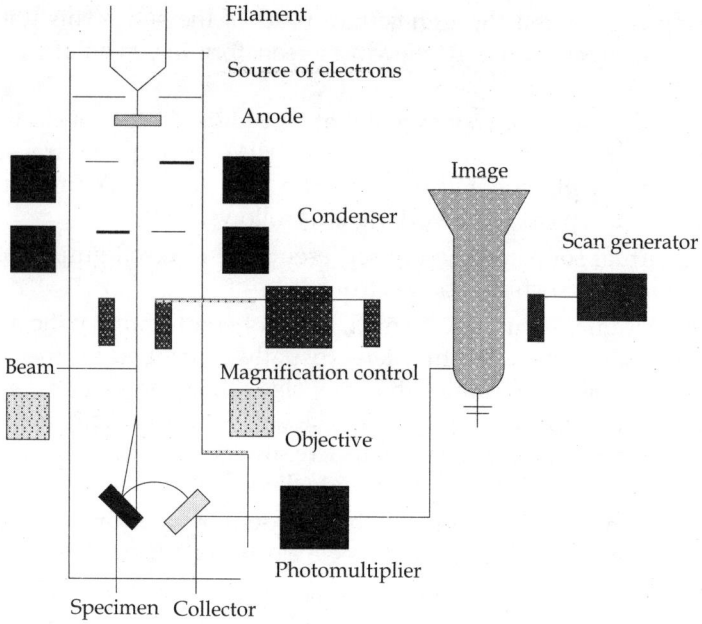

FIG. 2.9 Light traveling through the Scanning Electron Microscope

and limit the amount of current in the beam. It works in conjunction with the condenser aperture to eliminate the high-angle electrons from the beam.

3. The beam is then constricted by the condenser aperture (usually not user selectable), eliminating some high-angle electrons.

4. The second condenser lens forms the electrons into a thin, tight, coherent beam and is usually controlled by the 'fine probe current knob'.

5. A user selectable objective aperture further eliminates high-angle electrons from the beam.

6. A set of coils then 'scan' or 'sweep' the beam in a grid fashion (like a television), dwelling on points for a period of time determined by the scan speed (usually in the microsecond range).

7. The final lens, the objective, focuses the scanning beam onto the part of the specimen desired.

8. When the beam strikes the sample (and dwells for a few microseconds) interactions occur inside the sample and are detected with various instruments.

9. Before the beam moves to its next dwell point these instruments count the number of interactions and display a pixel on a CRT whose intensity is determined by this number (the more reactions the brighter the pixel).

10. This process is repeated until the grid scan is finished and then repeated, the entire pattern can be scanned 30 times per second.

Fluorescence Microscopy

Fluorescence microscopy is used to detect different structures, molecules or proteins within the cell. In fluorescence microscopy, the sample itself acts as the light source. This technique is used to study specimens, which can be made to fluoresce. The fluorescence microscope is based on the phenomenon that certain material emits energy detectable as visible light when irradiated with the light of a specific wavelength. The sample can either be fluorescing in its natural form like chlorophyll and some minerals, or treated with fluorescing chemicals. Fluorescent molecules absorb light at one wavelength and emit light at another, longer wavelength. When fluorescent molecules absorb a specific absorption wavelength for an electron in a given orbital, the electron rises to a higher energy level (the excited) state. Electrons in this state are unstable and will return to the ground state, releasing energy in the form of light and heat. This emission of energy is fluorescence. Because some energy is lost as heat, the emitted light contains less energy and therefore is a longer wavelength than the absorbed (or excitation) light.

In fluorescence microscopy, a cell is stained with a dye and the dye is illuminated with filtered light at the absorbing wavelength; the light emitted from the dye is viewed through a filter that allows only the emitted wavelength to be seen. The dye glows brightly against a dark background because only the emitted wavelength is allowed to reach the eyepieces or camera port of the microscope. Most microscopes are designed using epi-illumination. In epi-illumination excitation, light goes through the objective lens and illuminates the object. Light emitted from the specimen is collected by the same objective lens.

Fluorescence microscopy is a rapid expanding technique, both in the medical and biological sciences. The technique has made it possible to identify cells and cellular components with a high degree of specificity. For example, certain antibodies and disease conditions or impurities in inorganic material can be studied with the fluorescence microscopy.

AUTORADIOGRAPHY (FIG. 2.10)

It is a technique that uses X- ray film to visualize molecules or fragments of molecules that have been radioactively labeled by the incorporation of radioactive isotopes such as tritium (^{3}H), carbon-14 (^{14}C), phosphorus-32 (^{32}P) and sulphur-35 (^{35}S). An autoradiograph is an image produced on a photographic film by the radiation from a radioactive substance. In this technique, a compound (such as a metabolite or DNA) is radiolabeled by a substance that emits either beta radiation or gamma rays. A photographic film is then placed directly on the labeled tissue.

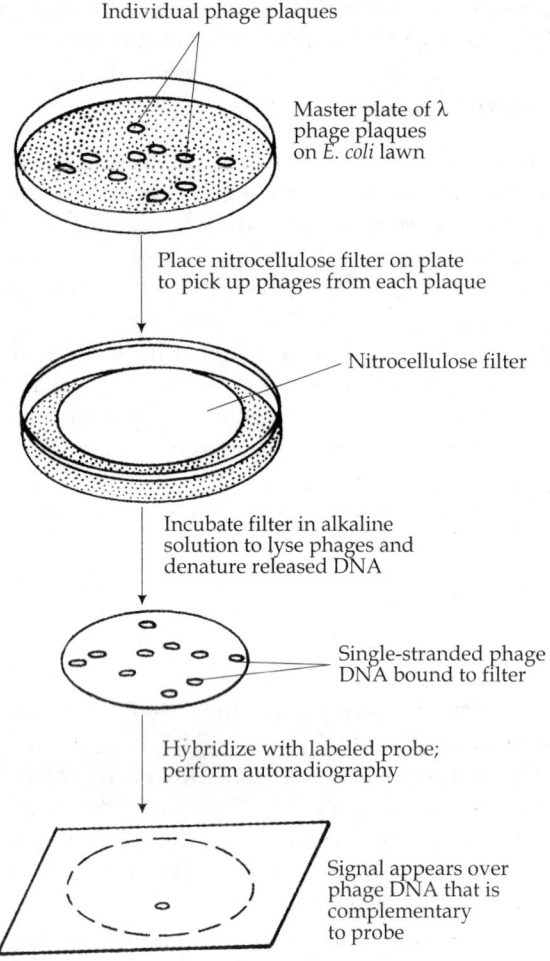

Individual phage plaques

Master plate of λ phage plaques on *E. coli* lawn

Place nitrocellulose filter on plate to pick up phages from each plaque

Nitrocellulose filter

Incubate filter in alkaline solution to lyse phages and denature released DNA

Single-stranded phage DNA bound to filter

Hybridize with labeled probe; perform autoradiography

Signal appears over phage DNA that is complementary to probe

FIG. 2.10 Autoradiography

Autoradiography has many applications in the laboratory, for example, it can be used to analyze the length and number of DNA fragments after they are separated from one another by a method called gel electrophoresis.

CELL FRACTIONATION (FIG. 2.11)

The eukaryotic cell contains many organelles, each of which performs one or more specialized functions; they are suspended within the cytoplasm and bounded by the plasma membrane. Each organelle has specific characteristics like size, shape and density which make it different from other organelles within the same cell. Because of these different features, individual types of organelles can be isolated from cells and studied. The technique of cell

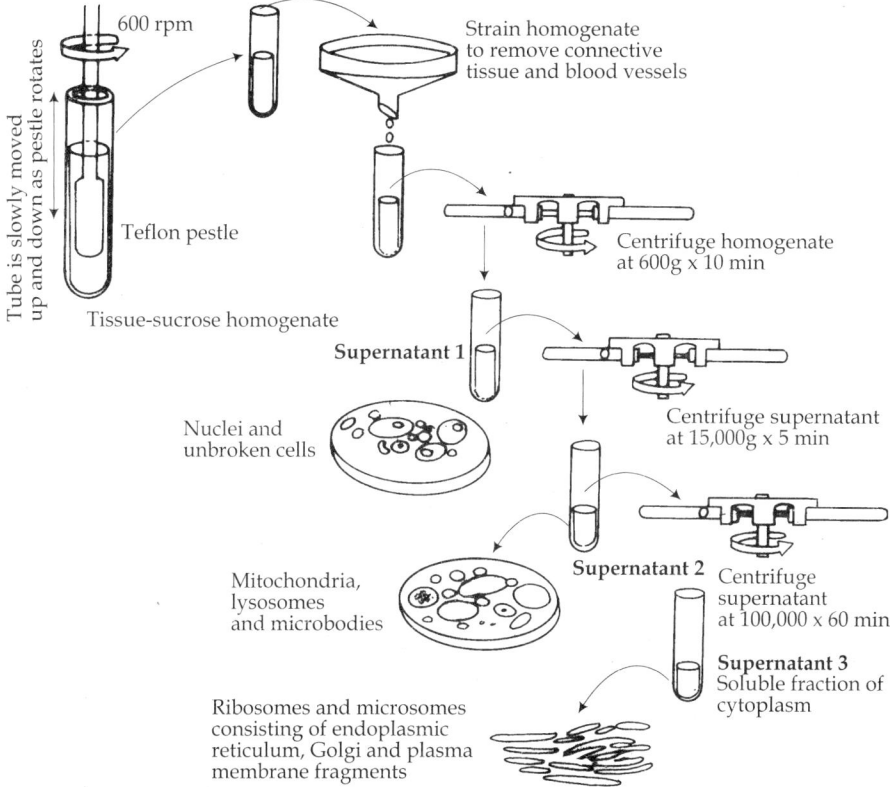

FIG. 2.11 Cell Fractionation

fractionation is employed in order to break up tissues and cells and isolate various organelles. The process of breaking the cells is called homogenization and the subsequent isolation of organelles is called fractionation. Isolation of organelles requires the use of physical chemistry techniques, and these techniques can range from the use of simple sieves, gravity sedimentation or differential precipitation, to ultracentrifugation of fluorescent labeled organelles in computer generated density gradients.

One technique for isolating organelles is to homogenize cells in a blender. Homogenization ruptures the cells freeing many of the organelles. To preserve the viability of the organelles, cells are homogenized in a phosphate buffered sucrose solution. Under carefully controlled conditions, organelles can continue to perform their functions outside of the cell for some time. In order to study specific organelles, homogenization is followed by some procedure that can isolate one type of organelle from the others. This technique is differential centrifugation, a process by which homogenized cells are centrifuged at increasingly higher speeds and for increasingly longer periods of time. Centrifugation tends to isolate the cellular components in order of density and, to some extent, size. The densest cell components and cell fragments settle out as a residue during the first centrifugation. Less dense organelles remain suspended in the buffer solution as the supernatant. With subsequent centrifugations, more and more organelles settle in layers in the residue, until only the least dense organelles remain in the supernatant.

Sedimentation Equilibrium

If a sample contains several different particles with different densities and sizes, the large particles will settle to the bottom of a tube faster than the smaller ones. If the relative centrifugal force is gradually increased, the time for the consequent separation of particles can be decreased.

By varying centrifugation force (speed) and time, while maintaining a continuous media density, different sizes of particles can be separated on the basis of their size. Large particles, such as whole cells and nuclei are sedimented at low speeds. Mitochondria and chloroplasts require higher speeds and/or longer times of centrifugation. Ribosomes require even greater forces and longer times. Thus, it is possible to design a protocol which first sediments large organelles, and then by increasing the centrifugation time or speed to sediment smaller particles from the same tube. This protocol is known as differential centrifugation, and the process makes use of both time and speed. Since the procedure sediments large organelles first, they are often contaminated by the smaller organelles which start at the bottom of the centrifuge tube.

At the beginning, the pellet area (bottom of tube) will contain both small and large randomly distributed organelles. As the centrifuge is run, larger particles move down the tube, but smaller ones do not move up; the process is based on sedimentation. As the process continues, the larger particles are removed and thus the smaller the particle, the purer the isolated fraction at later centrifugation steps. Of course, the smaller organelles are separated as both contaminants of the larger organelles, and as sediments of subsequent centrifugations. Thus, if you wish to maximize the collection of smaller organelles, or minimize the presence of smaller organelles in the large organelle fraction, it is necessary to re-centrifuge the larger fractions several times and to collect and pool the resulting smaller units.

It is possible, however, to sediment and float the particles simultaneously. If the particles to be separated have varying densities (gm/ml) they can be separated through a medium that allows particles of one density to 'float' and particles of higher density to sink to the bottom. Such media can be layered into the centrifuge tube, in step gradients or linear gradients.

In either gradient, the particles are centrifuged until they reach a density equal to the media, thus named as equilibrium density separations. This process has been greatly utilized in the analysis of molecular weights for proteins and nucleic acids, since to a great extent; the density of these molecules can be directly related to their size. Equilibrium density is also used to successfully isolate membranes and other high lipid-containing organelles.

CHROMATOGRAPHY

Chromatography is a separation method that is used to separate the molecules of different substances present in a solution. It involves passing the sample (the mobile phase) through a stationary phase which is an absorbing medium such as paper, gelatin, alumina or silica. The stationary phase retards the passage of the components in the sample. When components pass through the system at different rates they become separated in time. Each component has a characteristic time of passage through the system, called a 'retention time'.

Russian botanist, Mikhail Semyonovich Tsvet, invented the first chromatography technique in 1901 during his research on Chlorophyll. He used a liquid-adsorption column containing calcium carbonate to separate plant pigments. The technology of chromatography advanced rapidly throughout the 20th century. Researchers found that the principles underlying Tsvet's chromatography could be applied in many different ways, giving rise to the different varieties of chromatography, discussed below. Simultaneously, advances continually improved the technical performance of chromatography, allowing increasingly similar molecules to be resolved.

Paper Chromatography (Fig. 2.12)

In paper chromatography, compounds travel at different rates due to chemical interactions with the paper.

A small spot of solution containing the sample is applied to a strip of chromatography paper about one centimeter from the base. This sample is adsorbed onto the paper. This means that the sample will contact the paper

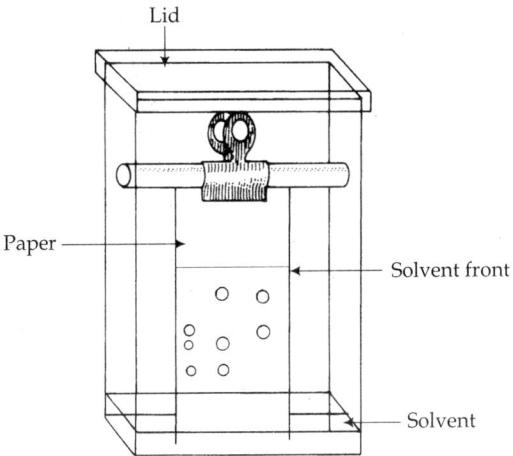

FIG. 2.12 Paper Chromatography

and may form interactions with it. Any substance that will react with (and thus bond to) the paper cannot be measured using this technique. The paper is then dipped into a suitable solvent (such as ethanol or water) and placed in a sealed container. As the solvent rises through the paper it meets the sample mixture which starts to travel up the paper with the solvent. Different compounds in the sample mixture travel different distances according to how strongly they interact with the paper. Paper chromatography takes some time and the experiment is usually left to complete for some hours.

The final chromatogram can be compared with other known mixture chromatograms to identify sample mixes. Two-way paper chromatography involves using two solvents and rotating the paper 90° in between. This is useful for separating complex mixtures of similar compounds.

Thin Layer Chromatography (TLC)

In thin layer chromatography or TLC the stationary phase consists of a thin layer of adsorbent like silica gel, alumina or cellulose on a flat carrier like a glass plate, a thick aluminum foil, or a plastic sheet. The process is similar to paper chromatography with the advantage of faster runs, better separations, and the choice between different adsorbents.

Column Chromatography (Fig. 2.13)

Column chromatography utilizes a vertical glass column filled with some form of solid support with the sample to be separated placed on top of this

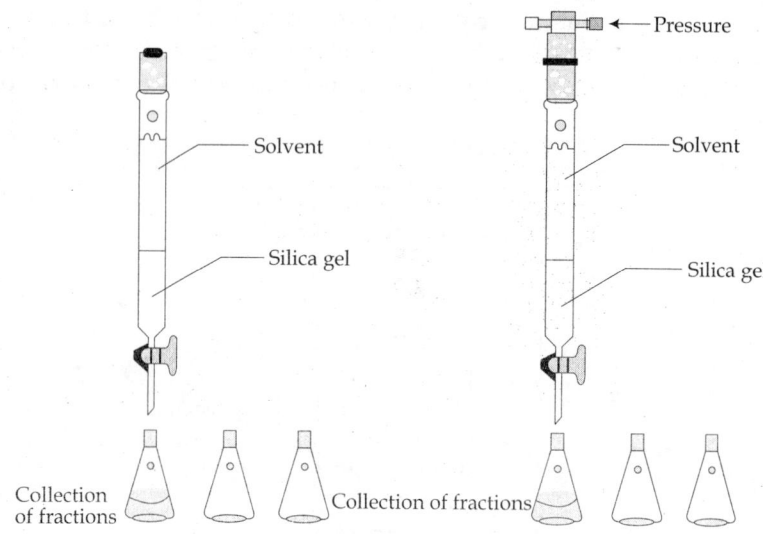

FIG. 2.13 Column Chromatography

support. The rest of the column is filled with a solvent which moves the sample through the column under the influence of gravity. Like other forms of chromatography, differences in rates of movement through the solid medium are translated to different exit times from the bottom of the column for the various elements of the original sample.

Gas-liquid Chromatography

Gas-liquid chromatography is based on a partition equilibrium of analyte between a liquid stationary phase and a mobile gas. It is useful for a wide range of non-polar analytes, but poor for thermally labile molecules.

Ion Exchange Chromatography (Fig. 2.14)

Ion exchange chromatography is a column chromatography that uses a charged stationary phase. It is used to separate charged compounds including

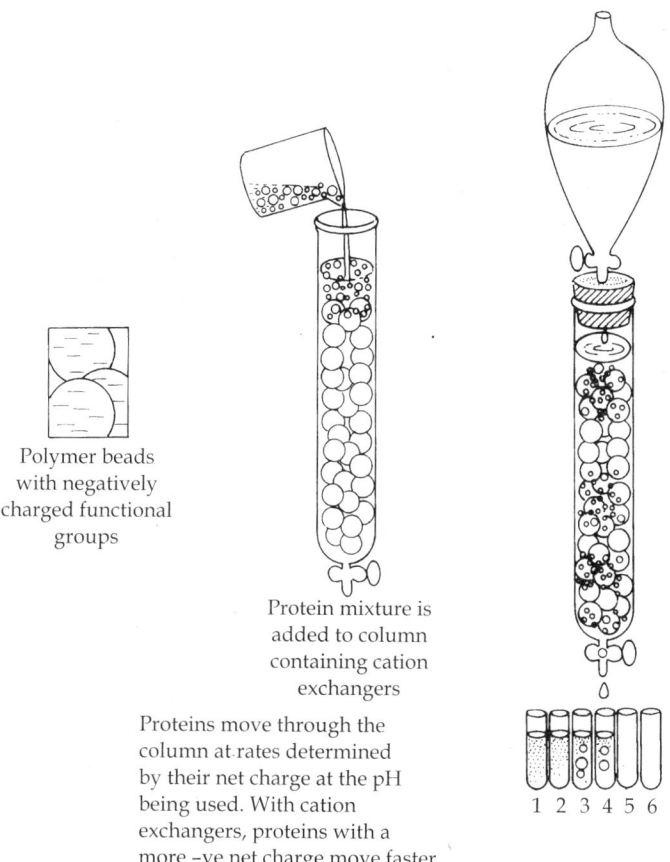

Polymer beads with negatively charged functional groups

Protein mixture is added to column containing cation exchangers

Proteins move through the column at rates determined by their net charge at the pH being used. With cation exchangers, proteins with a more –ve net charge move faster and elute earlier.

1 2 3 4 5 6

FIG. 2.14 Ion Exchange Chromatography

amino acids, peptides and proteins. The stationary phase is usually an ion exchange resin that carries charged functional groups which interact with oppositely charged groups of the compound to be retained:

Positively charged ion exchanger (*anion exchanger*) interacts with anions.

Negatively charged ion exchanger (*cation exchanger*) interacts with cations.

OBJECTIVE TYPE QUESTIONS

1. Who is regarded as the father of microscopy?
   - A. Janssen
   - B. Leeuwenhoek
   - C. Robert Hooke
   - D. Charles Spenser

2. _____ microscopy is useful for finding cells in suspension.
   - A. Light
   - B. Phase contrast
   - C. Dark field
   - D. Electron

3. Who developed the electron microscope?
   - A. Tsvet
   - B. Knoll and Ruska
   - C. Robert Hooke
   - D. Charles Spenser

4. The first chromatography technique was invented by _____
   - A. Tsvet
   - B. Janssen
   - C. Schwann
   - D. Charles Spenser

5. In ion exchange chromatography the anion exchanger interacts with _____
   - A. Cations
   - B. Anions
   - C. Both
   - D. None of the above

Answers

| 1. C | 2. C | 3. B | 4. A | 5. B |

SHORT ANSWER TYPE QUESTIONS

1. Who constructed the first advanced microscope in the late 17th and early 18th century?
2. Enlist the limitations of light microscopy.
3. Write a brief note on fluorescence microscopy.
4. Explain paper chromatography.
5. Define resolution.

LONG ANSWER TYPE QUESTIONS

1. Describe the parts of a light microscope.
2. Write a comparative account on transmission and scanning electron microscopy.

3. Write a note on phase contrast microscopy.
4. Explain chromatography and describe its different types.
5. Explain the process of cell fractionation.

Protoplasm

3

INTRODUCTION

The term protoplasm includes the substance present within the cell membrane and forms the 'living substance' of the cell. It can be differentiated into cytoplasm and the nucleoplasm. The term was coined by Purkinje in 1839. Huxley in 1868 defined protoplasm as 'the physical basis of life'.

PHYSICAL PROPERTIES OF PROTOPLASM

It is a transparent and jelly-like material. The consistency of protoplasm varies from slightly gelatinous white of a fresh egg to that of semi-solidified gelatin of jelly. If the protoplasm is more liquid it is termed as sol, but if it is more gelatinous, it is called a gel.

CHEMICAL PROPERTIES OF PROTOPLASM

The chemical nature of protoplasm can be divided into two categories of inorganic and organic substances.

Inorganic Substances

Inorganic substances are water, which make up 90% of the protoplasm, mineral salts, such as NaCl-salt, and gases like oxygen and carbon dioxide.

Organic Substances

Organic substances include carbohydrates, proteins, lipids, nucleic acids and enzymes.

CARBOHYDRATES

The term carbohydrate was originally used to describe compounds that were literally 'hydrates of carbon' because they had the empirical formula CH_2O. In recent years, carbohydrates have been classified on the basis of their structures. They are polyhydroxy aldehydes and polyhydroxy ketones. Among the compounds that belong to this family are cellulose, starch, glycogen, and most sugars.

There are three classes of carbohydrates: monosaccharides, disaccharides, and polysaccharides.

Monosaccharides

The monosaccharides are white, crystalline solids that contain a single aldehyde or ketone functional group. They are subdivided into two classes; aldoses and ketoses on the basis of whether they are aldehydes or ketones. They are also classified as a triose, tetrose, pentose, hexose, or heptose on the basis of whether they contain three, four, five, six, or seven carbon atoms.

If the monosaccharide has an aldehyde group $-\overset{\overset{\text{O}}{\|}}{\text{C}}-\text{H}$, it is called an aldose, e.g., glyceraldehyde (a triose), erythrose (a tetrose), ribose (a pentose), glucose (a hexose). If a keto group C=O is present in a monosaccharide, it is called a ketose, eg., dihydroxy acetone (a triose), erythrulose (a tetrose), ribulose (a pentose) and fructose (a hexose). Each of the carbon atoms of the chain is assigned a number beginning with the end that is closest to the carbon bearing the double-bonded oxygen. In the aldoses, the number 1 carbon forms the double bond with oxygen, whereas in the ketoses, it is usually the penultimate carbon.

| D-Erythrose | D-Ribose | D-Glucose | D-Galactose |

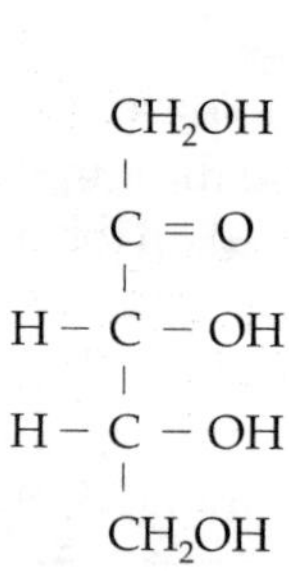

$$
\begin{array}{c}
CH_2OH \\
| \\
C = O \\
| \\
H - C - OH \\
| \\
H - C - OH \\
| \\
CH_2OH
\end{array}
$$

D-Ribulose

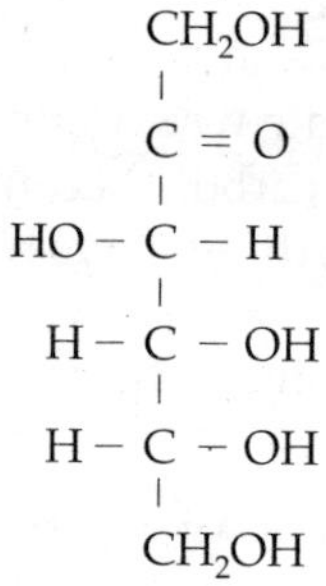

$$
\begin{array}{c}
CH_2OH \\
| \\
C = O \\
| \\
HO - C - H \\
| \\
H - C - OH \\
| \\
H - C - OH \\
| \\
CH_2OH
\end{array}
$$

D-Frutose

Chain and ring forms

Many simple sugars can exist in a chain form or a ring form. The ring form is favored in aqueous solutions, and the mechanism of ring formation is similar for most sugars. The glucose ring form is created when the oxygen on carbon number 5 links with the carbon comprising the carbonyl group (carbon number 1) and transfers its hydrogen to the carbonyl oxygen to create a hydroxyl group. The rearrangement produces alpha (α) glucose when the hydroxyl group is on the opposite side of the **-CH$_2$OH** group, or beta (β) glucose when the hydroxyl group is on the same side as the **-CH$_2$OH** group. Isomers, such as these, which differ only in their configuration about their carbonyl carbon atom are called anomers. The 'D' in the name derives from the fact that natural glucose is dextrorotatory, that is, it rotates polarized light to the right. Monosaccharides forming a five-sided ring, like ribose, are called furanoses. Those forming six-sided rings, like glucose, are called pyranoses. Thus, monosaccharides are optically active compounds. Although both D (dextrorotatory) and L (levorotatory) isomers are possible, most of the monosaccharides found in nature are in the D configuration.

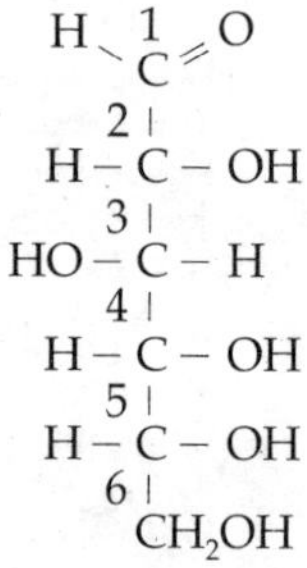

D-Glucose
(an aldose)

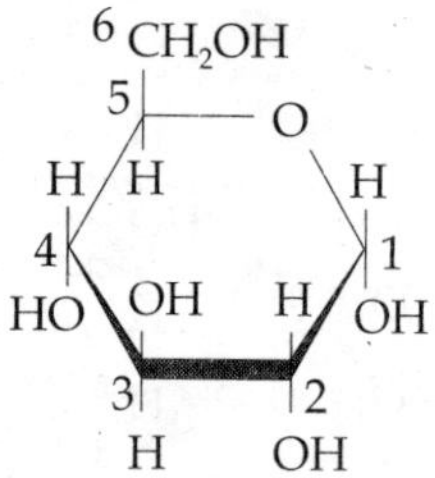

α-D-Glucose

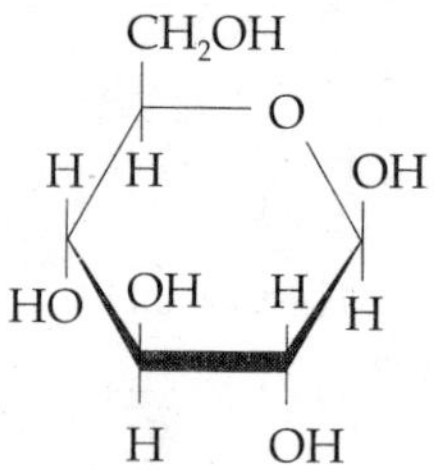

β-D-Glucose

Stereochemistry

Saccharides with identical functional groups but with different spatial configurations have different chemical and biological properties. Stereochemisty is the study of the arrangement of atoms in three-dimensional space. Stereoisomers are compounds in which the atoms are linked in the same order but differ in their spatial arrangement. Compounds that are mirror images of each other but are not identical are called *enantiomers*. The following structures illustrate the difference between β-D-Glucose and β-L-Glucose. Identical molecules can be made to correspond to each other by flipping and rotating. However, enantiomers cannot be made to correspond to their mirror images by flipping and rotating. Glucose is sometimes illustrated as a 'chair form' because it is a more accurate representation of the bond angles of the molecule. The 'boat' form of glucose is unstable.

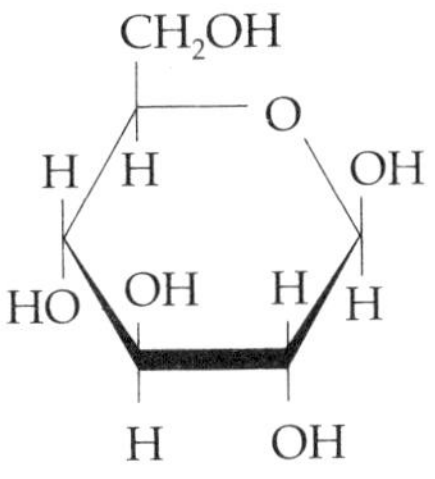

β-D-Glucose 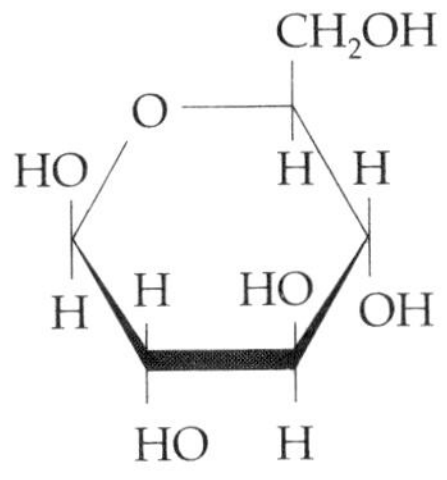   β-L-Glucose 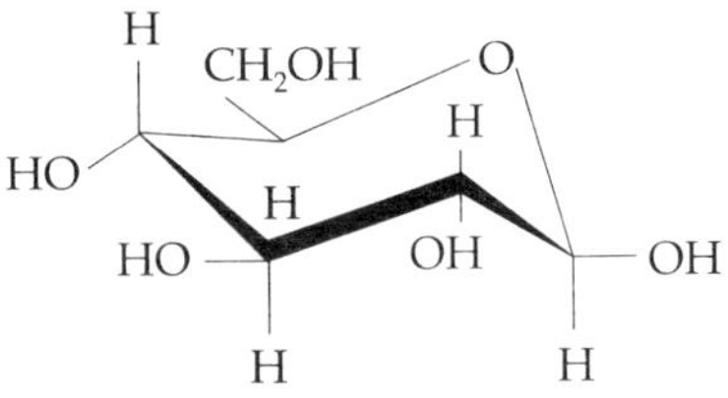   β-D-Glucose (chair form)

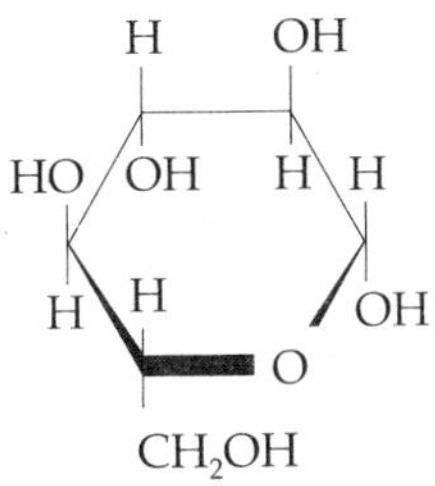

β-D-Glucose 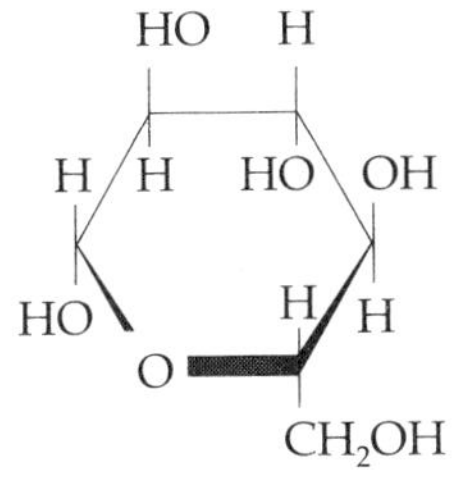   β-L-Glucose 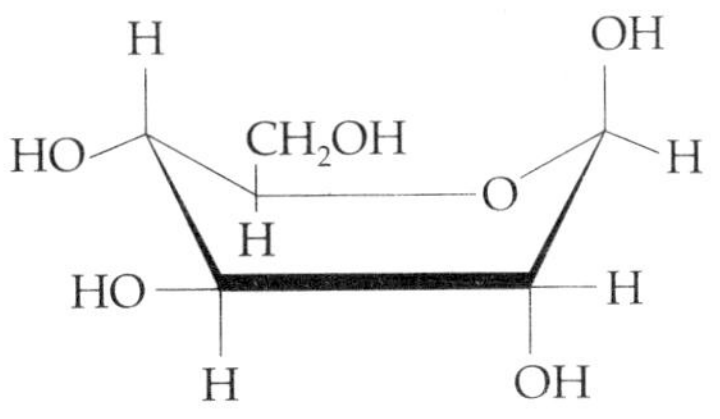   β-D-Glucose (boat form)

Disaccharides

Carbohydrates that can be hydrolyzed into two monosaccharides are called disaccharides, for example, sucrose, maltose, lactose.

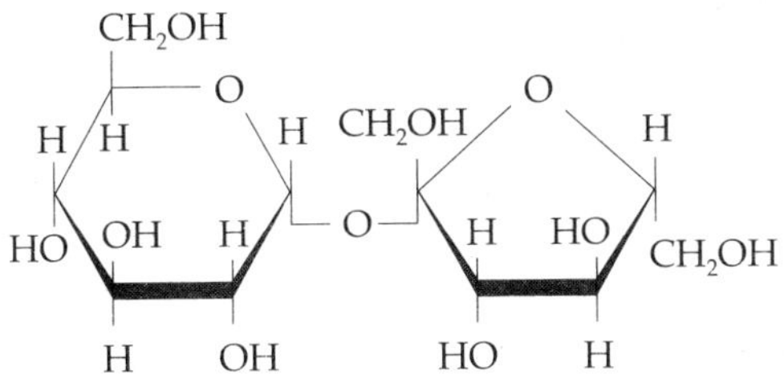

Sucrose

Lactose   Maltose

The common disaccharides have the general formula, $C_{12}H_{22}O_{11}$ and yield hexoses on hydrolysis:

$$C_{12}H_{22}O_{11} \quad + \quad H_2O \quad \rightarrow \quad 2C_6H_{12}O_6$$

The products of common disaccharides are:

Maltose + H_2O =   glucose + glucose
Lactose + H_2O =   glucose + galactose
Sucrose + H_2O =   glucose + fructose
Maltose + H_2O =   glucose + glucose

Sucrose, also called saccharose, is an ordinary table sugar refined from sugar cane or sugar beets. It is the main ingredient in turbinado sugar, evaporated or dried cane juice, brown sugar, and confectioner's sugar. **Lactose** has a molecular structure consisting of galactose and glucose. It is of interest because it is associated with **lactose intolerance** which is the intestinal distress caused by a deficiency of lactase, an intestinal enzyme needed to absorb and digest lactose in milk. Undigested lactose ferments in the colon and causes abdominal pain, bloating, gas, and diarrhea. Yogurt does not cause these problems because lactose is consumed by the bacteria that transform milk into yogurt. **Maltose** consists of two α-D-glucose molecules with the alpha bond at carbon 1 of one molecule attached to the oxygen at carbon 4 of the second molecule. This is called a 1α→4 glycosidic linkage. **Trehalose** has two α-D-glucose molecules connected through carbon number one in a 1α→1 linkage. **Cellobiose** is a disaccharide consisting of two β-D-glucose molecules that have a 1β→4 linkage as in cellulose. Cellobiose has no taste, whereas maltose and trehalose are about one-third as sweet as sucrose.

Trisaccharides

Raffinose, also called melitose, is a trisaccharide that is found in beans, cabbage, brussels sprouts, and broccoli. It consists of galactose connected to sucrose via a 1α→6 glycosidic linkage. Humans cannot digest this saccharide and it is fermented in the large intestine by gas-producing bacteria.

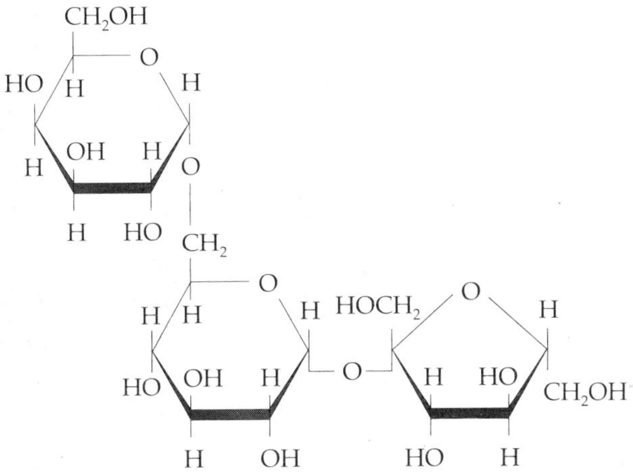

Raffinose

Polysaccharides

The carbohydrates that yield ten to many thousand monosaccharides upon hydrolysis are called polysaccharides. Their empirical formula is $(C_6H_{10}O_6)_n$. Polysaccharides can be divided into two groups:

Structural polysaccharides

These serve as extracellular or intracellular supporting elements, such as cellulose which is found in plant cell wall, mannan, a homopolymer of mannose found in yeast cell walls, chitin, present in the exoskeleton of arthropods, pectin, cementing material in the cell walls of all plant tissues, hyaluronic acid, found in connective tissues and peptidoglycans which are present in bacterial cell wall.

Nutrient polysaccharides

These serve as reserves of monosaccharides and are in continuous metabolic turnover. They include starch, glycogen, paramylum, inulin.

Further, polysaccharides can be classified as homo-polysaccharides (in which all the constituent sugars are same) and hetero-polysaccharides (in which constituent sugars are different).

Some important polysaccharides are described below:

Starch

Starch is the major form of stored carbohydrate in plants. Starch is composed of a mixture of two substances: amylose, an essentially linear polysaccharide, and amylopectin, a highly branched polysaccharide. Natural starches contain 10-20% amylose and 80-90% amylopectin. Amylose forms a colloidal dispersion in hot water whereas amylopectin is completely insoluble.

Amylose molecules consist typically of 200 to 20,000 glucose units which form a helix as a result of the bond angles between the glucose units.

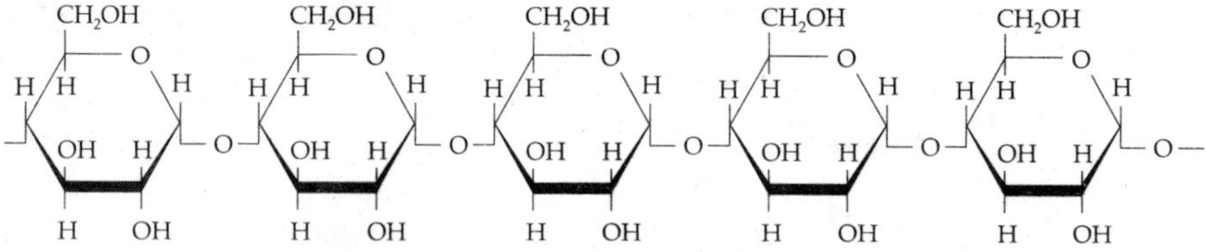

Amylose

Amylopectin differs from amylose in being highly branched. Short side chains of about 30 glucose units are attached with $1\alpha\rightarrow6$ linkages approximately every twenty to thirty glucose units along the chain. Amylopectin molecules may contain up to two million glucose units.

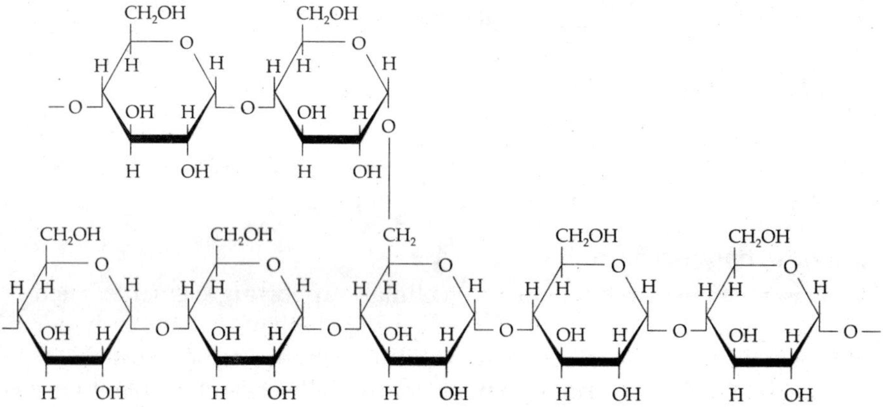

Amylopectin

Starches are transformed into many commercial products by hydrolysis using acids or enzymes as catalysts. Hydrolysis is a chemical reaction in which water is used to break long polysaccharide chains into smaller chains or into simple carbohydrates.

Glycogen

Glucose is stored as glycogen in animal tissues by the process of glycogenesis. When glucose cannot be stored as glycogen or used immediately for energy, it gets converted into fat. Glycogen is a polymer of α-D-Glucose identical to amylopectin, but the branches in glycogen tend to be shorter (about 13 glucose units) and more frequent. The glucose chains are organized globularly like branches of a tree originating from a pair of molecules of glycogenin, a protein with molecular weight of 38,000 that acts as a primer at the core of the structure. Glycogen is easily converted back to glucose to provide energy.

Dextran

Dextran is a polysaccharide similar to amylopectin, but the main chains are formed by 1α→6 glycosidic linkages and the side branches are attached by 1α→3 or 1α→4 linkages. Dextran is an oral bacterial product that adheres to the teeth, creating a film called plaque. It is also used commercially in confections, in lacquers, as food additive, and as plasma volume expander.

Dextran

Cellulose

Cellulose is a polymer of β-D-Glucose, which in contrast to starch, is oriented with $-CH_2OH$ groups alternating above and below the plane of the cellulose molecule thus producing long, unbranched chains. The absence of side chains allows cellulose molecules to lie close together and form rigid structures. Cellulose is the major structural material of plants. Wood is largely cellulose, and cotton is almost pure cellulose. Cellulose can be hydrolyzed to its constituent glucose units by microorganisms that inhabit the digestive tract of termites and ruminants. Cellulose may be modified in the laboratory by treating it with nitric acid (HNO_3) to replace all the hydroxyl groups with nitrate groups ($-ONO_2$) to produce cellulose nitrate (nitrocellulose or guncotton) which is an explosive component of smokeless powder. Partially nitrated cellulose, known as pyroxylin, is used in the manufacture of collodion, plastics, lacquers, and nail polish.

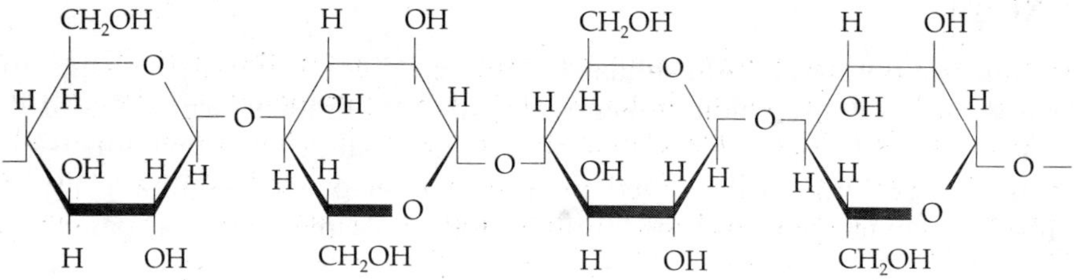

Cellulose

Inulin

Some plants store carbohydrates in the form of inulin as an alternative, or in addition, to starch. Inulins are polymers consisting of fructose units that typically have a terminal glucose. Inulins have a sweet taste and are present in many vegetables and fruits, including onions, garlic, bananas, etc.

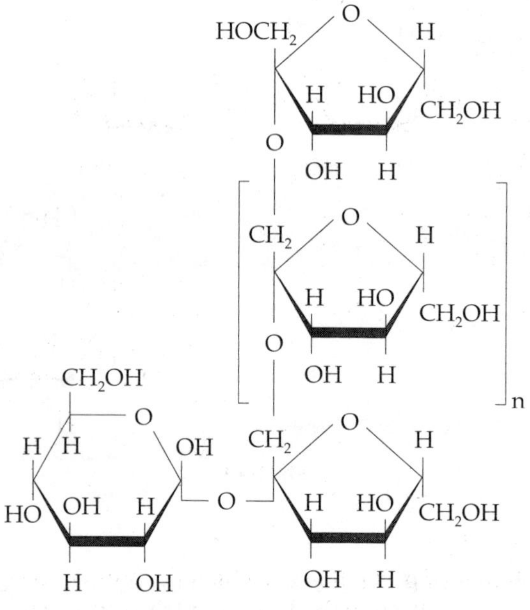

Inulin n = approx. 35

Chitin

Chitin is an unbranched polymer of N-Acetyl-D-glucosamine. It is found in fungi and is the principal component of arthropod and lower animal exoskeletons, for example, insect, crab, and shrimp shells. It may be regarded as a derivative of cellulose, in which the hydroxyl groups of the second carbon of each glucose unit have been replaced with acetamido (-NH(C=O)CH$_3$) groups.

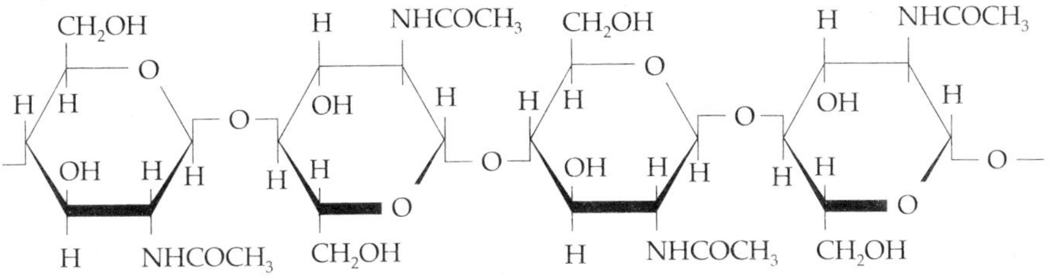

Chitin

PROTEINS

A protein (in Greek proteios = first element) is a complex, high molecular weight organic compound that consists of amino acids joined by peptide bonds. The term protein was coined by Mulder in 1840. Proteins are essential to the structure and function of all living cells and viruses and are the most abundant class of all biological molecules, comprising about 50 percent of cellular dry weight. They include enzymes, hormones, and antibodies. Certain food such as meat, fish, eggs, and beans are good sources of protein, necessary for the growth and repair of human tissue.

There are 20 different amino acids that make up essentially all proteins on earth. Each of these amino acids has a fundamental structure composed of a central carbon (also called the alpha carbon) bonded to:

- a hydrogen
- a carboxyl group
- an amino group
- a unique side chain or R-group

Thus, the characteristic that distinguishes one amino acid from another is its unique side chain, and it is the side chain that decides the chemical properties of amino acids.

All amino acids have the same general formula:

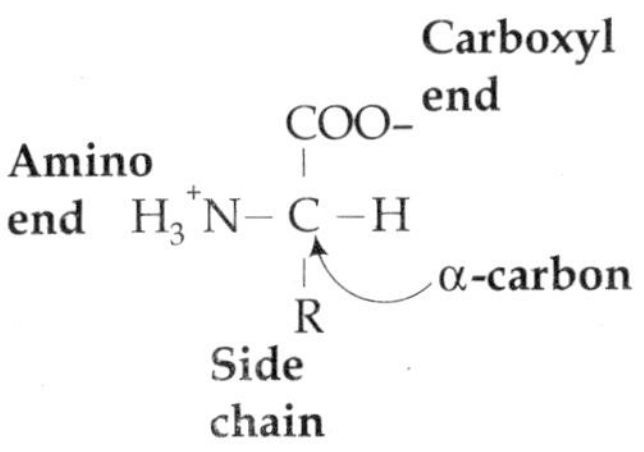

Except for glycine, which has hydrogen as its R-group, there is asymmetry about the alpha (α) carbon in all amino acids. Because of this, all amino acids except glycine can exist in either of two mirror-image forms. The two forms — called stereoisomers are referred to as D and L amino acids. With rare exceptions, all of the amino acids in proteins are L amino acids.

Naturally occurring amino acids, their abbreviations and structural formulae

Ala = alanine $CH_3CH(NH_2)COOH$	Arg = arginine $H_2N-(=NH)NHCH_2CH_2CH_2CH(NH_2)COOH$
Asn = asparagines $H_2N-C(=O)CH_2CH(NH_2)COOH$	Asp = aspartic acid $HOOC-CH_2CH(NH_2)COOH$
Cys = cysteine $HS\text{-}CH_2CH(NH_2)COOH$	Gln = glutamine $H_2N-C(=O)CH_2CH_2CH(NH_2)COOH$
Glu = glutamic acid $HOOC\text{-}CH_2CH_2CH(NH_2)COOH$	Gly = glycine $H_2N\text{-}CH_2COOH$
His = histidine* 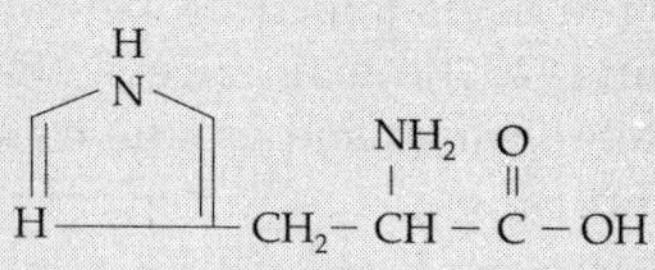	Phe = phenylalanine* 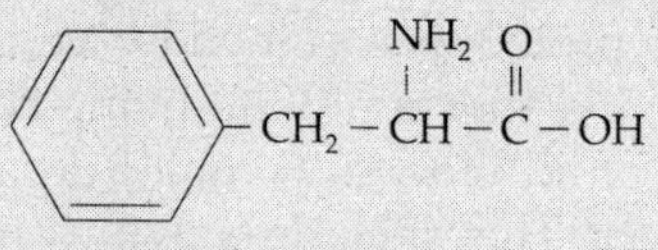

Ile = isoleucine* $CH_3CH_2CH(CH_3)CH(NH_2)COOH$	Lys = lysine * $H_2N-CH_2CH_2CH_2CH_2CH(NH_2)COOH$
Leu = leucine * $CH_3CH(CH_3)CH_2CH(NH_2)COOH$	Met = methionine* $CH_3-S-CH_2CH_2CH(NH_2)COOH$
Pro = proline 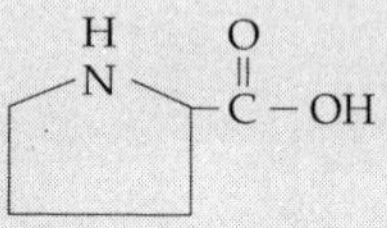	Trp = tryptophan* 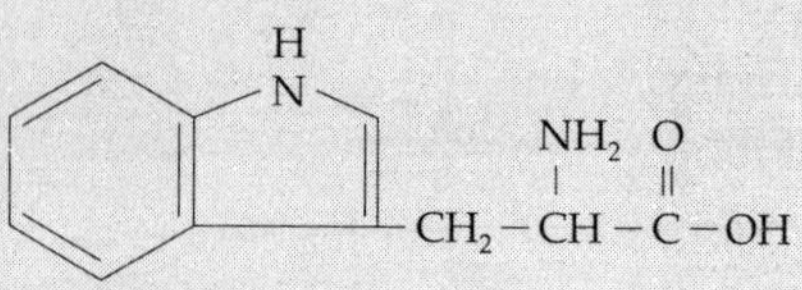

Ser = serine $HOCH_2CH(NH_2)COOH$	Thr = threonine* $CH_3CH(OH)CH(NH_2)COOH$
Tyr = tyrosine	Val = valine*

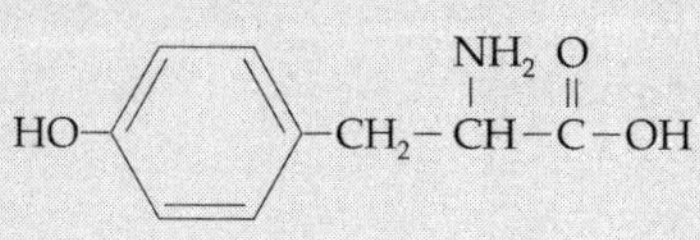

$CH_3CH(CH_3)CH(NH_2)COOH$

* Essential amino acids

The term 'essential amino acid' refers to an amino acid that is required to meet physiological needs and must be supplied in the diet. Arginine is synthesized by the body, but at a rate that is insufficient to meet growth needs. Methionine is required in large amounts to produce cysteine if the latter amino acid is not adequately supplied in the diet. Similarly, phenylalanine can be converted to tyrosine, but is required in large quantities when the diet is deficient in tyrosine. Tyrosine is essential for people with the disease phenylketonuria (PKU) whose metabolism cannot convert phenylalanine to tyrosine. Isoleucine, leucine, and valine are sometimes called 'branched-chain amino acids' because their carbon chains are branched.

Peptides and Proteins

Amino acids are covalently bonded together in chains by peptide bonds. If the chain length is short (say less than 30 amino acids) it is called a peptide; longer chains are called polypeptides or proteins. Peptide bonds are formed between the carboxyl group of one amino acid and the amino group of the next amino acid. Peptide bond formation occurs in a condensation reaction involving loss of a molecule of water.

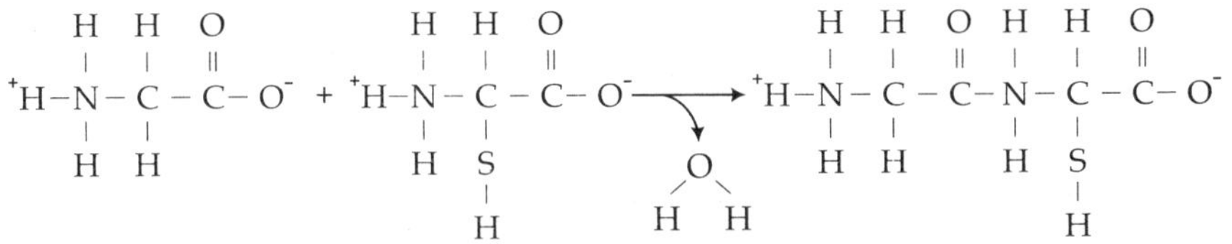

$$
\overset{H}{\underset{H}{\overset{|}{\underset{|}{H-N}}}} - \overset{H}{\underset{H}{\overset{|}{\underset{|}{C}}}} - \overset{O}{\overset{\|}{C}} - O^- \;+\; \overset{H}{\underset{H}{\overset{|}{\underset{|}{{}^+H-N}}}} - \overset{H}{\underset{S}{\overset{|}{\underset{|}{C}}}} - \overset{O}{\overset{\|}{C}} - O^- \longrightarrow \;\; {}^+H-N-C-C-N-C-C-O^-
$$

The head-to-tail arrangment of amino acids in a protein means that there is an amino group on one end (called the amino-terminus or N-terminus) and a carboxyl group on the other end (carboxyl-terminus or C-terminus).

Levels of Protein Structure

Structural features of proteins are usually described at four levels of complexity:

Primary structure

It is the linear arrangment of amino acids in a protein and the location of covalent linkages such as disulfide bonds between amino acids. For example, the linkages that may exist between the sulphur atoms of cystein amino acids located in the chain of protein insulin.

Secondary structure

It is formed when the simple long polymeric chains of protein molecules show areas of folding or coiling; examples include alpha helices and beta pleated sheets, which are stabilized by hydrogen bonding.

The Alpha Helix (Fig. 3.1)

The alpha helix is a periodic structure formed when main-chain atoms from residues spaced four residues apart hydrogen bond with one another. This gives rise to a helical structure, which in natural proteins is always right-handed. Each turn of the helix comprises 3.6 amino acids. Alpha helices are stiff, rod-like structures which are found in many unrelated proteins.

The amino acids in an α-helix are arranged in a helical structure, about 5Å wide. Each amino acid results in a 100° turn in the helix, and corresponds to a translation of 1.5Å along the helical axis. The helix is tightly packed; there is almost no free space within the helix. All amino acid side-chains are arranged at the outside of the helix. The N-H group of amino acid (n) can establish a hydrogen bond with the C=O group of amino acid (n+4).

Ordinarily, a helix has a buildup of positive charge at the N-terminal end and negative charge at the C-terminal end which is a destabilizing influence. As a result, α-helices are often capped at the N-terminal end by a negatively charged amino acid (like glutamic acid) in order to stabilise the helix dipole. Less common (and less effective) is C-terminal capping with a positively charged protein like lysine.

The Beta Sheet (Fig. 3.1)

The β sheet (also β-pleated sheet) is a commonly occurring form of regular secondary structure in proteins, first proposed by Linus Pauling and Robert Corey in 1951. It consists of two or more amino acid sequences within the same protein that are arranged adjacently and in parallel, but with alternating orientation such that hydrogen bonds can form between the two strands. The amino acid chain is almost fully extended throughout a β strand. The N-H groups in the backbone of one strand establish hydrogen bonds with the C=O

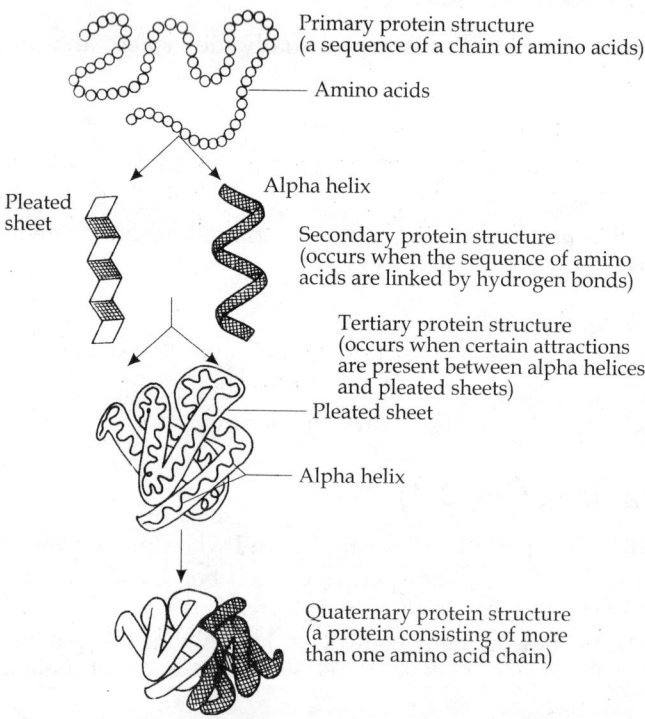

Primary protein structure
(a sequence of a chain of amino acids)

Amino acids

Alpha helix

Pleated sheet

Secondary protein structure
(occurs when the sequence of amino acids are linked by hydrogen bonds)

Tertiary protein structure
(occurs when certain attractions are present between alpha helices and pleated sheets)

Pleated sheet

Alpha helix

Quaternary protein structure
(a protein consisting of more than one amino acid chain)

FIG. 3.1 Levels of protein structure

groups in the backbone of the adjacent, parallel strand(s). The cumulative effect of such hydrogen bonds arranged in this way contributes to the sheet's stability and structural rigidity and integrity. The α-C atoms of adjacent strands stand 3.5Å apart.

The side chains from the amino acid residues found in a β sheet structure may also be arranged such that many of the adjacent sidechains on one side of the sheet are hydrophobic, while many of those adjacent to each other on the alternate side of the sheet are polar or charged (hydrophilic).

Some sequences involved in a β sheet, when traced along the backbone, take a *hairpin turn* in orientation (direction), sometimes through one or more prolines.

Tertiary structure

It is the name given to refer the overall shape of a single protein molecule. Although tertiary structure is sometimes described as being a result of interactions between amino acid residue side chains, a more correct understanding of tertiary structure is the interactions between elements of secondary protein structure, i.e. alpha-helices and beta-pleated sheets. Tertiary structure is often referred to as the 'fold structure' of a protein, since it is the result of the complex three-dimensional interplay of other structural and environmental elements. In globular proteins such as haemoglobin and myoglobin, the non-helical regions are engaged in folding by a variety of interactions between one part of the polypeptide chain and another and also between the polypeptide and neighbouring water molecules. Such interactions include a) ionic bonds, b) hydrogen bonds, c) hydrophobic bonds, and d) disulphide bonds. (Fig. 3.1)

Quaternary structure

It is the shape or structure that results from the union of more than one protein molecule, usually called protein subunits in this context, which function as part of the larger assembly or protein complex. Many globular proteins are made up of several polypeptide chains called sub-units stuck to each other by a variety of attractive forces but rarely by covalent bonds. The sub-units may be identical or different; an example is haemoglobin, a globular protein made up of four subunits. It is when polypeptide sub-units join together in large numbers they form supramolecular assemblies. (Fig. 3.1)

Protein Folding

Proteins are amino acid chains that acquire their biological and biochemical properties by fold into unique 3-dimensional structures. The shape into which a protein naturally folds is known as its native state, which is, in most cases, determined by its sequence of amino acids only. Protein folding is commonly a fast or very fast process, often but not always reversible, taking no more than a few milliseconds to occur. But this does not mean that it is a simple process. It can be viewed as a complex process between the different chemical interactions that can happen between the amino acid sidechains, the amidic backbone and the solvent. There are literally millions of possible three-dimensional configurations, often with minimal energetic differences between them.

Changing the Shape of Protein Molecules

Small changes to the shape of a protein can have a large effect on the way the protein behaves. Proteins, especially globular proteins in solution, may change shape in response to changes in their surroundings, such as changes in:

- pH
- temperature
- polarity of the solvent
- concentration of ions or molecules that can stick to the protein.

Small changes in pH can add or remove H^+ ions from side chain groups on the surface of a protein, without causing any permanent damage to the conformation. At a certain pH, called the isoelectric point, the protein molecule will have no overall ionic charge and so will have its minimum solubility in water. Different proteins have different isoelectric points. Chemists can use this to precipitate one protein from a mixture of proteins in solution by adjusting the pH to the isoelectric point of that protein.

Urea (NH_2CONH_2) is one of several small molecules which, at high concentrations, can weaken the non-covalent forces keeping the secondary and tertiary structure intact. The protein in this state has none of its original biological properties. If the urea is removed, denaturation may be reversed as the protein can slowly coil and fold back into its original conformation, regaining all its biological properties.

Nomenclature

Classes of proteins

Based on solubility of proteins there are two classes, viz., Simple and Complex

SIMPLE PROTEINS

The simple proteins include:

1.	Albumins	soluble in water (distilled), globular, most enzymes
2.	Globulins	soluble in dilute aqueous solutions; insoluble in pure distilled water
3.	Prolamines	insoluble in water; soluble in 50% to 90% simple alcohols
4.	Glutelins	insoluble in most solvents; soluble in dilute acids/bases
5.	Protamines	not based upon solubility; includes small molecular weight proteins with 80% Arginine and no Cysteine

6. Histones   90% Arg, Lys, or His

7. Scleroproteins   insoluble in most solvents, fibrous structure cartilage and connective tissue

COMPLEX PROTEINS

The complex proteins include:
1. Lipoproteins   (Proteins + lipids) blood, membrane and transport proteins
2. Glycoproteins   (Proteins + carbohydrates) antibodies, cell surface proteins
3. Nucleoproteins   (Proteins + nucleic acids) ribosomes and organelles

Classification of Proteins Based on Shape

Based on shape, proteins can be classified into two types, viz., Fibrous and Globular.

1. **Fibrous proteins:** These are thread like proteins having secondary protein structure. They are water soluble and occur in structures such as collagen, elastin, keratin, fibrin and myosin.

2. **Globular proteins:** These have no systematic structures. There may be single chains, two or more chains which interact in the usual ways or there may be portions of the chains with helical structures, pleated structures, or completely random structures. Globular proteins are relatively spherical in shape as the name implies. Common globular proteins include egg albumin, hemoglobin, myoglobin, insulin, serum globulins in blood, and many enzymes.

Haemoglobin (Fig. 3.2(a))

Haemoglobin is a heterotetramer, consisting of two alpha subunits and two beta subunits, the alpha subunits (141 residues in human hemoglobin) and

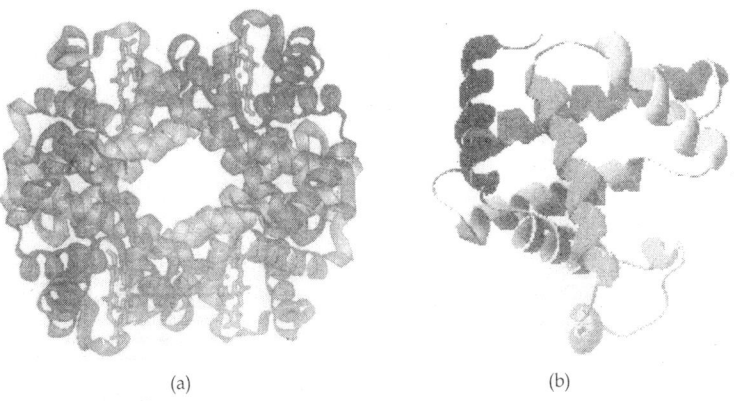

(a)   (b)

FIG. 3.2 (a) 3-D Structure of Hemoglobin (b) 3-D Structure of Myoglobin

beta subunits (146) being homologous but the alpha chains have one fewer helix than the beta chains. This form is called hemoglobin A.

Hemoglobin is found in erythrocytes where it is responsible for binding oxygen in the lung and transporting the bound oxygen throughout the body where it is used in aerobic metabolic pathways. Each subunit of a hemoglobin tetramer has a heme prosthetic group (Fig. 3.3). The common peptide subunits are designated a, b, g and d which are arranged into the most commonly occurring functional hemoglobins.

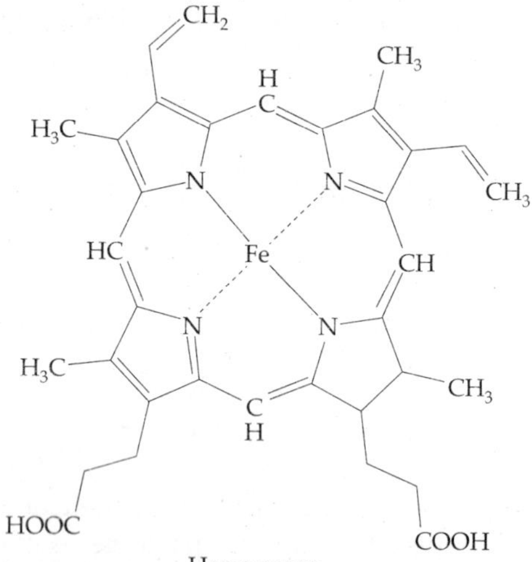

Heme group

FIG. 3.3 Heme Group

Although the secondary and tertiary structure of various hemoglobin subunits are similar, reflecting extensive homology in amino acid composition, the variations in amino acid composition that do exist impart marked differences in oxygen carrying properties of haemoglobin. In addition, the quaternary structure of haemoglobin leads to physiologically important allosteric interactions between the subunits, a property lacking in monomeric myoglobin which is otherwise very similar to the β-subunit of haemoglobin.

Myoglobin (Fig. 3.2(b))

Myoglobin and hemoglobin are hemeproteins whose physiological importance is principally related to their ability to bind molecular oxygen. Myoglobin is a monomeric heme protein found mainly in muscle tissue where it serves as an intracellular storage site for oxygen. During periods of oxygen deprivation oxymyoglobin releases its bound oxygen which is then used for metabolic purposes.

The tertiary structure of myoglobin is that of a typical water soluble globular protein. Its secondary structure is unusual in that it contains a very high proportion (75%) of α-helical secondary structure. A myoglobin polypeptide is comprised of 8 separate right handed α-helices, designated A through H that is connected by short non-helical regions. Amino acid R-groups packed into the interior of the molecule are predominantly hydrophobic in character while those exposed on the surface of the molecule are generally hydrophilic, thus making the molecule relatively water soluble.

Each myoglobin molecule contains one heme prosthetic group inserted into a hydrophobic cleft in the protein. Each heme residue contains one central coordinately bound iron atom that is normally in the Fe^{2+}, or ferrous, oxidation state. The oxygen carried by hemeproteins is bound directly to the ferrous iron atom of the heme prosthetic group. Oxidation of the iron to the (Fe^{3+}) ferric oxidation state renders the molecule incapable of normal oxygen binding. Hydrophobic interactions between the tetrapyrrole ring and hydrophobic amino acid R groups on the interior of the cleft in the protein strongly stabilize the heme protein conjugate. In addition a nitrogen atom from a histidine R-group located above the plane of the heme ring is coordinated with the iron atom further stabilizing the interaction between the heme and the protein. In oxymyoglobin the remaining bonding site on the iron atom (the 6th coordinate position) is occupied by the oxygen, whose binding is stabilized by a second histidine residue.

Carbon monoxide also binds coordinately to heme iron atoms in a manner similar to that of oxygen, but the binding of carbon monoxide to heme is much stronger than that of oxygen. The preferential binding of carbon monoxide to heme iron is largely responsible for the asphyxiation that results from carbon monoxide poisoning.

Classification of Proteins Based on Functions

On the basis of functions proteins can be classified into a number of types as stated below:

1. **Enzymes**: catalytic activity and function
2. **Transport Proteins**: bind and carry ligands
3. **Storage Proteins**: ovalbumin, gluten, casein, ferretin
4. **Contractile (motor) proteins**: can contract, change shape, elements of cytoskeleton (actin, myosin, tubulin)
5. **Structural (support) proteins**: collagen of tendons and cartilage, elastin of ligaments (tropoelastin), keratin of hair, feathers, and nails, fibroin of silk and webs
6. **Defensive (protect) proteins**: antibodies (IgG), fibrinogen and thrombin, snake venoms, bacterial toxins
7. **Regulatory (signal) proteins**: regulate metabolic processes, hormones, transcription factors and enhancers, growth factor proteins
8. **Receptors (detect stimuli) proteins**: light and rhodopsin, membrane receptor proteins and acetylcholine or insulin

ENZYMES

Introduction

An enzyme is a macromolecule that catalyzes, or speeds up, a chemical reaction. Most enzymes are proteins and the word 'enzyme' is often used to mean a protein enzyme, but some RNA molecules also have catalytic activity, and to differentiate them from protein enzymes, they are referred to as RNA enzymes or ribozymes. The word comes from the Greek word *énsymo*, which comes from *én* ('at' or 'in') and *simo* (leaven or yeast) and was coined in 1878 by Kuhne. In 1897, Hans and Eduard Buchner inadvertently used yeast extracts to ferment sugar, despite the absence of living yeast cells. They were interested in making extracts of yeast cells for medical purposes, and as one possible way of preserving them, they added large amounts of sucrose to the extract. To their surprise, they found that the sugar was fermented, even though there were no living yeast cells in the mixture. The term 'enzyme' was used to describe the substance(s) in yeast extract that brought about the fermentation of sucrose. In 1926 the first enzyme was obtained in pure form by Prof. J. B. Sumner. He isolated the enzyme 'urease' in crystalline form from jack beans by means of acetone.

Enzymes are essential to sustain life because most chemical reactions in biological cells would occur too slowly, or would lead to different products without enzymes. A malfunction (mutation, overproduction, underproduction or deletion) of a single critical enzyme can lead to a severe disease. For example, the most common type of Phenylketonuria is caused by a single amino acid mutation in the enzyme Phenylalanine hydroxylase, which catalyzes the first step in the degradation of Phenylalanine. The resulting build-up of phenylalanine and related products can lead to mental retardation if the disease is untreated.

Like all catalysts, enzymes work by providing an alternate pathway of lower activation energy of a reaction, thus allowing the reaction to proceed much faster. Enzymes may speed up reactions by a factor of many millions. An enzyme, like any catalyst, remains unaltered by the completed reaction and can therefore continue to function. Because enzymes do not affect the relative energy between the products and reagents, they do not affect equilibrium of a reaction. However, the advantage of enzymes compared to most other catalysts is their stereo-, regio- and chemoselectivity and specificity.

Enzyme activity can be affected by other molecules. Inhibitors are naturally occurring or synthetic molecules that decrease or abolish enzyme activity. Activators are molecules that increase activity. Some irreversible inhibitors bind enzymes very tightly, effectively inactivating them. Many drugs and poisons act by inhibiting enzymes. Aspirin inhibits the Cyclooxygenase 1 and 2 (COX-1 and COX-2) enzymes that produce the inflammation messenger prostaglandin, thus suppressing pain and inflammation. The poison cyanide inhibits Cytochrome c oxidase, which effectively blocks cellular respiration.

While all enzymes have a biological role, some enzymes are used commercially for other purposes. Many household cleaners use enzymes to speed up chemical reactions (*e.g.*, breaking down protein or starch stains in clothes).

Nomenclature

More than 5,000 enzymes are known. Typically the suffix *-ase* is added to the name of the substrate (*e.g.*, lactase is the enzyme that catalyzes the cleavage of lactose) or the type of reaction (*e.g.*, DNA polymerase catalyzes the formation of DNA polymers). However, this is not always the case, especially when enzymes modify multiple substrates. For this reason, Enzyme Commission or EC numbers are used to classify enzymes based on the reactions they catalyze. Even this is not a perfect solution, as enzymes from different species or even very similar enzymes in the same species may have identical EC numbers.

Except for some of the originally studied enzymes such as pepsin, rennin, and trypsin, most enzyme names end in 'ase'. The International Union of Biochemistry (I.U.B.) initiated standards of enzyme nomenclature which recommend that enzyme names indicate both the substrate acted upon and the type of reaction catalyzed. Under this system, the enzyme uricase is called urate: O_2 oxidoreductase, while the enzyme glutamic oxaloacetic transaminase (GOT) is called L-aspartate: 2-oxoglutarate aminotransferase.

Classification

Enzymes can be classified by the kind of chemical reaction catalyzed.

1. Addition or removal of water:

   Hydrolases: these include esterases, carbohydrases, nucleases, deaminases, amidases, and proteases

   Hydrases: such as fumarase, enolase, aconitase and carbonic anhydrase

2. Transfer of electrons:

   Oxidoreductases: They bring about the main energy yielding reactions of living tissue and act by transferring electrons and hydrogen ions. They include oxidases and dehydrogenases.

3. Transfer of a radical:

   Transglycosidases: transfer of monosaccharides

   Transphosphorylases and phosphomutases: transfer of a phosphate group

   Transaminases: transfer of amino group

   Transmethylases: transfer of a methyl group

   Transacetylases: transfer of an acetyl group

4. Splitting or forming a C−C bond:

   Desmolases: These are the enzymes that break linkages not attacked by water. For example, decarboxylases that break C−C linkage and transaminases that break C−N linkages.

5. Changing geometry or structure of a molecule:

   Isomerases: catalyse reactions that bring about intramolecular rearrangement of atoms in substrates.

6. Joining two molecules through hydrolysis of pyrophosphate bond in ATP or other tri-phosphate:

   Ligases: catalyse reactions in which the pyrophosphate bond of ATP is broken down and linkage is formed between two molecules. For example, the formation of bonds like C−C, C−S, C−N and C−O.

Specificity of Enzymes

One of the properties of enzymes that makes them as important as diagnostic and research tools is the specificity they exhibit relative to the reactions they catalyze. A few enzymes exhibit absolute specificity; that is, they will catalyze only one particular reaction. Other enzymes will be specific for a particular type of chemical bond or functional group. In general, there are four distinct types of specificity:

1. Absolute specificity: the enzyme will catalyze only one type of reaction.
2. Group specificity: the enzyme will act only on molecules that have specific functional groups, such as amino, phosphate and methyl groups.
3. Linkage specificity: the enzyme will act on a particular type of chemical bond regardless of the rest of the molecular structure.
4. Stereochemical specificity: the enzyme will act on a particular steric or optical isomer.

   Though enzymes exhibit great degrees of specificity, co-factors may serve many apoenzymes, in case of enzymes that consist of two parts; a protein part called apoenzyme and a non-protein part called the co-factor. The combination of two can be referred to as the enzyme system. The enzyme activity can take place when both components are present together.

   For example, nicotinamide adenine dinucleotide (NAD) is a coenzyme for a great number of dehydrogenase reactions in which it acts as a hydrogen acceptor. Among them are the alcohol dehydrogenase, malate dehydrogenase and lactate dehydrogenase reactions.

The Lock and Key Theory (Fig. 3.4)

The specific action of an enzyme with a single substrate can be explained using a Lock and Key analogy first postulated in 1894 by Emil Fischer. In this

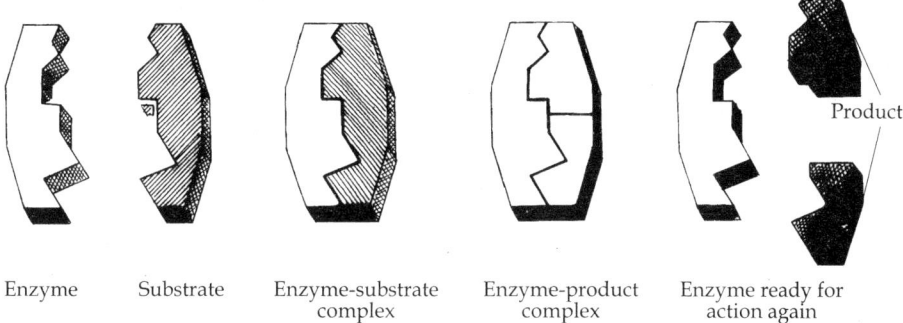

| Enzyme | Substrate | Enzyme-substrate complex | Enzyme-product complex | Enzyme ready for action again |

FIG. 3.4   Lock and Key Model of Enzyme Action

analogy, the lock is the enzyme and the key is the substrate. Only the correct sized key (substrate) fits into the key hole (active site) of the lock (enzyme).

Smaller keys, larger keys, or incorrectly positioned teeth on keys (incorrectly shaped or sized substrate molecules) do not fit into the lock (enzyme). Only the correctly designed key opens a particular lock.

Induced-fit Theory (Fig. 3.5)

In 1958, Daniel Koshland suggested a modification to the 'lock and key' model. Enzymes are rather flexible structures. The active site of an enzyme can be modified as the substrate interacts with the enzyme. The side chains of amino acids which make up the active site are molded into a precise shape which enables the enzyme to perform its catalytic function. In some cases the substrate molecule changes shape slightly as it enters the active site.

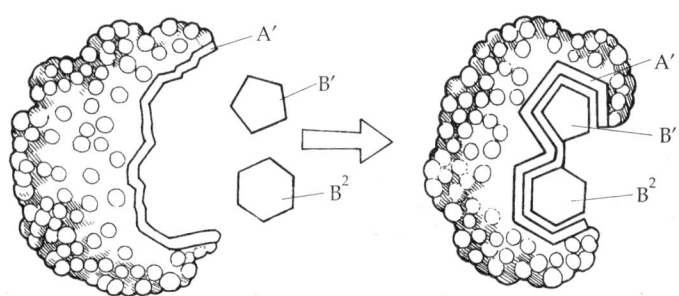

FIG. 3.5   Induced Fit Model of Enzyme Action

Functional Properties of Enzymes

1. An enzyme is a protein macromolecule that catalyzes, or speeds up, a chemical reaction.
2. The enzyme works at extremely small concentrations.
3. Thermodynamically, enzyme speeds up a reaction by lowering the activation energy.

4. An enzyme loses its catalytic properties if subjected to agents like heat, strong acids or bases and organic solvents.

Factors Affecting Enzyme Activity

Temperature

The temperature of a system is, to some extent, a measure of the kinetic energy of the molecules in the system. Thus, lower the kinetic energy, the lower the temperature of the system and, likewise, the higher the kinetic energy, the greater the temperature of the system. This has several effects on the rates of reactions.

(1) More energetic collisions

When molecules collide, the kinetic energy of the molecules can be converted into chemical potential energy of the molecules. If the chemical potential energy of the molecules becomes great enough, the activation energy of an exergonic reaction can be achieved and a change in chemical state will result. Thus, the greater the kinetic energy of the molecules in a system, the greater is the resulting chemical potential energy when two molecules collide. As the temperature of a system is increased it is possible that more molecules per unit time will reach the activation energy. Thus the rate of the reaction may increase.

(2) The number of collisions per unit time will increase

In order to convert substrate into product, enzymes must collide with and bind to the substrate at the active site. Increasing the temperature of a system will increase the number of collisions of enzyme and substrate per unit time. Thus, within limits, the rate of the reaction will increase.

(3) The heat of the molecules in the system will increase

As the temperature of the system is increased, the internal energy of the molecules in the system will increase. The internal energy of the molecules may include the translational energy, vibrational energy and rotational energy of the molecules, the energy involved in chemical bonding of the molecules as well as the energy involved in nonbonding interactions. Some of this heat may be converted into chemical potential energy. If this chemical potential energy increase is great enough some of the weak bonds that determine the three dimensional shape of the active proteins may be broken. This could lead to a thermal denaturation of the protein and thus inactivate the protein. Thus too much heat can cause the rate of an enzyme catalyzed reaction to decrease because the enzyme or substrate becomes denatured and inactive.

Given the above considerations, each enzyme has a temperature range in which a maximal rate of reaction is achieved. This maximum is known as the temperature optimum of the enzyme.

pH

The pH of a solution can have several effects of the structure and activity of enzymes. For example, pH can have an effect of the state of ionization of acidic or basic amino acids. Acidic amino acids have functional carboxyl groups in their side chains. While basic amino acids have amine functional groups in their side chains. If the state of ionization of amino acids in a protein is altered then the ionic bonds that help to determine the 3-D shape of the protein can be altered. This can lead to altered protein recognition or an enzyme might become inactive.

Changes in pH may not only affect the shape of an enzyme but it may also change the shape or charge properties of the substrate so that either the substrate cannot bind to the active site or it cannot undergo catalysis.

In general, enzymes have a pH optimum. However the optimum is not the same for each enzyme.

Enzyme-concentration

If enzyme is not present in sufficient quantity, the reaction rate will be slow because there is not enough enzyme for all of the reactant molecules. As the amount of enzyme is increased, the rate of reaction increases. If there are more enzyme molecules than are needed, adding additional enzyme will not increase the rate. Reaction rate therefore increases as enzyme concentration increases but then it levels off.

Substrate-concentration

At lower concentrations, the active sites on most of the enzyme molecules are not filled because there is not much substrate. But higher concentrations cause more collisions between the molecules so that enzymes are more likely to encounter reactant molecules.

The velocity of a reaction is reached maximum when the active sites are almost continuously filled. Increased substrate concentration after this point will not increase the rate. Reaction rate therefore increases as substrate concentration is increased but it levels off.

Kinetics

In 1913, Leonor Michaelis and Maud Menten proposed a quantitative theory of enzyme kinetics, which is referred to as Michaelis-Menten kinetics. Their work was further developed by G. E. Briggs and J. B. S. Haldane, who derived numerous kinetic equations that are still widely used today.

Enzymes can perform up to several million catalytic reactions per second; to determine the maximum speed of an enzymatic reaction, the substrate concentration is increased until a constant rate of product formation is achieved. This is the maximum velocity (V_{max}) of the enzyme. In this state, all enzyme active sites are saturated with substrate. However, V_{max} is the only one kinetic parameter for which biochemists are interested. The amount of substrate needed to achieve a given rate of reaction is also of interest. This can be expressed by the Michaelis-Menten constant (K_m), which is the substrate concentration required for an enzyme to reach one half its maximum velocity. Each enzyme has a characteristic K_m for a given substrate.

The efficiency of an enzyme can be expressed in terms of k_{cat}/K_m. The quantity k_{cat}, also called the Turnover number, incorporates the rate constants for all steps in the reaction, and is the quotient of V_{max} and the total enzyme concentration. k_{cat}/K_m is a useful quantity for comparing different enzymes against each other, or the same enzyme with different substrates, because it takes both affinity and catalytic ability into consideration. The theoretical maximum for k_{cat}/K_m, called the diffusion limit, is about 10^8 to 10^9 (M^{-1} s^{-1}). At this point, every collision of the enzyme with its substrate will result in catalysis and the rate of product formation is not limited by the reaction rate but by the diffusion rate.

Enzymes that reach this k_{cat}/K_m value are called catalytically perfect or kinetically perfect. Example of such enzymes are Triose phosphate isomerase, Carbonic anhydrase, Acetyl cholinesterase, Catalase, Fumarase, β-lactamase, and Superoxide dismutase.

Competitive Inhibition (Fig. 3.6)

Each enzyme is designed to bind only its intended substrate. However, it is possible in many cases that a different molecule which is sufficiently similar to the substrate binds to the active site of the enzyme. If this similar molecule is present in a concentration comparable to the concentration of the substrate, it will compete with the substrate for bonding sites on the enzyme and it will interfere with the catalytic action of the enzyme and the substrate. This phenomenon is known as competitive inhibition because the enzyme is

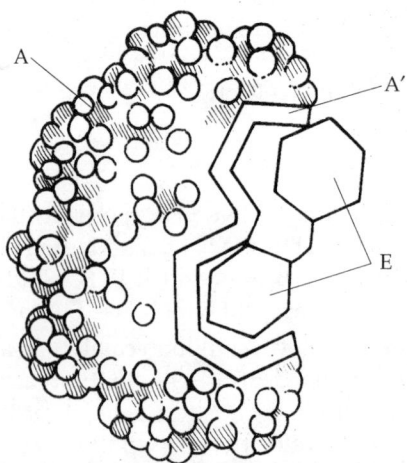

FIG. 3.6 Competitive Inhibition

inhibited by the inactive substrate, or competitor, so called because it competes with the real substrate for the active site.

Non-Competitive Inhibition (Fig. 3.7)

Non-competitive inhibitors never bind to the active center, but to other parts of the enzyme that can be far away from the substrate binding site, consequently, there is no competition between the substrate and inhibitor for the enzyme. The extent of inhibition depends entirely on the inhibitor concentration and will not be affected by the substrate concentration.

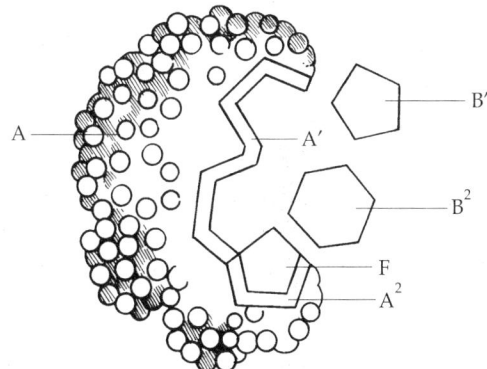

[F-Non-competitive inhibitor]
[A^2-Non-competitive inhibitor binding site]

FIG. 3.7 Non-Competitive Inhibition

Feedback Inhibition (Fig. 3.8)

Another kind of inhibition is called feedback inhibition. In feedback inhibition, there is a second binding site on the enzyme where the inhibitor binds, so that the inhibitor is not necessarily similar in structure to the substrate.

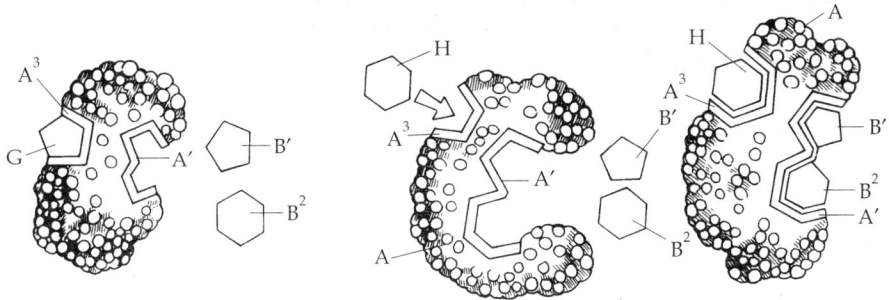

[G-Negative regulator] [A^3-Allosteric site] [H-Positive regulator]

FIG. 3.8 Allosteric Regulation

The absence or presence of the inhibitor at this second binding site activates or deactivates the enzyme, presumably by changing the conformation of the enzyme so that the active site is made available or unavailable to the substrate. The inhibitor is usually the product of a reaction farther down the metabolic pathway.

Uses of Inhibitors

Many therapeutic drugs are enzyme inhibitors. Important examples are penicillin, which inhibits an enzyme necessary for bacterial cell wall synthesis, and aspirin, an inhibitor of the synthesis of molecules that mediate pain and swelling. More recent examples are drugs used in the treatment of human immunodeficiency virus (HIV) and acquired immunodeficiency syndrome (AIDS) that prevent maturation of the virus by inhibiting the HIV protease and drugs that lower cholesterol by inhibiting a key step in cholesterol biosynthesis.

Among the irreversible inhibitors are organophosphorus compounds, which inhibit the enzyme acetylcholinesterase and similar enzymes. Organophosphorous compounds include nerve gases (such as sarin), that work on the human nervous system, and insecticides like malathion.

LIPIDS

Lipids are a class of hydrocarbon-containing organic compounds essential for the structure and function of living cells. Lipids are characterized by being water-insoluble and soluble in non-polar organic solvents such as ether, chloroform, benzene and acetone. Usually, they are aliphatic but they can have benzene rings in their structure. Although the word lipid is commonly used as a synonym to fat, the latter is a subgroup of triglyceride lipids.

Lipids are generally non-polar; however some of them have polar characters in addition to being largely nonpolar. In those cases, the bulk of their structure is nonpolar or hydrophobic (water-fearing), meaning that it does not interact well with polar solvents like water, and part of their structure is polar or hydrophilc (water-loving) which tends to associate with polar solvents like water. This makes them amphiphatic molecules (having both hydrophobic and hydrophilic portions).

Types of Lipids

Lipids are usually classified by the kind and number of carbon chains, but can be categorised as:
1. Fatty acids
   (i) Saturated
   (ii) Unsaturated (eicosanoids)
2. Glycerides

(i) Neutral
- Monoglycerides
- Diglycerides
- Triglycerides (fats)

(ii) Phosphoglycerides

3. Nonglycerides
    (i) Sphingolipids
    (ii) Steroids
    (iii) Waxes

4. Complex lipids
    (i) Lipoproteins
    (ii) Glycolipids

Fatty Acids

Fatty acids consist of the elements carbon (C), hydrogen (H) and oxygen (O) arranged as a carbon chain skeleton with a carboxyl group ($-COOH$) at one end. Chemically, fatty acids can be described as long chain monocarboxylic acids, and have a general structure of $CH_3(CH_2)_nCOOH$. The length of the chain usually ranges from 12 to 24, always with an even number of carbons. When the carbon chain contains no double bonds, it is called saturated. If it contains one or more such bonds, it is unsaturated. The presence of double bonds generally reduces the melting point of fatty acids. Furthermore, unsaturated fatty acids can occur either in *cis* or *trans* geometric isomers. In nature, almost all double bonds in fatty acids are found in the *cis* configuration.

Chemical Names and Descriptions of some Common Fatty Acids

Common Names	Carbon Atoms	Double Bonds	Scientific Name	Sources
Butyric acid	4	0	butanoic acid	butterfat
Caproic Acid	6	0	hexanoic acid	butterfat
Caprylic Acid	8	0	octanoic acid	coconut oil
Capric Acid	10	0	decanoic acid	coconut oil
Lauric Acid	12	0	dodecanoic acid	coconut oil
Myristic Acid	14	0	tetradecanoic acid	palm kernel oil
Palmitic Acid	16	0	hexadecanoic acid	palm oil
Palmitoleic Acid	16	1	9-hexadecenoic acid	animal fats
Stearic Acid	18	0	octadecanoic acid	animal fats
Oleic Acid	18	1	9-octadecenoic acid	olive oil
Vaccenic Acid	18	1	11-octadecenoic acid	butterfat
Linoleic Acid	18	2	9,12-octadecadienoic acid	grape seed oil

Alpha-Linolenic Acid	18	3	9,12,15-octadecatrienoic acid	flaxseed
Gamma-Linolenic Acid	18	3	6,9,12-octadecatrienoic acid	borage oil
Arachidic Acid	20	0	eicosanoic acid	peanut oil
Gadoleic Acid	20	1	9-eicosenoic acid	fish oil
Arachidonic Acid (AA) EPA	20	4	5,8,11,14-eicosatetraenoic acid	liver fats
	20	5	5,8,11,14,17-eicosa-pentaenoic acid	fish oil
Behenic acid	22	0	docosanoic acid	rapeseed oil
Erucic acid	22	1	13-docosenoic acid	rapeseed oil
DHA	22	6	4,7,10,13,16,19-docosahexaenoic acid	fish oil
Lignoceric acid	24	0	tetracosanoic acid	small amounts in most fats

Saturated and Unsaturated Fatty Acids

Saturated fatty acids (SFAs) have all the hydrogen that the carbon atoms can hold, and therefore, have no double bonds between the carbons. Monounsaturated fatty acids (MUFAs) have only one double bond. Polyunsaturated fatty acids (PUFAs) have more than one double bond. Fats, which are mostly from animal sources, have all single bonds between the carbons in their fatty acid tails, thus all the carbons are also bonded to the maximum number of hydrogens possible. Since the fatty acids in these triglycerides contain the maximum possible amount of hydrogens, these would be called saturated fats. The hydrocarbon chains in these fatty acids are, thus, fairly straight and can pack closely together, making these fats solid at room temperature. Oils, mostly from plant sources, have some double bonds between some of the carbons in the hydrocarbon tail, causing bends or 'kinks' in the shape of the molecules. Because some of the carbons share double bonds, they are not bonded to as many hydrogens as they could if they were not double bonded to each other. Therefore these oils are called unsaturated fats. Because of the kinks in the hydrocarbon tails, unsaturated fats cannot pack as closely together, making them liquid at room temperature. Many people have heard that the unsaturated fats are 'healthier' than the saturated ones. Hydrogenated vegetable oil (as in shortening and commercial peanut butters where a solid consistency is sought) started out as 'good' unsaturated oil. However, this commercial product has had all the double bonds artificially broken and hydrogens artificially added (in a chemistry lab-type setting) to turn it into saturated fat that bears no resemblance to the original oil from which it came (so it is solid at room temperature).

Examples of some saturated fatty acids are:

Butyric acid: $CH_3(CH_2)_2COOH$

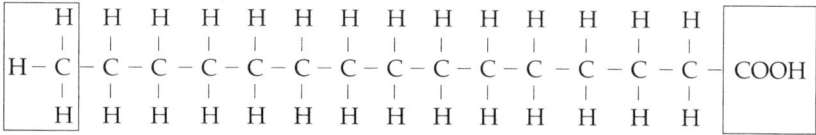

Butyric acid

Lauric acid (dodecanoic acid): $CH_3(CH_2)_{10}COOH$

Myristic acid (tetradecanoic acid): $CH_3(CH_2)_{12}COOH$

Methyl	**LAURIC ACID**	Carboxyl
OIL Solube	12 carbons	Water-Solube
OMEGA (ω) END		DELTA (Δ) END

Methyl	**MYRISTIC ACID**	Carboxyl
OIL Solube	14 carbons	Water-Solube
OMEGA (ω) END		DELTA (Δ) END

Palmitic acid (hexadecanoic acid): $CH_3(CH_2)_{14}COOH$

Stearic acid (octadecanoic acid): $CH_3(CH_2)_{16}COOH$

Methyl	**PALMITIC ACID**	Carboxyl
OIL Solube	16 carbons	Water-Solube
OMEGA (ω) END		DELTA (Δ) END

Methyl	**STEARIC ACID**	Carboxyl
OIL Solube	18 carbons	Water-Solube
OMEGA (ω) END		DELTA (Δ) END

Arachidic acid (eicosanoic acid): $CH_3(CH_2)_{18}COOH$

In unsaturated fatty acids, the pieces of the hydrocarbon tail can be arranged around a C=C double bond in two ways. In *cis* bonds, the two pieces of the carbon chain on either side of the double bond are either 'up' or 'down', such that both are on the same side of the molecule. In *trans* bonds, the two pieces of the molecule are on opposite sides of the double bond, that is, one 'up' and one 'down' across from each other. Naturally occurring unsaturated vegetable oils have almost all *cis* bonds, but using oil for frying causes some of the *cis* bonds to convert to *trans* bonds. If oil is used only once like when you fry an egg, only a few of the bonds do this so it is not too bad for health. However, if oil is constantly reused, more and more of the *cis* bonds are changed to *trans* until significant numbers of fatty acids with *trans* bonds build up. The reason this is of concern is that fatty acids with *trans* bonds are carcinogenic, or cancer-causing.

Examples of unsaturated fatty acids are:

Alpha-linolenic acid: $CH_3CH_2CH=CHCH_2CH=CHCH_2CH=CH(CH_2)_7COOH$

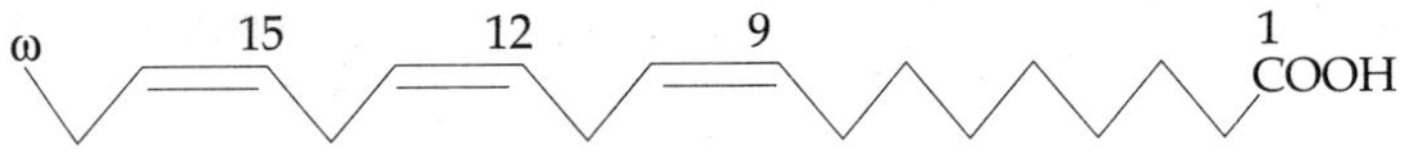

Alpha-Linolenic Acid (omega-3)

Linoleic acid: $CH_3(CH_2)_4 CH = CHCH_2CH = (CH_2)_7COOH$

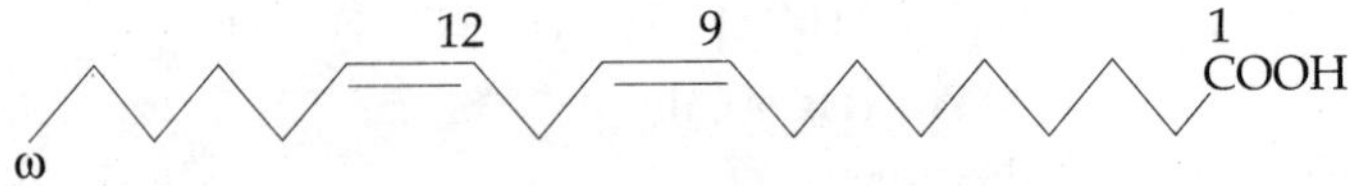

Linoleic Acid (omega-6)

Arachidonic acid: $CH_3(CH_2)_4CH = CHCH_2CH = CHCH_2CH = CHCH_2CH = CH(CH_2)_3COOH$

Oleic acid: $CH_3(CH_2)_7 CH = CHCH_2CH(CH_2)_7COOH$

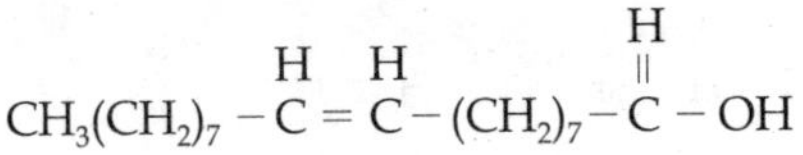

Oleic acid

Erucic acid: $CH_3(CH_2)_7 CH = CH(CH_2)_{11}COOH$

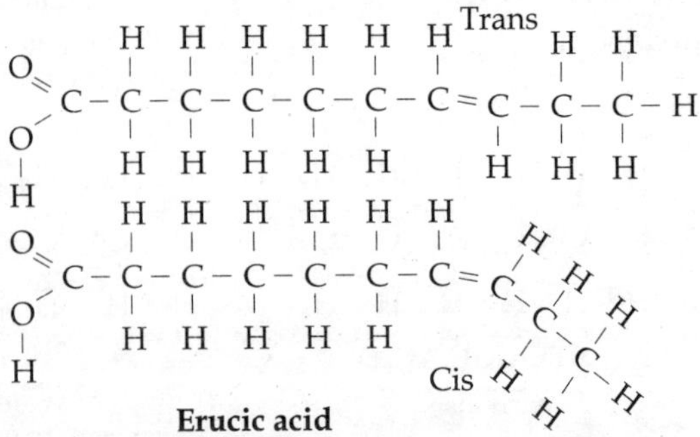

Erucic acid

We need fats in our bodies and in our diet. Animals generally use fat for energy storage because fat stores 9 KCal/g of energy. Plants, which do not move around, can afford to store food for energy in a less compact but more easily accessible form, so they use starch (a carbohydrate) for energy storage. Carbohydrates and proteins store only 4 KCal/g of energy, so fat stores over twice as much energy/gram as the same amount of proteins or carbohydrates. By the way, this is also related to the idea behind some of the high-carbohydrate weight loss diets. The human body burns carbohydrates and fats for fuel in a given proportion to each other. The theory behind these diets is that if they supply carbohydrates but not fats, then it is hoped that the fat needed to balance with the sugar will be taken from the dieter's body stores. Fat is also is used in our bodies to (a) cushion vital organs like the kidneys and (b) serve as insulation, especially just beneath the skin.

Numeric Designations

The numeric designations used for fatty acids are made according to the number of carbon atoms, followed by the number of sites of unsaturation (for example, palmitic acid is a 16-carbon fatty acid with no unsaturation and is designated by 16:0). The site of unsaturation in a fatty acid is indicated by the symbol Δ and the number of the first carbon of the double bond (for example, palmitoleic acid is a 16-carbon fatty acid with one site of unsaturation between carbons 9 and 10, and is designated by $16:1^{\Delta 9}$).

Saturated fatty acids of less than eight carbon atoms are liquid at physiological temperature, whereas those containing more than eight are generally solid. The presence of double bonds in fatty acids significantly lowers the melting point relative to a saturated fatty acid.

The majority of body fatty acids are acquired in the diet. However, the lipid biosynthetic capacity of the body (fatty acid synthase and other fatty acid modifying enzymes) can supply the body with all the various fatty acid structures needed. Two key exceptions to this are the highly unsaturated fatty acids known as linoleic acid and linolenic acid, containing unsaturation sites beyond carbons 9 and 10. These two fatty acids cannot be synthesized from precursors in the body, and are thus considered the essential fatty acids, that is, they must be provided in the diet. Since plants are capable of synthesizing linoleic and linolenic acid, humans can aquire these fats by consuming a variety of plants or else by eating the meat of animals that have consumed these plant fats.

Glycerides

Glycerides are esters formed from glycerol and fatty acids. Glycerol has three hydroxyl functional groups and which can be esterified with one, two or three fatty acids to form monoglycerides, diglycerides and triglycerides.

Vegetable oils and animal fats contain mostly triglycerides, but are broken down by natural enzymes (lipases) into mono- and di-glycerides and free fatty acids. Soaps often contain glycerides. Glycerol is a product that can soften dehydrated skin. It can absorb moisture from the air. If 100% glycerol is exposed to air, in 10 to 12 hours it would absorb moisture and become 80% glycerol and 20% water.

Monoglycerides

A monoglyceride is a glyceride consisting of one fatty acid chain covalently bonded to a glycerol molecule through an ester linkage.

Diglycerides

A diglyceride is a glyceride consisting of two fatty acid chains covalently bonded to a glycerol molecule through ester linkages.

Triglycerides

A triglyceride is a glyceride consisting of three fatty acid chains covalently bonded to a glycerol molecule through ester linkages. Triglycerides are the main constituents of vegetable oils and animal fats. Triglycerides have lower densities than water (they float on water), and at normal room temperatures may be solid or liquid. When solid, they are called 'fats' or 'butters' and when liquid they are called 'oils'.

$$CH_3(CH_2)_7 - \overset{H}{C} = \overset{H}{C} - (CH_2)_7 - \overset{\overset{O}{\|}}{C} - OH$$

Oleic acid

$$HO - CH_2$$
$$HO - CH$$
$$HO - CH_2$$

Glycerol or Glycerin

Phospholglycerides

Phospholglycerides are esters of only two fatty acids, phosphoric acid and a trifunctional alcohol — glycerol (IUPAC name is 1,2,3-propantriol). The fatty acids are attached to the glycerol at the 1 and 2 positions on glycerol through ester bonds. There may be a variety of fatty acids, both saturated and unsaturated, in the phospholipids.

The third oxygen on glycerol is bonded to phosphoric acid through a phosphate ester bond (oxygen-phosphorus double bond oxygen). In addition, there is usually a complex amino alcohol also attached to the phosphate through a second phosphate ester bond. The complex amino alcohols include choline, ethanolamine, and the amino acid-serine.

The properties of a phospholipid are characterized by the properties of the fatty acid chain and the phosphate/amino alcohol. The long hydrocarbon chains of the fatty acids are of course non-polar. The phosphate group has a negatively charged oxygen and a positively charged nitrogen to make this group ionic. In addition there are other oxygens of the ester groups, which make on whole end of the molecule strongly ionic and polar. Phospholipids are major components in the lipid bilayers of cell membranes.

There are two common phospholipids, lecithin and cephalins.

Lecithin: contains the amino alcohol, choline.

Cephalins: contain the amino alcohols serine or ethanolamine

Lecithin

Lecithin is probably the most common phospholipid. It is found in egg yolks, wheat germ, and soybeans. Lecithin is extracted from soybeans for use as an emulsifying agent in foods. Lecithin is an emulsifier because it has both polar and non-polar properties, which enable it to cause the mixing of other fats and oils with water components.

Cephalins

Cephalins are phosphoglycerides that contain ethanolamine or the amino acid serine attached to the phosphate group through phosphate ester bonds. A variety of fatty acids make up the rest of the molecule. Cephalins are found in most cell membranes, particularly in brain tissues. They are also important in the blood clotting process as they are found in blood platelets.

Sphingolipids

Sphingolipids are composed of a backbone of sphingosine which is derived from glycerol. Sphingosine is N-acetylated by a variety of fatty acids generating a family of molecules referred to as ceramides. Sphingolipids predominate in the myelin sheath of nerve fibers. Sphingomyelin is an abundant sphingolipid generated by transfer of the phosphocholine moiety of phosphatidylcholine to a ceramide, thus sphingomyelin is a unique form of phospholipid.

The other major class of sphingolipids is the glycosphingolipids generated by substitution of carbohydrates to the *sn1* carbon of the glycerol backbone of a ceramide. There are four major classes of glycosphingolipids:

Cerebrosides: contain a single moiety, principally galactose

Sulfatides: sulfuric acid esters of galactocerebrosides

Globosides: contain two or more sugars

Gangliosides: similar to globosides except that it also contains sialic acid

Steroids

Sterols, such as cholesterol, are alcohols with the cyclopentanophenanthrene ring system (atoms 1 through 17 in the structure below). This substructure is also found in steroid hormones such as testosterone, progesterone, and cortisol. Cholesterol is classified as an alcohol because it has a hydroxyl group (**-OH**) in position 3 of the ring system. Cholesterol is produced by the liver and is found in all body tissues where it helps to organize cell membranes and control their permeability. Cholesterol derivatives in the skin are converted to vitamin D when the skin is exposed to sunlight. Vitamin D_3 mediates intestinal calcium absorption and bone calcium metabolism. A high level of cholesterol in the blood is considered to be a risk factor for cardiovascular diseases.

Cholesterol (a sterol)

Vitamin D$_3$
(cholecalciferol)

Testosterone
(a steroid hormone)

Waxes

Wax is an ester of ethylene glycol (ethane-1,2-diol) and two fatty acids, as opposed to a fat which is an ester of glycerin (propane-1,2,3-triol) and three fatty acids. It may also be a combination of other fatty alcohols with fatty acids. It is a type of lipid.

Lipoproteins

Lipoproteins are clusters of proteins and lipids all tangled up together. These act as a means of carrying lipids, including cholesterol, around in our blood. There are two main categories of lipoproteins distinguished by their density. LDL or low density lipoprotein is the 'bad protein', being associated with deposition of 'cholesterol' on the walls of arteries. HDL or high density lipoprotein is the 'good protein', being associated with carrying 'cholesterol' out of the blood system, and is more dense/more compact than LDL.

Glycolipids

Glycolipids are the lipids that contain one or more carbohydrate groups. Their role is to provide energy and also serve as markers for cellular recognition. They occur where a carbohydrate chain is associated with phospholipids in the cell surface membrane.

OBJECTIVE TYPE QUESTIONS

1. Which functional group is found in amino acids?
   A. Carboxyl   B. Amino
   C. Ester   D. Both A and B

2. Which functional group is found in simple sugars called reducing sugars?
   A. Carboxyl group   B. Amino group
   C. Ester group   D. Aldehyde

3. Which of these would be found in triglycerides?
   A. Amino groups   B. Aldehyde groups
   C. Ester linkages   D. Peptide group

4. Which of these is the organic acid functional group?
   A. Ketone   B. Aldehyde
   C. Ester   D. Carboxyl

5. The bond that joins amino acids together is called a:
   A. Glyceride bond   B. Peptide bond
   C. Carboxyl bond   D. Transamination bond

6. Enzymes speed reactions by:
   A. Increasing the activation energy
   B. Decreasing the activation energy
   C. Making the free-energy change of the reaction more negative
   D. Increasing the entropy of the reaction

7. Substrates of an enzyme normally fit into:
   A. The active site   B. The inactive site
   C. The allosteric site
   D. For some enzymes, into the active site, and for others into the allosteric site

8. Phospholipids are important because
    A. They make up DNA
    B. They are found in phosphorylated adipose deposits
    C. They are found in membranes
    D. They are lipids
9. Hemoglobin is a _____
    A. Monomer   B. Dimer
    C. Tetramer   D. Pentamer
10. Which major class of organic molecules is the most hydorphobic?
    A. Proteins   B. Lipids
    C. Carbohydrates   D. None of the above

Answers

1. D	2. D	3. C	4. D	5. B
6. B	7. A	8. C	9. C	10. B

SHORT ANSWER TYPE QUESTIONS

1. Name the different kinds of elements found in carbohydrates.
2. Name three major functions of lipids.
3. Name the sub-units of proteins and state how many different kinds of molecules are found in proteins.
4. Which major class of organic compounds contains the phospholipids.
5. What two major classes of organic compounds are always found in cell membranes?

LONG ANSWER TYPE QUESTIONS

1. What are carbohydrates? Discuss the classification and types.
2. What are proteins? Describe the primary, secondary and tertiary structures?
3. What are enantiomers? Give an example.
4. Explain the structures of starch and glycogen.
5. Write a note on the structure of myoglobin.
6. What are enzymes? Enlist the functional properties of enzymes.
7. Write a note on competitive inhibition.
8. Differentiate between saturated and unsaturated fatty acids.
9. Write an account on glycerides.
10. Which major class of organic compounds has the highest energy content? Explain why?

Nucleic Acid

4

INTRODUCTION

A nucleic acid is a complex, high-molecular weight biochemical macromolecule composed of nucleotide chains that transfer genetic information. The most common nucleic acids are deoxyribonucleic acid (DNA) and ribonucleic acid (RNA). Nucleic acids are found in all living cells and viruses.

History

Nucleic acids were discovered in 1868 by a Swiss biochemist Friedrich Miescher (1844-1895), when he isolated a cellular substance containing nitrogen and phosphorus. Miescher thought it to be a phosphorus-rich nuclear protein and named it Nuclein.

Later, German biochemist Albrecht Kossel in 1880 discovered that it was actually a protein plus nucleic acid. He also isolated two purines (adenine and guanine) and three pyrimidines (thymine, cytosine, and uracil), as well as carbohydrates.

The American biochemist Phoebus Levene, who had once studied with Kossel, identified two nucleic acid sugars — ribose (in 1909) and deoxyribose (in 1929). This meant that there were two nucleic acids, one named for each type of sugar. Levene also defined a nucleic acid's main unit as a phosphate-base-sugar nucleotide. The exact linking of nucleotides into a linear polymer chain was discovered in the 1940s by the British organic chemist Alexander Todd.

In the early 1950s, Rosalind Franklin (Fig. 4.1), an Englishwoman, was studying the scattered patterns made by the X-rays bouncing off the crystals of various substances including DNA. Other people, like Linus Pauling, were also attempting to figure out the structure of DNA.

James Deway Watson (1928-till date)
Francis Harry Compton Crick (1916-2004)

Rosalind Franklin
(1920-1958)

FIG. 4.1

James Watson, a young American scientist was in England working with Francis Crick, another young researcher (Fig. 4.1). On seeing Franklin's photographs of DNA X-ray crystallography they were able to determine that the structure of DNA was organized into a double spiral or double helix. Based on Franklin's data, in 1953, Watson and Crick published a paper in which they proposed and described a hypothetical structure for DNA. For their discovery, Watson and Crick received the Nobel Prize jointly with Maurice Wilkins, who was also working on the same aspect, in 1962. Unfortunately, in the intervening time, Rosalind Franklin had died in 1958 because of ovarian cancer, probably due in large part to her work with X-rays.

Chemical Structure of Nucleic Acids

The term nucleic acid is the generic name of a family of biopolymers. The monomers are called nucleotides. Each monomer consists of three components: a nitrogenous heterocyclic base, either a Purine or a Pyrimidine; a pentose sugar, and a phosphate group. The nitrogenous bases possible in the two nucleic acids are different: Adenine, Cytosine, and Guanine are present in both RNA and DNA, while Thymine is possible only in DNA and Uracil is present only in RNA.

Nucleic acids may be single-stranded or double-stranded. A double-stranded nucleic acid consists of two single-stranded nucleic acids hydrogen-bonded together. RNA is usually single-stranded, but any given strand is likely to fold back upon itself to form double-helical regions. DNA is usually double-stranded, though some viruses have single-stranded DNA as their genome. The sugars and phosphates in nucleic acids are connected to each other in an alternating chain, linked by shared oxygens, forming a phosphodiester functional group. In conventional nomenclature, the carbons to which the phosphate groups are attached are the 3' and the 5' carbons. The bases extend from a glycosidic linkage to the 1' carbon of the pentose ring.

DEOXYRIBONUCLEIC ACID (DNA)

DNA is a nucleic acid, usually in the form of a double helix, that contains the genetic instructions specifying the biological development of all cellular forms of life and most viruses.

The two strands of the DNA double helix are antiparallel, which means that they run in opposite directions. The sugar-phosphate backbone is on the outside of the helix, and the bases are on the inside. The backbone can be thought of as the sides of a ladder, whereas the bases in the middle form the rungs of the ladder. Each rung is composed of two base pairs. Either an adenine-thymine pair that form a two-hydrogen bond together, or a cytosine-guanine pair that form a three-hydrogen bond. The base pairing is thus restricted.

The average molecular weight of a nucleotide is 300. The diameter of the DNA molecule is about 20 Å, the base pair is about 3.4 Å thick (Fig. 4.2).

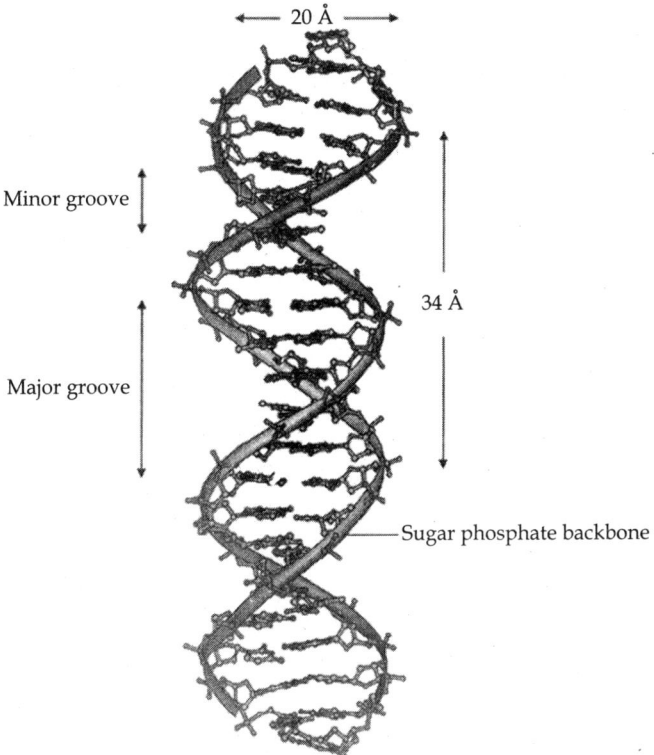

FIG. 4.2 Double Helical DNA Molecule

Nitrogenous Bases

Nucleobases are the parts of RNA and DNA that are involved in pairing up. These include cytosine, guanine, adenine, thymine. These are abbreviated as C, G, A and T, respectively. Adenine and guanine belong to the double-ringed class of molecules called purines (abbreviated as R). Cytosine, thymine, and uracil are all pyrimidines (abbreviated as Y).

Cytosine: It is a pyrimidine derivative, with a heterocyclic aromatic ring and two substituents attached (an amine group at position 4 and a keto group at position 2). In DNA and RNA, cytosine is paired with guanine (Fig. 4.3(b)). However, it is inherently unstable, and can change into uracil (spontaneous deamination). This can lead to a point mutation if not repaired by the DNA repair enzymes. Cytosine can also be methylated into 5-methylcytosine by an enzyme called DNA methyltransferase.

Guanine: With the formula $C_5H_5N_5O$, guanine is a derivative of purine, consisting of a fused pyrimidine-imidazole ring system with conjugated double bonds. Being unsaturated, the bicyclic molecule is planar.

Adenine: It is one of the two purine nucleobases. In DNA, adenine binds to thymine via two hydrogen bonds to assist in stabilizing the nucleic acid structures (Fig. 4.3(a)).

Thymine: It is a pyrimidine nucleobase which is also known as 5-methyluracil. As the name implies, thymine may be derived by methylation of uracil at the 5th carbon. In DNA, thymine (T) binds to adenine (A) via two hydrogen bonds to assist in stabilizing the nucleic acid structures (Fig. 4.3(a)).

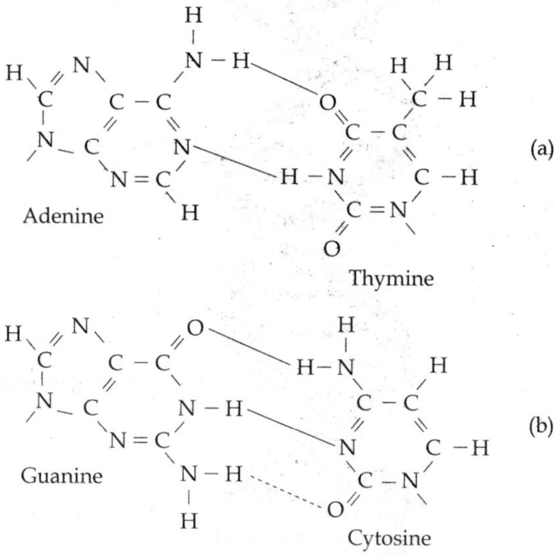

FIG. 4.3 (a) The pairing of adenine and thymine by two hydrogen bonds (b) The pairing of guanine and cytosine by three hydrogen bonds

Nucleosides

A nucleobase covalently bound to the 1' carbon of a ribose or deoxyribose is called a nucleoside. If the sugar molecule involved is ribose, it is called ribonucleoside and if the sugar is deoxyribose, it is deoxyribonucleoside. The glycosidic bonds are developed between sugar molecules and nitrogenous bases. Examples of these include cytidine, uridine, adenosine, guanosine, thymidine and inosine.

Nucleotides

Nucleosides can be phosphorylated by specific kinases in the cell, producing nucleotides, which are the molecular building blocks of DNA and RNA. Nucleotide names are abbreviated into standard three- or four-letter codes. The first letter is lower case and indicates whether the nucleotide in question is a deoxyribonucleotide (denoted by a 'd') or a ribonucleotide (no letter). The second letter indicates the nucleoside corresponding to the nucleobase:

G- Guanine

A- Adenine

T- Thymine

C- Cytosine

U- Uracil (*not present in DNA, but takes the place of Thymine in RNA*)

The third and fourth letters indicate the length of the attached phosphate chain (Mono-, Di-, Tri-) and the presence of a phosphate (P).

For example, deoxy-cytidine-triphosphate is abbreviated as dCTP.

Polynucleotides

Polynucleotides are formed by the condensation of two or more nucleotides. The condensation most commonly occurs between the alcohol of a 5'-phosphate of one nucleotide and the 3'-hydroxyl of a second, with the elimination of H_2O, forming a phosphodiester bond. The formation of phosphodiester bonds in DNA and RNA exhibits directionality. The primary structure of DNA and RNA (the linear arrangement of the nucleotides) proceeds in the 5' → 3' direction. The common representation of the primary structure of DNA or RNA molecules is to write the nucleotide sequences from left to right synonymous with the 5' → 3' direction.

Chargaff's Rule

In 1951, Chargaff pointed out that DNA contains equal proportion of purine and pyrimidine bases.

- The amount of 'A' base found in the DNA of a cell equals 'T' and the amount of 'C' found in a cell equals the amount of 'G'.

- The structure proposed was that of a double stranded molecule in the shape of a helix, that turned to the right, had a major and minor groove, was antiparallel, and had 'A' forming two hydrogen bonds with 'T', and had 'C' forming three hydrogen bonds with 'G'.

Summary of DNA Structure

1. DNA is made up of subunits called nucleotides.
2. Each nucleotide is made up of a sugar, a phosphate and a base.
3. There are 4 different bases in a DNA molecule:
   adenine (a purine)
   cytosine (a pyrimidine)
   guanine (a purine)
   thymine (a pyrimidine)
4. The number of purine bases equals the number of pyrimidine bases.
5. The number of adenine bases equals the number of thymine bases.
6. The number of guanine bases equals the number of cytosine bases.
   The basic structure of the DNA molecule is helical, with the bases being stacked on top of each other.

Thermal Properties of DNA

As cells divide it is a necessity that the DNA is also copied (replicated), in such a way that each daughter cell acquires the same amount of genetic material. In the process of DNA replication two strands of the helix must first be separated, the process is termed denaturation. This process can also be carried out *in vitro*. If a solution of DNA is subjected to high temperature, the H-bonds between bases become unstable and the strands of the helix separate in a process of thermal denaturation.

The base composition of DNA varies widely from molecule to molecule and even within different regions of the same molecule. Regions of the duplex that have predominantly A-T base pairs will be less thermally stable than those rich in G-C base-pairs. In the process of thermal denaturation, a point is reached at which 50% of the DNA molecule exists as single strands. This point is the melting temperature (T_M), and is characteristic of the base composition of that DNA molecule. The T_M depends upon several factors in addition to the base composition. These include the chemical nature of the solvent and the identities and concentrations of ions in the solution.

When thermally melted DNA is cooled, the complementary strands again re-form the correct base pairs, in a process termed annealing or hybridization. The rate of annealing is dependent upon the nucleotide sequence of the two strands of DNA.

Forms of DNA

Z-DNA

Z-DNA is a form of DNA in which the double helix winds to the left in a zig-zag pattern (instead of to the right, like the more common B-DNA form). Z-DNA was the first crystal structure of a DNA molecule to be solved. It was solved by Alexander Rich and co-workers in 1979. Z-DNA is quite different from the right-handed forms. This unique type of DNA can form alternating purine-pyrimidine tracts under very specific conditions. These conditions include high salt, the presence of some cations, and DNA supercoiling.

B-DNA

This is the common form of DNA proposed by Watson and Crick. The molecule is a right-handed double-helix. The backbone chains are antiparallel and the base pairs are centered on the helix axis. On an average, each base pair is rotated 35.6 degrees from the adjacent base pair. The length of the helix is about 34 Å. It contains 10 pairs of mononucleotide units.

A-DNA

The A form appears likely to occur only in dehydrated samples of DNA, such as those used in crystallographic experiments, and possibly in hybrid pairings of DNA and RNA strands.

Circular DNA

When the ends of a piece of double-helical DNA are joined so that it forms a circle, as in plasmid DNA, the strands are topologically knotted. This means they cannot be separated by gentle heating or by any process that does not involve breaking a strand. The task of unknotting topologically linked strands of DNA falls to enzymes known as topoisomerases. Some of these enzymes unknot circular DNA by cleaving two strands so that another double-stranded segment can pass through. Unknotting is required for the replication of circular DNA as well as for various types of recombination in linear DNA.

Properties of different helical forms

Geometry attribute	A-form	B-form	Z-form
Helix sense	right-handed	right-handed	left-handed
Repeating unit	1 bp	1 bp	2 bp
Rotation/bp	33.6°	35.9°	60°/2
Mean bp/turn	10.7	10.4	12
Inclination of bp to axis	+19°	−1.2°	−9°
Rise/bp along axis	0.23 nm	0.332 nm	0.38 nm
Pitch/turn of helix	2.46 nm	3.32 nm	4.56 nm
Mean propeller twist	+18°	+16°	0°

Sugar pucker	C3'-endo	C2'-endo	C: C2'-endo, G: C2'-exo
Diameter	2.55 nm	2.37 nm	1.84 nm

RIBONUCLEIC ACID (RNA)

RNA is a nucleic acid polymer consisting of nucleotide monomers. RNA nucleotides contain ribose rings and uracil unlike deoxyribonucleic acid (DNA), which contains deoxyribose and thymine. RNA is primarily made up of four different bases: adenine, guanine, cytosine, and uracil. The first three are the same as those found in DNA, but in RNA uracil replaces thymine as the base complementary to adenine. This base is also a pyrimidine and is very similar to thymine. Uracil is energetically less expensive to produce than thymine, which may account for its use in RNA. In DNA, however, uracil is readily produced by chemical degradation of cytosine, so having thymine as the normal base makes detection and repair of such incipient mutations more efficient. Thus, uracil is appropriate for RNA, where quantity is important but lifespan is not, whereas thymine is appropriate for DNA where maintaining sequence with high fidelity is more critical (Fig. 4.4).

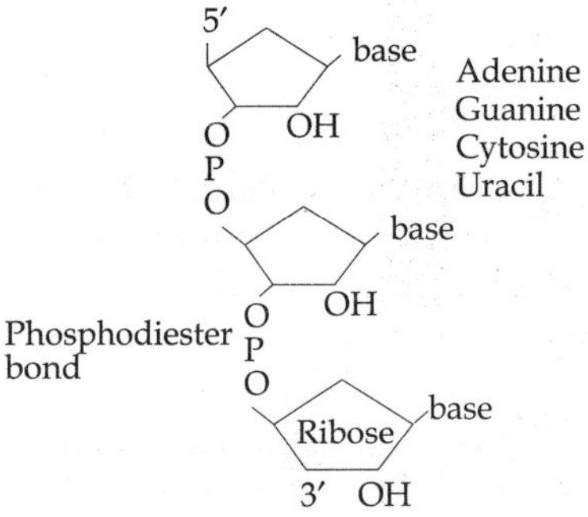

FIG. 4.4 Ribonucleic acid

Comparison with DNA

Unlike DNA, RNA is almost always a single-stranded molecule and has a much shorter chain of nucleotides. RNA contains ribose, rather than the deoxyribose found in DNA (there is a hydroxyl group attached to the pentose ring in the 2' position whereas DNA has a hydrogen atom rather than a hydroxyl group). This hydroxyl group makes RNA less stable than DNA because it is more prone to hydrolysis. Several types of RNA (tRNA, rRNA) contain a great deal of secondary structure, which promote stability.

TYPES OF RIBONUCLEIC ACID (RNA)

Messenger Ribonucleic Acid (mRNA)

The mRNA is RNA that encodes and carries information from DNA during transcription to sites of protein synthesis to undergo translation in order to yield a gene product. The brief life of an mRNA molecule begins with transcription and ultimately ends in degradation. During its life, an mRNA molecule may also be processed, edited, and transported prior to translation. Eukaryotic mRNA molecules often require extensive processing and transport, while prokarotic molecules do not.

Structure

A fully processed mRNA includes a 5′ cap, 5′ UTR, coding region, 3′ UTR, and poly(A) tail (Fig. 4.5).

5′ cap: A *5′ cap*, also termed as RNA cap, RNA 7-methylguanosine cap or RNA m⁷G cap, is a modified guanine nucleotide that has been added to the 'front' or 5′ end of the messenger RNA shortly after the start of transcription. The 5′ cap consists of a terminal 7-methylguanosine residue which is linked through a 5′-5′-triphosphate bond to the first transcribed nucleotide. Its presence is critical for recognition by the ribosome and protection from RNases.

Cap addition is coupled to transcription, and occurs co-transcriptionally, such that each influences the other. Shortly after the start of transcription, the 5′ end of the mRNA being synthesized is bound by a cap-synthesizing complex associated with RNA polymerase. This enzymatic complex catalyzes the chemical reactions that are required for mRNA capping. Synthesis proceeds as a multi-step biochemical reaction.

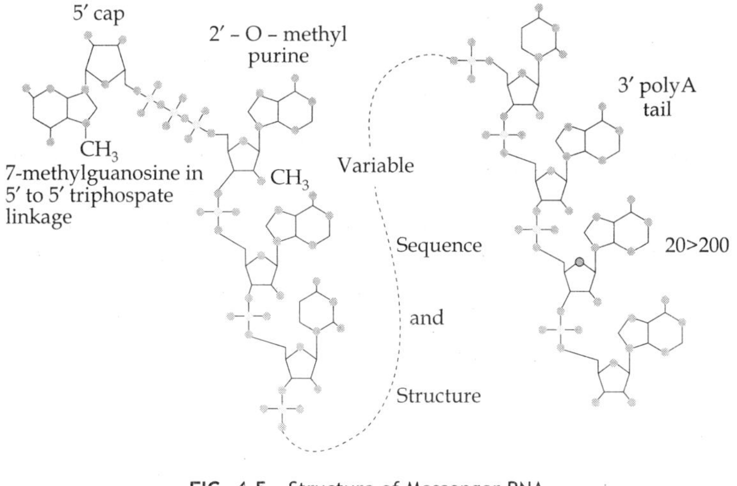

FIG. 4.5 Structure of Messenger RNA

Coding regions: Coding regions are composed of codons, which are decoded and translated into protein by the ribosome. Coding regions begin with the start codon and end with the one of three possible stop codons. In addition to protein-coding, portions of coding regions may also serve as regulatory sequences as exonic splicing enhancers or exonic splicing silencers.

Untranslated regions: Untranslated regions (UTRs) are sections of the RNA before the start codon and after the stop codon that are not translated, termed the five prime untranslated region (5' UTR) and three prime untranslated region (3' UTR), respectively. These regions are transcribed as part of the same transcript as the coding region. Several roles in gene expression have been attributed to the untranslated regions, including mRNA stability, mRNA localization, and translational efficiency. The ability of a UTR to perform these functions depends on the sequence of the UTR and can differ between mRNAs.

Stability of mRNAs may be mediated by the 5' UTR and 3' UTR due to varying affinity for certain RNA degrading enzymes called ribonucleases, which can promote or inhibit the relative stability of the RNA molecule. The greater the stability of an mRNA, the more protein that may be produced from that transcript.

Translational efficiency, and even inhibition of translation altogether, can be mediated by UTRs. Proteins that bind to either the 3' or 5' UTR may affect translation by interfering with the ability of ribosome to bind to the mRNA.

3' poly(A): The 3' poly(A) tail is a long sequence of adenine nucleotides (often several hundred) added to the 'tail' or 3' end of the pre-mRNA through the action of an enzyme, polyadenylate polymerase. The poly(A) tail is added on to the transcripts that contain a specific sequence, the AAUAAA signal. The importance of the AAUAAA signal is demonstrated by a mutation in the human alpha 2-globin gene which mutates the original sequence AATAAA into AATAAG, which can lead to hemoglobin deficiencies.

Transfer Ribonucleic Acid (tRNA)

The tRNA is a small RNA chain (74-93 nucleotides) that transfers a specific amino acid to a growing polypeptide chain at the ribosomal site of protein synthesis during translation. It has sites for amino-acid attachment and codon (a particular sequence of 3 bases) recognition. The codon recognition is different for each tRNA and is determined by the anticodon region, which contains the complementary bases to the ones encountered on the mRNA. Each tRNA molecule binds only one type of amino acid, but because the genetic code is degenerate, more than one codon exists for each amino acid.

Transfer RNA is the 'adaptor' molecule hypothesized by Francis Crick, which mediates recognition of the codon sequence in mRNA and allows its translation into the appropriate amino acid.

Structure

tRNA has primary structure (the order of nucleotides from 5' to 3'), secondary structure, usually visualized as the *cloverleaf structure*, (Fig. 4.6), and tertiary structure (all tRNAs have a similar L-shaped 3D structure that allows them to fit into the P and A sites of the ribosome). The primary structure was reported in 1969 by Robert W. Holley. The secondary and tertiary structures were derived from X-ray crystallographic studies reported independently in 1974 by American and British research groups headed, respectively, by Alexander Rich and Aaron Klug.

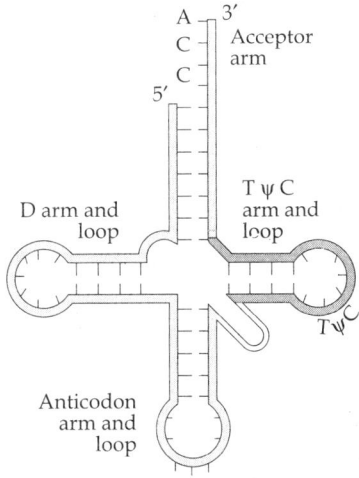

FIG. 4.6 Clover Leaf Model of Transfer RNA

There are at least 32 different kinds of tRNA in an eukaryotic cell. They are relatively small molecules; each one is made up of only 73-93 ribonucleotides. Although tRNA is a single stranded RNA, it bends around in certain places resulting in some ribonucleotides pairing up with others in the same chain, forming three loops. Each tRNA molecule has one amino acid attached to its 3' end. Since there are only 20 amino acids and around 32 different kinds of tRNAs, some amino acids are carried by more than one type of tRNA on one of the three loops called an anticodon. Anticodons are made up of three bases and are involved in translation. The particular amino acid attached to a tRNA molecule is determined by its anticodon sequence. The tRNA molecule contains:

1. The 5'-terminal phosphate.
2. The acceptor stem (also called the *amino acid stem*) is a 7-bp stem that incudes the 5'-terminal nucleotide and the 3'-terminal nucleotide with the 3'-terminal OH group (which can bind the amino acid). The acceptor stem may contain non-Watson-Crick base pairs.

3. The CCA tail is a CCA sequence at the 3' end of the tRNA molecule. This sequence is important for the recognition of tRNA by enzymes critical in translation. In prokaryotes, the CCA sequence is transcribed. In eukaryotes, the CCA sequence is added during processing.
4. The D arm is a 4-bp stem ending in a loop that often contains dihydrouridine.
5. The anticodon arm is a 5-bp stem containing the anticodon.
6. The T arm is a 5-bp stem containing the sequence TΨC.
7. Modified forms of the standard adenine, guanine, cytosine, and uracil bases are present in tRNA.

Ribosomal Ribonucleic Acid (rRNA)

Ribosomal RNA (rRNA) is the primary constituent of ribosomes. Ribosomes are the protein-manufacturing organelles of cells and exist in the cytoplasm. Like all RNAs, rRNA is also transcribed from DNA. Ribosomal proteins are transported into the nucleus and assembled together with rRNA before being transported through the nuclear membrane. This type of RNA makes up the vast majority of RNA found in a typical cell. While proteins are also present in the ribosomes, solely rRNA is able to form peptides. Therefore ribosome is often referred to as ribozyme.

Mammalian cells have two mitochondrial (23S and 16S) rRNA molecules and four types of cytoplasmic rRNA (28S, 5.8S, 5S present in the large subunit of ribosome, and 18S present in the small subunit). 28S, 5.8S and 18S rRNAs are encoded by a *single transcription unit* organized into 5 clusters (each has 30-40 repeats) on the 13, 14, 15, 21 and 22 chromosomes. These are transcribed by RNA polymerase I. 5S occurs in tandem arrays (~200-300 true 5S genes and many dispersed pseudogenes), the largest one on the chromosome 1q41-42. 5S rRNA is transcribed by RNA polymerase III.

Cytoplasmic rRNA genes are highly repetitive because of huge demand of ribosomes for protein synthesis (gene dosage) in the cell.

NON-CODING RNA

A non-coding RNA (ncRNA) is any RNA molecule that functions without being translated into a protein. The DNA sequence from which a non-coding RNA is transcribed as the end product is often called an RNA gene or non-coding RNA gene. The most prominent examples of non-coding RNAs are transfer RNA (tRNA) and ribosomal RNA (rRNA), both of which are involved in the process of translation and gene expression. However, since the late 1990s, many new non-coding RNAs have been found.

Small Nuclear RNA

Small nuclear RNA (snRNA) is a class of small RNA molecules that are found within the nucleus of eukaryotic cells. They are transcribed by RNA polymerase II or RNA polymerase III. They are involved in a variety of important processes such as RNA splicing (removal of introns from hnRNA), regulation of transcription factors (7SK RNA) or RNA polymerase II (B2 RNA), and maintaining the telomeres. They are always associated with specific proteins, and the complexes are referred to as small nuclear ribonucleoprotein (snRNP) or sometimes as snurps.

Small Nucleolar RNAs

Small nucleolar RNAs (snoRNAs) are a class of small RNA molecules that guide chemical modifications (methylation or pseudouridylation) of ribosomal RNAs (rRNAs) and other RNA genes. These modifications enhance the function of the mature RNA. They are frequently encoded in the introns of ribosomal proteins and are synthesized by RNA polymerase II, but can also be transcribed as independent (sometimes polycistronic) transcriptional units. snoRNAs are a component in the small nucleolar ribonucleoprotein (snoRNP), which contains snoRNA and proteins.

MicroRNA

The microRNA (also miRNA) are single stranded RNA molecules that consist of 21-23 nucleotides, which regulate the expression of other genes.

Efference RNA

Efference RNA (eRNA) is derived from intron sequences of genes or from non-coding DNA. The function is assumed to be regulation of translational activity by interference with the transcription apparatus or target proteins of the translated peptide in question, or by providing a concentration-based measure of protein expression.

OBJECTIVE TYPE QUESTIONS

1. Nucleic acids are composed of three types of molecules. Which of the following is not part of a nucleic acid?
   A. A carbohydrate
   B. A heterocyclic aromatic ring
   C. A lipid
   D. A phosphate ester

2. Which of the following bases occurs only in RNA?
   A. Adenine
   B. Uracil
   C. Guanine
   D. Thymine

3. Which of the following statements about nucleosides is false?
   A. Nucleosides are N-glycosides.
   B. The carbohydrate is a furanose ring.
   C. The anomeric C is attached to a N atom in the base.
   D. The base is trans to the $-CH_2OH$ group of the furanose ring (i.e., stereochemistry).

4. Which of the following statements about Polynucleotides is false?

    A. A phosphodiester joins the 3′ oxygen of one nucleotide to the 5′ oxygen of another.

    B. The sequence of nucleotides is written from the 5′ end to 3′ end.

    C. The two most important polynucleotides are RNA and DNA.

    D. Oligonucleotides have approximately 50 nucleotides or less.

5. Which of the following statements about the secondary structure of DNA is false?

    A. The helix has two identical grooves that allow access to the base pairs.

    B. The secondary structure is an antiparallel double helix.

    C. There are three forms of the double helix, the more common B-DNA, as well as A-DNA and Z-DNA.

    D. The two strands are held together by hydrogen bonding between complimentary base pairs, and base-stacking.

6. Which of the following process only involves DNA?

    A. Replication   B. Expression

    C. Transcription   D. Translation

7. mRNA is primarily used for?

    A. Storing genetic information.

    B. Transcription of DNA.

    C. Translation of DNA.

    D. The addition of individual amino acids to the protein.

8. The DNA in the form of a long strand is called _____.

    A. Monomer   B. Dimer

    C. Polymer   D. Tetramer

9. The building blocks of nucleic acids are

    A. Amino acids   B. Nucleotides

    C. Pentose sugars   D. Phosphate groups

10. A nucleotide may contain a(n)

    A. Base   B. 5-carbon sugar

    C. Phosphate group   D. All of these

Answers

1. C	2. B	3. D	4. B	5. A
6. A	7. B	8. C	9. B	10. D

SHORT ANSWER TYPE QUESTIONS

1. The DNA double helix looks like a twisted ladder. What makes up each rung of the ladder? What holds the rungs together at the sides?
2. Does the 'free arm' of deoxyribose (the carbon that is not a member of the pentose ring) point in the direction in which the coding strand is read, or against it?
3. In DNA, each strand is made up of two zones or regions. One zone of each strand is made up of identical repeating units, while another zone is made up of differing units. What are these zones of each strand called?
4. How many kinds of 5-membered rings are there in DNA?
5. Write note on Z-DNA.

LONG ANSWER TYPE QUESTIONS

1. Explain in detail the structure of mRNA.
2. Which DNA double helix do you think would be harder to separate into two strands: DNA composed predominantly of AT base pairs, or of GC base pairs? Why?
3. Write an account on transfer RNA.
4. What is Chargaff's rule?
5. Write a note on different forms of DNA.

Cell Boundaries

5

CELL MEMBRANE

Introduction

The cell membrane is defined as the external limiting membrane of a cell that separates the cell wall and the cytoplasm, and regulates the flow of material into and out of the cell forming a selective permeability barrier. It is composed of lipids (fat molecules), proteins and some carbohydrates; the lipids have a hydrophobic end (insoluble in water) and a hydrophilic end (water-soluble) and are arranged in bilayers in which the lipids line up into two layers with the hydrophobic ends facing each other and the hydrophilic ends facing the outside and the inside of the cell to face extracellular and intracellular fluids, respectively. Cell surface membranes often contain receptor proteins and cell adhesion proteins. There are also other proteins with a variety of functions.

Cell membrane is also known as plasma membrane or plasmalemma. The term plasma membrane was suggested by C. Nageli and C. Cramer in 1855 while Q. Plowe in 1933 coined the term plasmalemma.

CHEMICAL COMPOSITION OF CELL MEMBRANE

Chemically, cell membrane is composed of 40% lipids and 60% proteins. In some cases carbohydrates may be associated with proteins or lipids giving rise to glycoproteins or glycolipids, respectively.

Lipids

The major membrane lipids are phospholipids, the fatty acid chains are in the range of 16-18 carbons long as the chains with fewer than 12 carbons cannot form a stable bilayer. Phospholipid chains are amphipathic molecules, that is, the head end has a negatively-charged (polar) region, while the remainder of the molecule, the tail, consists of two (non-polar) long fatty acid chains. The phospholipids in cell membranes self-organize into a bimolecular layer, with the non-polar fatty acid chains in the middle while the polar regions are oriented toward the membrane surfaces due to their attraction to the polar water molecules in the extracellular and cytosolic fluids.

The plasma membrane also contains other lipids. For example, cholesterol, a steroid lipid fills in small gaps in the phospholipid structure, thus improving membrane impermeability to small water-soluble molecules like glucose. Cholesterol also acts as a membrane antifreeze agent, decreases bilayer fluidity at higher temperatures (e.g., raising lipid bilayer 'melting point') and preventing hydrocarbon chains of phospholipids from aggregating at lower temperatures (e.g., lowering membrane 'freezing point'). Plasma membranes may contain up to ~1 cholesterol molecule for each phospholipid molecule. The precise lipid composition of plasma membranes varies from one cell type to another, and also varies among the membranes of organelles within each cell type.

There are ~5×10^6 lipid molecules in a 1 μ^2 area of lipid bilayer or ~2.5 bilayer lipid pairs/nm^2 of cell membrane surface. Thus, the plasma membrane of a typical 20 μ human tissue cell contains ~10 billion lipid molecules. Phospholipids are not covalently bound to each other; therefore each lipid molecule is free to move independently, resulting in considerable random lateral movement parallel to the bilayer surfaces. The long fatty acid chains, each include one unsaturated bond, producing a kink in the otherwise straight chain that prevents close packing (and solidification). The chains also wiggle back and forth, so the lipid bilayer has fluid like characteristics much like a layer of oil on a water surface.

The different types of lipids are present in the cell membranes. The ten main types of lipids are:

  (i) Cholesterol

  (ii) Glycolipids

  (iii) Phosphatidylcholine

  (iv) Sphingomyelin

  (v) Phosphatidylethnolamine

  (vi) Phosphatydilinositol

 (vii) Phosphatidylserine

(viii) Phosphatidylglycerol

  (ix) Diphosphatidylglycerol (Cardiolipin)

  (x) Phosphatidic acid

Proteins

Membrane proteins are embedded in the lipid bilayer plasma membrane. Indeed, it has been said that the lipid bilayer serves as a 'solvent' for membrane proteins. The plasma membrane of a typical 20 μ human tissue cell contains ~0.1 billion protein molecules.

There are two classes of membrane proteins: Integral (intrinsic) membrane proteins and peripheral (extrinsic) membrane proteins.

Integral membrane proteins are closely associated with membrane lipids and cannot be extracted from the membrane without disrupting the lipid bilayer. Like phospholipids, integral proteins are amphipathic. Polar amino acid side chains lie in one region of the molecule and non-polar side chains are in a separate region. Thus integral proteins vertically align with the amphipathic lipids in the plasma membrane — protein polar regions position themselves at the surfaces in association with polar water molecules, while the protein non-polar regions are attracted to the interior in association with the non-polar fatty acid chains at the center of the lipid bilayer membrane (Fig. 5.1).

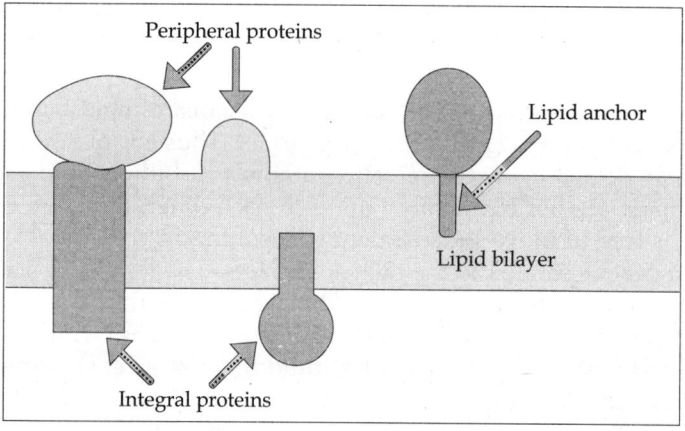

FIG. 5.1 Peripheral and Integral Proteins

Peripheral membrane proteins are bound to the hydrophilic regions of integral membrane proteins or to the hydrophilic heads of membrane lipids by weak electrostatic forces. Most peripheral proteins are located near the cytoplasmic surface of the plasma membrane rather than on the extracellular surface and mediate such properties as cell shape and motility. Peripheral proteins are not amphipathic and do not associate with the hydrophobic regions of the lipids in the membrane interior (Fig. 5.1).

Carbohydrates

The plasma membrane also contains small amounts of carbohydrate. This carbohydrate is covalently linked to some of the membrane lipids and proteins. Carbohydrate portions of the membrane glycoproteins are always located at the extracellular surface, forming the glycocalyx (together with collagen proteins and glycosaminoglycans). A small proportion of membrane carbohydrate is glycolipids, but most is in the form of glycoproteins. The sugar units are usually short oligosaccharide chains attached to serine, threonine, or asparagine side chains.

STRUCTURAL MODELS OF CELL MEMBRANE

In order to explain the physical and biological features of the cell membrane a number of models have been suggested from time to time.

The Lipid Bilayer Model

In 1895, Overton observed that the fat soluble substances could pass easily through the cell membrane. He, therefore, concluded that cell membrane contains lipids.

Phospholipids are the most common molecules in membranes. They consist of two parts, a hydrophilic (polar) head group and a hydrophobic (lipophilic) fatty acid chain. This structure lends a tendency to form organized structures and monomolecular layers whose degree of organization is dependent on the length of the fatty acid residues and the number of double bonds to phospholipids. The Dutch scientists E. Gorter and F. Grendel in 1925 recognized that two such layers of opposite orientations together would produce a fine membrane model and thus suggested the first ever model of cell membrane (Fig. 5.2).

Protein-lipid-protein (Sandwich Model)

This model was proposed in 1934, which is known as **Davson-Danielli model** (after the scientists who described it). In the simplified 'sandwich model' of a cell membrane it was suggested that a phospholipid bilayer is sandwiched between two layers of protein (Fig. 5.2(a)). Having phospholipids (with phosphates) rather than ordinary lipids is essential because the lipid layer is permeable to polar water molecules. Chemical analyses also revealed a great deal of protein in membranes. Though membranes were more permeable to lipids than to water, they still absorbed water faster than a pure phospholipid layer should have. Since most proteins are water absorbent, a new model was developed (lipid membrane model of Gorter and Grandel). It suggested that the phospholipid bilayer was coated on both sides with water-soluble proteins, in a kind of lipoprotein sandwich (Fig. 5.2(b)).

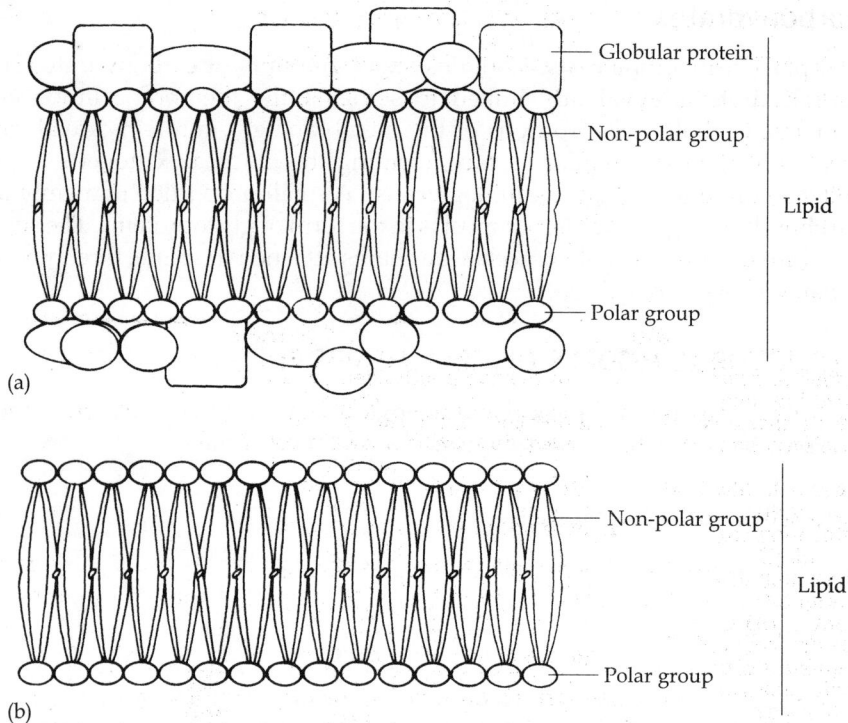

FIG. 5.2 (a) The Davson and Danielli model of cell membrane (b) The lipid membrane model of Gorter and Grendel

Microsurgical methods reinforced the idea that the membrane is a fluid. If a cell membrane is pushed with a probe, it bends like the surface of a balloon, and springs back when released. If it is penetrated, however, the membrane simply conforms around the probe. When the probe is withdrawn, the membrane reseals as if it is a liquid flowing into itself.

New chemical methods, especially methods for analysing the structure of proteins, revealed that the proteins of membranes were highly variable in both quantity and type and also showed that the proteins in membranes, rather than being essentially hydrophilic, were largely lipophilic and hydrophobic.

The Unit Membrane Model

In 1957, J. D. Robertson proposed a modified version of the membrane model, based primarily on electron microscope studies, which he called the 'unit membrane'. He suggested that under the high magnification of the transmission electron microscope, membranes have a characteristic 'trilaminar' appearance consisting of two darker outer lines and a lighter inner region. According to the unit membrane model, the two outer, darker lines are the protein layers and the inner region the lipid bilayer.

Fluid Mosaic Model

This model, known as the fluid mosaic model, was proposed by biochemists S. J. Singer and Garth L. Nicolson in 1972. The model retains the basic lipid bilayer structure first proposed by Gorter and Grendel and modified by Danielli and Davson and Robertson. The proteins, however, are thought to be globular and to float within the lipid bilayer rather than forming the layers of the sandwich-type model.

The hydrophobic tails of the phospholipids, the major lipid component of the membrane, face inward, away from the water. The hydrophilic heads of the phospholipids are on the outside where they interact with water molecules in the fluid environment of the cell. Floating within this bilayer are the proteins, some of which span the entire bilayer and may contain channels, or pores, to allow passage of molecules through the membrane. The entire membrane is fluid — the lipid molecules move within the layers of the bilayer while the 'floating' proteins also freely move within the bilayer (Fig. 5.3).

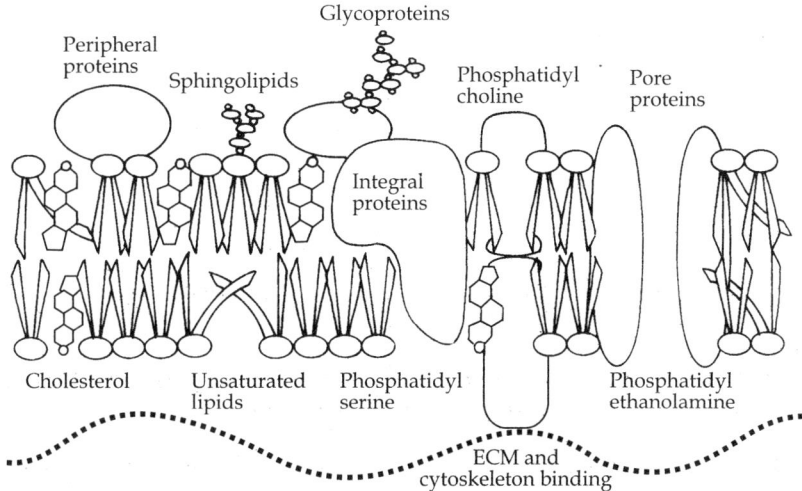

FIG. 5.3 The Fluid Mosaic Model

According to the Fluid Mosaic Model, the basic structure of the membrane is provided by the phospholipid molecules. Varying the lipid tails can give membranes different properties. For example, unsaturated fatty tails make a membrane more liquid, while the addition of cholesterol to the fatty layer makes them more viscous and more repellent to water.

Proteins are responsible for the special characteristics of different types of membranes, controlling their ability to transport molecules, to receive chemical messages, to attach to adjacent cells, etc.

The nature of these membrane proteins was studied by Unwin and Henderson (1984). They found that the portion of the protein that spans the lipid bilayer is hydrophobic in nature (i.e., similar to the lipids forming the bilayer) and arranged in a three-dimensional shape, often in the form of an alpha helix.

Since 1972, the fluid mosaic model has been modified because of more recent discoveries, such as those of Unwin and Henderson, but it still remains the model preferred by biologists today. It explains the current knowledge of membrane structure and also serves as the basis for our understanding of how membranes function.

FUNCTIONS OF THE CELL MEMBRANE

1. Transport

Diffusion

Diffusion is the movement of particles from an area of higher concentration to an area of lower concentration. The movement is due to collisions, which occur more frequently in areas of higher concentration (Fig. 5.4).

Diffusion

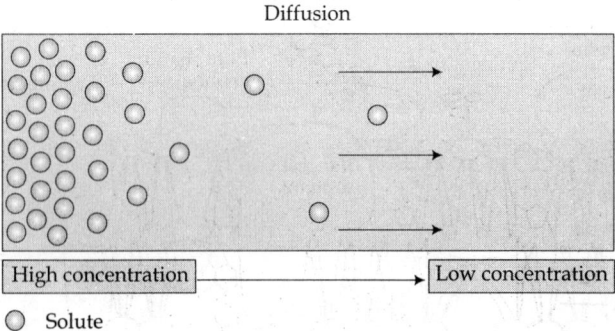

High concentration — Low concentration

○ Solute

FIG. 5.4 Solute transport is from the left to the right: movement of the solutes is due to the concentration gradient (dC/dx).

Water, carbon dioxide, and oxygen are among the few simple molecules that can cross the cell membrane by diffusion (or a type of diffusion known as osmosis). Diffusion is one principle method of movement of substances within cells, as well as the method for essential small molecules to cross the cell membrane. Gas exchange in gills and lungs operates by this process. Carbon dioxide is produced by all cells as a result of cellular metabolic processes. Since the source is inside the cell, the concentration gradient is constantly being replenished, thus the net flow of CO_2 is out of the cell. Metabolic processes in animals and plants usually require oxygen, which is in greater concentration inside the cell, thus the net flow of oxygen is into the cell.

The plasma membrane is differentially permeable because some particles can pass through, others cannot. It can control the extent to which certain substances pass through.

Non-polar molecules pass through cell membranes more readily than polar molecules because the center of the lipid bilayer (the fatty acid tails) is non-polar and does not readily interact with polar molecules.

The following substances can pass through the cell membrane:

(i) Non-polar molecules (for example, lipids)

(ii) Small polar molecules such as water

The substances that cannot pass through the cell membrane are:

(i) Ions and charged molecules (for example, salts dissolved in water)

(ii) Large polar molecules (for example, glucose)

(iii) Macromolecules

Osmosis

Osmosis is the diffusion of water across a differentially permeable membrane. It occurs when a solute (for example, salt, sugar, protein, etc.) cannot pass through a membrane but the solvent (water) can pass through it. Water always moves from its higher concentration (has less solute) to its lower concentration. In general, water moves toward the area with a higher solute concentration because it has a lower water concentration.

Osmotic pressure is a hydrostatic pressure, on the side of membrane with higher solute concentration, produced by water diffusing to that side of membrane, for example,

(1) A differentially permeable membrane separates two solutions.

(2) One side of the beaker has more water (lower percentage of solute) and the other side has less water (higher percentage of solute).

(3) The membrane does not permit passage of the solute.

(4) Membrane permits passage of water with net movement of water from one to the other side of the beaker.

(5) Osmotic pressure allows liquid increase on side of the membrane with greater percentage of solute (Fig. 5.5).

Osmotic pressure is the pressure that develops in a system due to osmosis. Osmosis is a constant process in life: for example, water is absorbed in large intestine, retained by kidneys, and taken up by blood.

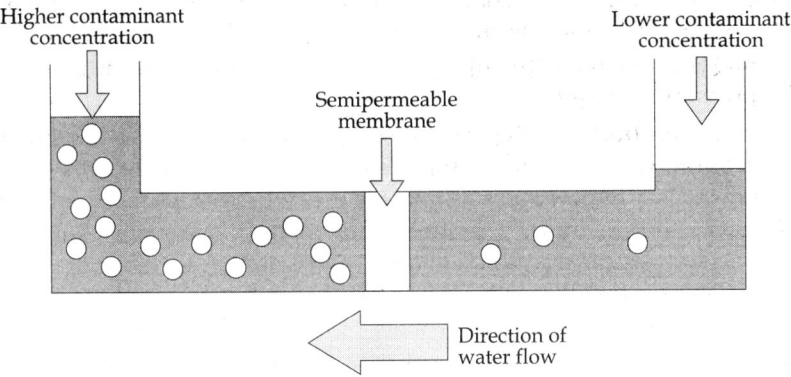

FIG. 5.5 Osmosis

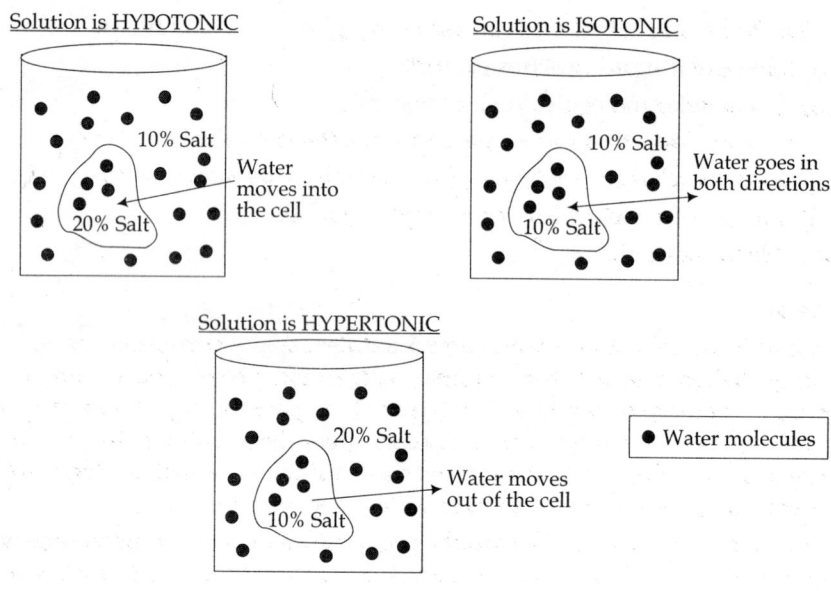

FIG. 5.6 Tonicity

Tonicity

Tonicity refers to the relative concentration of solute on either side of a membrane (Fig. 5.6).

Isotonic: In an isotonic solution, the concentration of solute is same on both sides of the membrane (inside the cell and outside). A cell placed in an isotonic solution neither gains nor loses water. Most cells in the body are in an isotonic solution.

Hypotonic: A hypotonic solution is one that has less solute (more water). Cells in hypotonic solution tend to gain water. Animal cells can *lyse* (rupture) in a hypotonic solution due to the osmotic pressure. Freshwater organisms live in a hypotonic solution and have a tendency to gain water. The contractile vacuole in freshwater protozoans removes water that enters the cell. The cell wall of plant cells prevents the cell from rupturing. The osmotic pressure, called *turgor pressure,* helps support the cell. A cell in which the contents are under pressure is *turgid.*

Hypertonic solution: A hypertonic solution is one that has a high solute concentration. Cells in a hypertonic solution will lose water. The marine environment is a hypertonic solution for many organisms. They often have mechanisms to prevent dehydration or to replace lost water. Animal cells placed in a hypertonic solution will undergo crenation, a condition where the cell shrivels up as it loses water. Plant cells placed in a hypertonic solution will undergo *plasmolysis,* a condition where the plasma membrane pulls away from the cell wall as the cell shrinks. The cell wall is rigid and does not shrink.

Facilitated diffusion

Facilitated diffusion involves the use of a protein to facilitate the movement of molecules across the membrane. In some cases, molecules pass through channels within the protein. In other cases, the protein changes shape, allowing molecules to pass through. In this case additional energy is not required because the molecule is traveling down a concentration gradient (high concentration to low concentration). The energy of movement comes from the concentration gradient (Fig. 5.7).

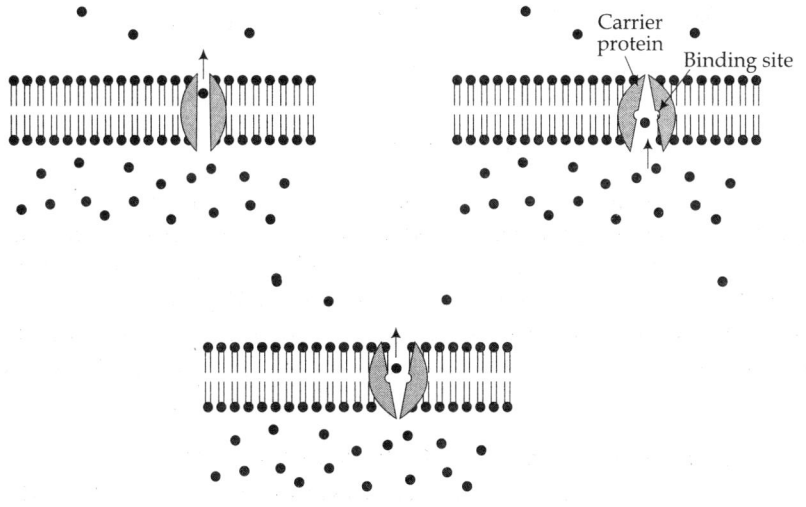

Carrier protein

Binding site

FIG. 5.7 Facilitated Diffusion

Active transport

Active transport is used to move ions or molecules *against* a concentration gradient (low concentration to high concentration). Active transport is like a water pump; it uses energy to pump water uphill where a siphon cannot. Facilitated diffusion is like a siphon in which additional energy is not required but it can only allow movement downhill.

Movement against a concentration gradient requires energy. The energy is supplied by ATP which is released by breaking a phosphate bond to produce ADP:

$$ATP = ADP + P_i + energy$$

Cells that use a lot of active transport have many mitochondria to produce the ATP needed (Fig. 5.8).

The sodium-potassium pump

The sodium-potassium pump is observed in human nerve and muscle cells. Na^+ is maintained at low concentrations inside the cell and K^+ is at higher concentrations. The reverse is the case on the outside of the cell. When a nerve

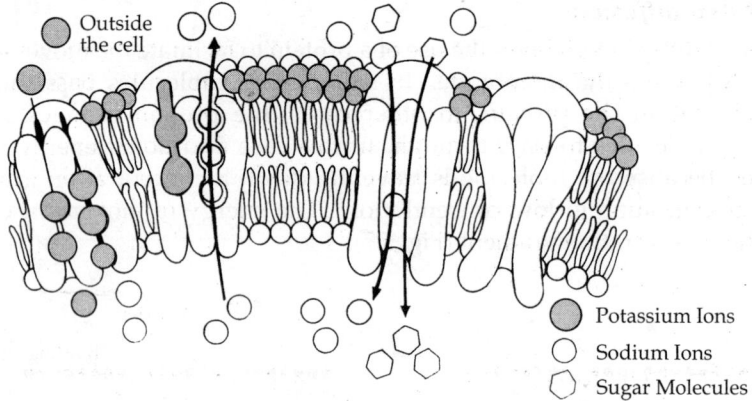

FIG. 5.8 Active Transport

message is propagated, the ions pass across the membrane, thus sending the message. After the message has passed, the ions must be actively transported back to their starting positions across the membrane. Up to one-third of the ATP used by a resting animal is used to reset the Na-K pump.

Thus, the sodium-potassium pump uses active transport to move three sodium ions to the outside of the cell for each two potassium ions that it moves in.

Mechanism of operation of the sodium-potassium pump

The diagrams below illustrate the mechanism of operation of the sodium-potassium pump. In these diagrams, the structure in the centre of the

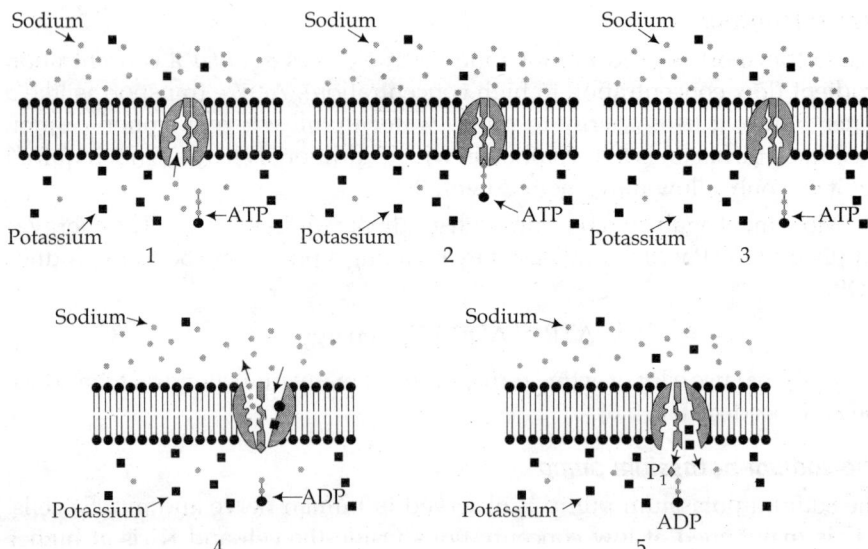

FIG. 5.9 Sodium-potassium Pump

membrane is used to represent the pump protein. Circles are used to represent sodium ions and squares are used to represent potassium ions. Notice that the pump protein has three sodium binding sites and two potassium binding sites (Fig. 5.9).

Three sodium ions enter the pump. ATP bonds to the pump. Now, one phosphate bond in the ATP molecule breaks, releasing its energy to the pump protein. The pump protein changes shape, releasing the sodium ions to the outside. The two potassium binding sites are also exposed to the outside, allowing two potassium ions to enter the pump.

When the phosphate group detaches from the pump, the pump returns to its original shape. The two potassium ions leave and three sodium ions enter. The cycle then repeats itself.

Vesicle-mediated transport

Vesicles and vacuoles that fuse with the cell membrane may be utilized to release or transport chemicals out of the cell or to allow them to enter a cell. Exocytosis is the term applied when transport is out of the cell.

Endocytosis is the case when a molecule causes the cell membrane to bulge inward, forming a vesicle. Phagocytosis is the type of endocytosis where an entire cell is engulfed. Pinocytosis is when the external fluid is engulfed (Fig. 5.10). Receptor-mediated endocytosis occurs when the material to be transported binds to certain specific molecules in the membrane. Examples include the transport of insulin and cholesterol into animal cells.

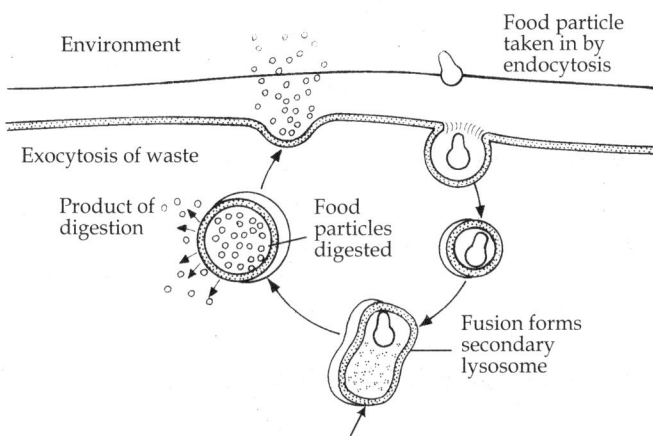

FIG. 5.10 Endocytosis and Exocytosis

Receptor-mediated endocytosis

This occurs when specific macromolecules bind to plasma membrane receptors. This allows cells to receive specific molecules and then sort them within the cell. A macromolecule that binds to receptor is called a ligand; binding of ligands to receptors causes receptors to gather at one location. This location is a coated pit with a layer of fibrous protein called clathrin, on cytoplasmic side. Pits appear associated with exchange of substances between cells (e.g., maternal and fetal blood).

2. Cell Identification Markers

Lipids and proteins within the membrane may have a carbohydrate chain attached. These glycolipids and glycoproteins often function as cell identification markers, allowing cells to identify other cells. This is particularly important in the immune system where cells patrolling the body's tissues identify and destroy foreign invaders such as bacteria or viruses.

3. Cell Adhesion-Junctions

Proteins associated with the cell membranes of animal cells may bind to proteins of adjacent cells. These connections, called junctions may serve to bind cells together to prevent the movement of material between the cells, or to allow cells to communicate with each other.

4. Receptors

Receptors enable cells to detect hormones and a variety of other chemicals in their environment. The binding of a molecule and a receptor initiates a chemical change within the cell. Hormones are molecules that cells use to communicate with one another. For example, cells in the pancreas produce the hormone insulin when glucose levels in the blood become elevated. The hormone travels within the blood to other parts of the body. It stimulates liver and muscle cells to begin removing the glucose and storing it as glycogen.

5. Vesicle Trafficking

Vesicles may follow microtubules to their destination. Proteins within the membrane of the vesicle recognize and attach to proteins in other membranes. This allows vesicles to attach to the membranes of other organelles such as the endoplasmic reticulum, Golgi apparatus, or lysosomes.

EXTRACELLULAR MATRIX OF ANIMAL CELLS

1. Extracellular matrix is meshwork of insoluble proteins with carbohydrate chains that are produced and secreted by animal cells; fills spaces between animal cells.
2. This matrix most likely influences the development, migration, shape and function of cells.
3. Collagen gives the matrix strength and elastin gives it resilience.
4. Fibronectins and laminins bind to membrane receptors; permit communication between matrix and cytoplasm.

5. Fibronectins and laminins form pathways that direct the migration of cells during development.

6. Proteoglycans are glycoproteins that provide a packing gel that joins the various proteins in matrix and most likely regulate signaling proteins that bind to receptors in the plasma protein.

JUNCTIONS BETWEEN CELLS

Cell junctions are points of contact that physically link neighboring cells or provide functional links; animal cells have three types: adhesion junctions, tight junctions, and gap junctions.

Adhesion junctions *(macula adherens)* — Adhesion junctions (desmosomes) are molecular complexes of cell adhesion proteins and linking proteins that attach the cell surface adhesion proteins to intracellular keratin cytoskeleton filaments. The cell adhesion proteins of the desmosome are members of the cadherin family of cell adhesion molecules. They are transmembrane proteins that bridge the space between adjacent epithelial cells by way of homophilic binding of their extracellular domains to other desmosomal cadherins on the adjacent cell. The desmosomal linking proteins such as desmoplakin bind to the intracellular domain of cadherins and form a connecting bridge to the cytoskeleton (for example, in heart, stomach, bladder).

Tight Junctions *(zonula occludens)* — This is an occluding type of junction, where the plasma membranes of adjacent cells pinch tightly together. This creates a selective barrier between the spaces, allowing only certain materials to pass. Found often between cells of animal intestine, tight junctions assure that the correct chemical environment of intestinal lumen is maintained (Fig. 5.11).

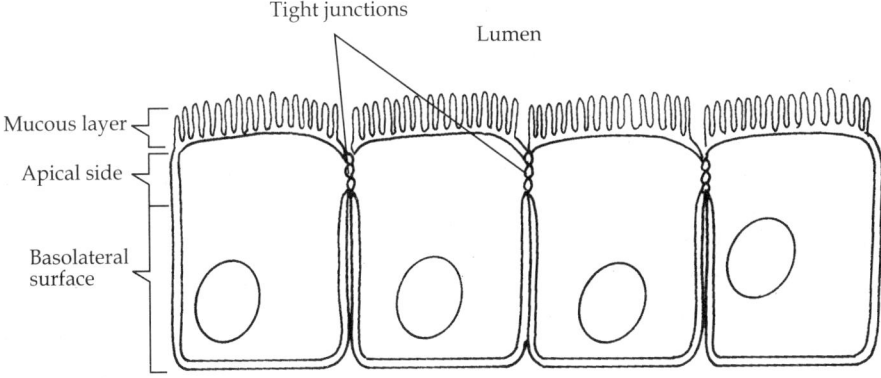

FIG. 5.11 Tight Junctions

Gap junction *(nexus)* — A gap junction allows cells to communicate.

(i) They are formed by the joining of two identical plasma membrane channels; channel of each cell is lined by six plasma proteins.

(ii) They provide strength to the cells involved and allow the movement of small molecules and ions from the cytoplasm of one cell to the cytoplasm of the other cell.

(iii) Gap junctions are important to function of heart muscle and smooth muscle because they permit diffusion of ions required for cells to contract (Fig. 5.12).

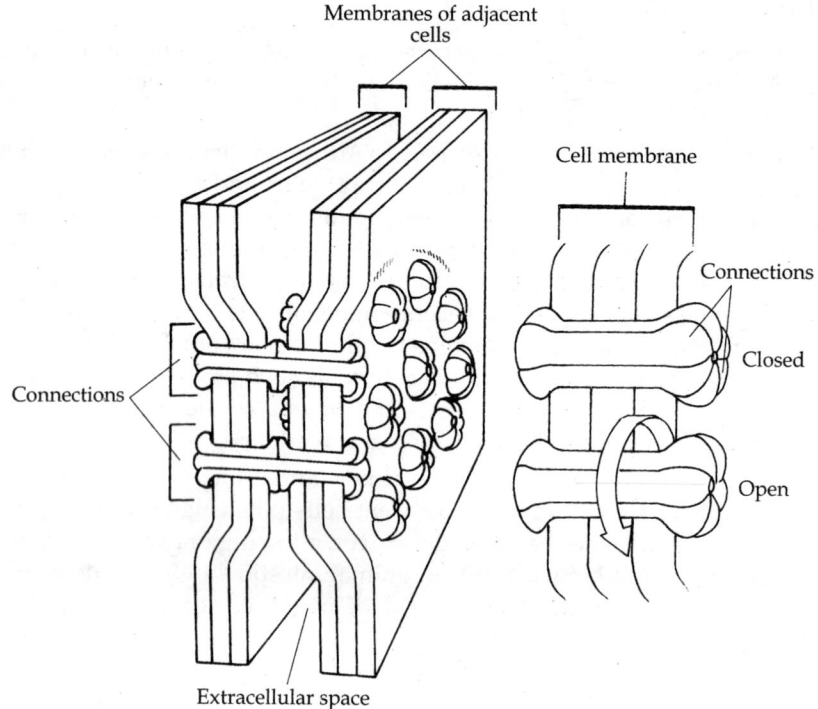

FIG. 5.12 Gap Junction

Hemidesmosomes—When visualized by electron microscopy, they are similar in appearance to desmosomes. Rather than linking two cells, hemidesmosomes attach one cell to the extracellular matrix. Rather than using cadherin, hemidesmosomes use integrin cell adhesion proteins. Hemidesmosomes are assymetrical and are found in epithelial cells connecting the basal face to other cells.

GLYCOCALYX

It is the carbohydrate-rich zone on the cell surface. Although most of the carbohydrate is attached to intrinsic plasma membrane molecules, the glycocalyx usually also contains both glycoproteins and proteoglycans that have been secreted into the extracellular space and then adsorbed onto the cell surface.

Functions

1. It protects the plasma membrane from physical and chemical injury.
2. It provides immunity by enabling the immune system to recognize and selectively attack foreign organisms.
3. It forms the basis for compatibility of blood transfusions, tissue grafts and organ transplants.
4. It helps the cells to adhere together.
5. It aids in fertilization by enabling sperm to recognize and bind to eggs.
6. It helps in embryonic development by guiding embryonic cells to their destinations in the body.

CELL WALL

Introduction

A cell wall is a more or less solid layer surrounding a cell, a rigid boundary consisting of peptidoglycans in prokaryotic cells and cellulose in plant cells. It is found in bacteria, archea, fungi, plants, and algae. Animals and most other protists have cell membranes but the surrounding cell wall is absent. When a cell wall is removed using cell wall degrading enzymes, the remaining part of the cell and its surrounding plasma membrane is called protoplast.

The cell wall is differentiated into three layers (Fig. 5.13):

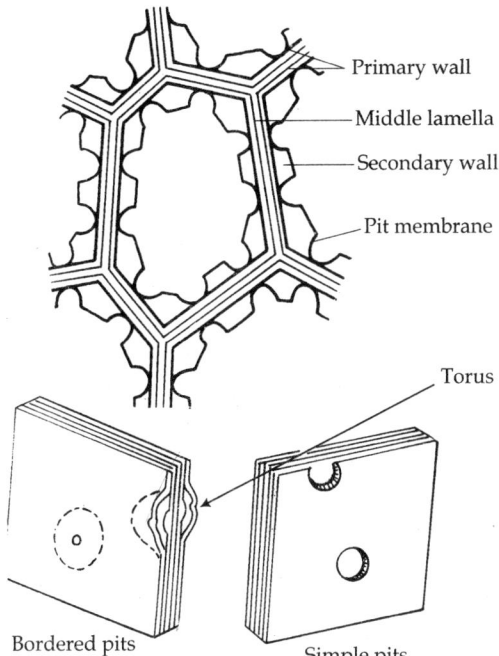

FIG. 5.13 Cell wall

1. **Middle lamella:** It is the outermost layer, viscous jelly like glue that binds adjacent cells, composed primarily of pectic polysaccharides.

2. **Primary wall:** This wall is deposited by cells before and during active growth. The primary wall is comprised of pectic polysaccharides, cross-linking glycans (hemicellulose), cellulose and protein. The actual content of the wall components varies with the species and age. All plant cells have a middle lamella and primary wall.

3. **Secondary wall:** Some cells deposit additional layers inside the primary wall. This occurs after growth stops or when the cells begin to differentiate. The secondary wall is mainly for support and is comprised primarily of cellulose and lignin. Often it can be distinguished into distinct layers, S1, S2 and S3, which differ in the orientation, or direction, of the cellulose microfibrils.

4. **Tertiary wall:** In some cells, a tertiary wall is formed on the inner surface of the secondary wall. It is primarily composed of xylan.

CHEMICAL COMPOSITION

The main ingredient of cell wall is polysaccharide (or complex carbohydrates or complex sugars) which are built from monosaccharides (or simple sugars). Carbohydrates are good building blocks because they can produce a nearly infinite variety of structures. The other components in the wall include proteins and lignin.

Cellulose: Cellulose is a polymer, composed of linear chains of covalently linked glucose residues. It is chemically very stable and is extremely insoluble. The primary cell wall consists of one glucose polymer of roughly 6000 glucose units and in the secondary wall their number increases to 13 - 16000 units. Cellulose readily forms hydrogen bonds with itself (intramolecular H-bonds) and with other cellulose chains (intermolecular H-bonds). A cellulose chain will form hydrogen bonds with about 36 other chains to yield a microfibril. Cellulose chains form crystalline structures called microfibrils. A microfibril with a diameter of 20 - 30 nm contains about 2000 molecules. This is somewhat analogous to the formation of a thick rope from thin fibers. Microfibrils are 5-12 nm wide and give the wall strength—they have a tensile strength equivalent to steel. Some regions of the microfibrils are highly crystalline while others are more 'amorphous' (Fig. 5.14).

Cross-linking glycans (Hemicellulose): This includes diverse group of carbohydrates that are characterized by being soluble in strong alkali. They are linear (straight), flat, with a β-1,4 backbone and relatively short side chains. Two common types include xyloglucans and glucuronarabino xylans. Other less common ones include glucomannans, galactoglucomannans, and galactomannans. The main feature of this group is that they do not form microfibrils. However, they form hydrogen bonds with cellulose and hence the reason they are called 'cross-linking glycans'.

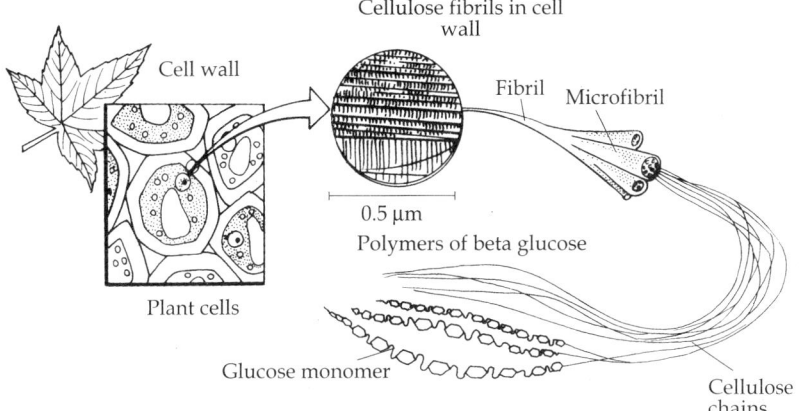

FIG. 5.14 Cellulose in plant cell walls

Pectic polysaccharides: They are the easiest constituents to remove from the wall. They form gels (*i.e.,* used in jelly making). They are the polysaccharides that are particularly rich in galacturonic acid (galacturonans - pectic acids). Polymers of primarily β 1, 4 galacturonans (polygalacturonans) are called homogalacturons (HGA) and are particularly common. These are helical in shape. Divalent cations, like calcium, also form cross-linkages to join adjacent polymers creating a gel. Pectic polysaccharides can also be cross-linked by dihydrocinnamic or diferulic acids. Other pectic acids include Rhamnogalacturonan II (RGII) which features rhamnose and galacturonic acid in combination with a large diversity of other sugars in varying linkages. Dimers of RGII can be cross-linked by boron atoms linked to apiose sugars in a side chain. Although most pectic polysaccharides are acidic, others are composed of neutral sugars including arabinans and galactans. The pectic polysaccharides serve a variety of functions including determining wall porosity, providing a charged wall surface for cell-cell adhesion (middle lamella), cell-cell recognition, pathogen recognition and others.

Protein: Wall proteins are typically glycoproteins (polypeptide backbone with carbohydrate side-chains). The proteins are particularly rich in the amino acids hydroxyproline (hydroxyproline-rich glycoprotein, HPRG), proline (proline-rich protein, PRP), and glycine (glycine-rich protein, GRP). These proteins form rods (HRGP, PRP) or beta-pleated sheets (GRP). Extensin is a well-studied HPRG. HPRG is induced by wounding and pathogen attack. The wall proteins also have a structural role since: (1) the amino acids are characteristic of other structural proteins such as collagen and gelatin; and (2) to extract the protein from the wall requires destructive conditions. Protein appears to be cross-linked to pectic substances and may have sites for lignification. The proteins may serve as the scaffolding used to construct the other wall components. Another group of wall proteins are heavily glycosylated with arabinose and galactose. These arabinogalactan proteins, or AGP's, seem to be tissue specific and may function in cell signaling. They may be important in embryogenesis, and growth and guidance of the pollen tube.

Lignin: It is a polymer of phenolics, especially phenylpropanoids. Lignin is primarily a strengthening agent in the wall. It also resists fungal/pathogen attack.

Suberin, wax, cutin: A variety of lipids are associated with the wall for strength and waterproofing.

Water: The wall is largely hydrated and comprised of between 75-80% water.

ALGAL CELL WALLS

Sendbusch in 2003 studied the cell walls in algae. Algal cell walls contain cellulose and a variety of glycoproteins. The presence of additional polysaccharides in algal cell walls is used for the study of algal taxonomy.

- Manosyl form microfibrils in the cell walls of a number of marine green algae including those from the genera, *Codium, Dasycladus,* and *Acetabularia* as well as in the walls of some red algae, like *Porphyra* and *Bangia.*
- Xylanes are polymers where the β-D-xylosyl residues are linked *via* 1→ 3 and 1→4 glycosidic bonds.
- Alginic acid is a common polysaccharide in the cell walls of brown algae.
- Sulfonated polysaccharides occur in the cell walls of most algae; those common in red algae include agarose, porphyran, funoran, etc.

Other compounds that may accumulate in algal cell walls include sporopollenin and calcium.

PROKARYOTIC CELL WALLS

Cell walls of bacteria are primarily used for protection against hostile environments or, in the case of pathogenic bacteria, against the immune system of the host. They contain peptidoglycan, which can be made visible by Gram-staining in Gram-positive bacteria . The cell walls of bacteria are also vital for containing the high osmotic pressure inside bacterial cells caused by the high concentration of solutes in the cytoplasm.

FUNGAL CELL WALLS

Not all the species of fungi have cell walls but in those where the cell wall is present it is composed of cellulose, glucosamine, and chitin.

FUNCTIONS OF THE CELL WALL

The cell wall serves a variety of functions including:

1. It helps in maintaining the cell shape.
2. It gives support and mechanical strength (i.e., allows plants to grow tall, hold out thin leaves to obtain light).

3. It prevents the cell membrane from bursting in a hypotonic medium (i.e., resists water pressure).

4. It controls the rate and direction of cell growth and regulates cell volume.

5. It is ultimately responsible for controlling plant morphogenesis.

6. It has a metabolic role (i.e., some of the proteins in the wall are enzymes for transport, secretion).

7. It acts as a physical barrier to: (a) pathogens; and (b) water in suberized cells. However, the wall is very porous and allows the free passage of small molecules, including proteins up to 60,000 MW.

8. The wall serves as a storage for carbohydrates.

9. It performs signaling by means of fragments of wall, called oligosaccharins that act as hormones. Oligosaccharins, result from normal development or pathogen attack, they serve a variety of functions including: (a) stimulate ethylene synthesis; (b) induce phytoalexin synthesis; (c) induce chitinase and other enzymes; (d) increase cytoplasmic calcium levels and (d) cause an "oxidative burst". This burst produces hydrogen peroxide, superoxide and other active oxygen species that attack the pathogen directly or cause increased cross-links in the wall making the wall harder to penetrate.

10. Recognition responses, for example, (a) the wall of roots of legumes is important in the nitrogen-fixing bacteria colonizing the root to form nodules; and (b) pollen-style interactions are mediated by wall chemistry.

11. Economic products: cell walls are important for products such as paper, wood, fiber, energy, shelter, and even roughage in our diet.

CELL WALL FORMATION

The cell wall is made during cell division when the cell plate is formed between daughter cell nuclei. The cell plate forms from a series of vesicles produced by the Golgi apparatus. The vesicles migrate along the cytoskeleton and move to the cell equator. The vesicles coalesce and dump their contents. The membranes of the vesicle become the new cell membrane. The Golgi synthesizes the non-cellulosic polysaccharides. At first, the Golgi vesicles contain mostly pectic polysaccharides that are used to build the middle lamella. As the wall is deposited, other non-cellulosic polysaccharides are made in the Golgi and transported to the growing wall.

Cellulose is made at the cell surface. The process is catalyzed by the enzyme cellulose synthase that occurs in a complex in the membrane.

Exactly how the wall components join together to form the wall once they are in place, is not completely understood. Two methods seem likely:

1. Self assembly: This means that the wall components spontaneously aggregate; and

2. Enzymatic assembly: various enzymatic reactions are designed for wall assembly. For example, one group of enzymes stitches xylans together in the wall to form long chains. Oxidases may catalyze additional cross-linking between wall components and pectin methyl esterase may also play an important role.

PLASMODESMATA

Plasmodesmata are narrow channels that act as intercellular cytoplasmic bridges to facilitate communication and transport of materials between plant cells; they are specialized channels that allow the intercellular movement of water, various nutrients, and other molecules (including signalling molecules). Plasmodesmata are located in narrow areas of cell walls called primary pit fields, and they are so dense in these areas (up to one million per square millimeter) that they make up one percent of the entire area of the cell wall (Fig. 5.15).

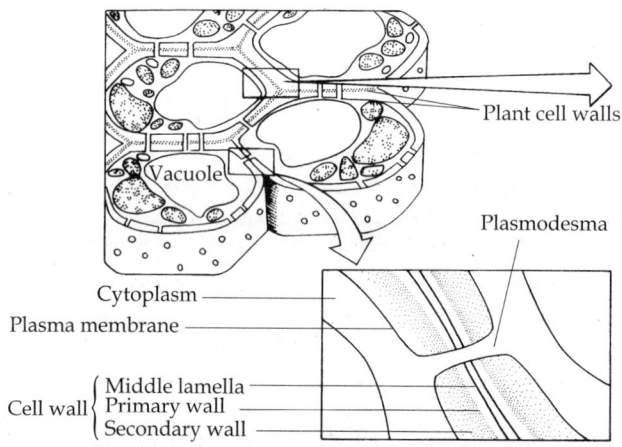

FIG. 5.15 Plasmodesmata

It has been demonstrated that the plasma membrane is continuous between cells, the outer leaflet is continuous with the cell wall and the inner leaflet is continuous with the plasmodesmal pore. Within the plasmodesmal pore, a tightly wound cylinder of membrane termed the desmotubule runs the length of the plasmodesma. The desmotubule provides a rigid stability to plasmodesmata and confers a fixed diameter and pore size to the plasmodesmal canal, much like a cytoskeletal structure.

The space between the plasmalemma and the desmotubule is the cytoplasmic sleeve or cytoplasmic annulus, and transport through plasmodesmata has been proposed to occur either through the lipid portions of the desmotubule or this cytoplasmic sleeve or both.

In some plasmodesmata, there is a region at each end of the plasmodesmal channel that is constricted and termed the neck region, where the plasma membrane component of the plasmodesmata closely associates with the central desmotubule. The neck region is proposed to contain several, spoke-like protein subunits that are located both extracellularly and between the desmotubule and the plasmalemma (linking these two structures), and these proteins can then act as a sphincter to regulate the passage of materials through the plasmodesmata, much like gap junctions in animal cells.

OBJECTIVE TYPE QUESTIONS

1. What do scientists mean when they say that the cell membrane is selectively permeable?
   A. It is impervious to most substances
   B. It only allows certain molecules to pass through
   C. It is a very rigid structure
   D. It is impervious to most substances

2. The cell membrane consists mainly of
   A. Water and carbon dioxide   B. Cellulose and pectins
   C. Phospholipids and proteins   D. Actin and myosin

3. All cell walls found in prokaryotes contain
   A. Peptidoglycans   B. Lipoproteins
   C. Lipopolysaccharides   D. Phospholipids

4. The primary cell wall is the part of a eukaryotic cell wall which is
   A. Closest to the inside of the cell
   B. In the middle of the cell wall
   C. Farthest outside of the cell
   D. The strongest of the three layers

5. Which of the layer in a eukaryotic cell wall contains pectins?
   A. The middle lamella   B. The secondary cell wall
   C. The primary cell wall   D. None of the above

6. Simple diffusion is a way of moving through the cell membrane utilized mainly by
   A. Very large polar molecules
   B. Very large nonpolar molecules
   C. Ions
   D. Small nonpolar molecules

7. A method of passive transport which utilizes special proteins called carrier proteins is called
   A. Facilitated diffusion   B. Simple diffusion
   C. Active transport   D. Protein channels

8. Active transport is different from simple diffusion and passive transport in that it
   A. Does not require an input of energy
   B. Involves the expenditure of energy
   C. Can be used to transport molecules across the cell membrane
   D. Cannot be used to transport molecules across the cell membrane

9. The plasma membrane does all of these except _____
   A. Contains the hereditary material
   B. Acts as a boundary or border for the cytoplasm
   C. Regulates passage of material in and out of the cell
   D. Functions in the recognition of self

10. Cell walls are found in all kingdoms, except _____
    A. Plants   B. Animals
    C. Bacteria   D. Fungi

Answers

1. B	2. C	3. A	4. A	5. A
6. D	7. A	8. B	9. A	10. B

SHORT ANSWER TYPE QUESTIONS

1. What are four kinds of molecules found in the cell membrane? How does each contribute to the functioning of the cell membrane?
2. Write a short note on osmosis.
3. Write a brief account on plasmodesmata.
4. Briefly state the chemical constituents of cell wall.
5. Differentiate between active and passive transport.

LONG ANSWER TYPE QUESTIONS

1. Write a note on chemical composition of plasma membrane.
2. Describe the 'fluid mosaic model' of the plasma membrane.
3. Describe the structure and functions of the cell wall.
4. Write a note on the functions of cell membrane.
5. Why do certain proteins 'float' in the cell membrane?

Mitochondria

6

INTRODUCTION

Mitochondria (Gr., *mito* = thread, *chondrion* = granule) are small, elongated, sac-like structures bounded by a double membrane. They are the sites of aerobic respiration, and generally are the major energy production center in eukaryotes; the energy packets are in the form of ATP molecules and mitochondria are thus referred to as the 'power houses' of cells.

HISTORY

The history of mitochondria dates back to nearly 150 years ago. In 1857, Kölliker discovered the mitochondria in muscle cells. In 1882, Flemming coined the term *Fila* to describe these structures. In 1890, Richard Altmann developed a technique to dye mitochondria and postulate their metabolic and genetic autonomy. He named them as *Bioplasts*. The name Mitochondria was coined by Benda in 1898. In 1912, Otto Warburg hypothesized the existence of a respiratory enzyme in mitochondria that activates the oxygen and can be inhibited by cyanide. Keilin, in 1923, showed the variation of redox state of cytochrome during respiration. In 1929, Fiske and Subbarow isolated ATP. Later, in 1933, Keilin isolated the cytochrome c and reconstituted the electron transfer into homogenate of myocardial tissue. Kalckar and Belitser, in 1937, made the first studies of oxidative phosphorylation. Claude, in 1940, isolated mitochondria from liver cells. In 1948, Kennedy and Lehninger showed that tricarboxylic acid cycle, β-oxidation and oxidative phosphorylation take place in mitochondria. In 1951, Lehninger observed the coupling between oxidative phosphorylation and the transfer of electrons in the respiratory chain of mitochondria. In 1965, Mitchell and Moyle showed the mitochondrial proton translocation. Chappell, in 1968, obtained the evidence for a number of transport systems in which anions are involved. In 1974, Nicholls found the control of heat production in brown fat mitochondria by the modulation of their proton conductance. In 1992, Wallace reported mtDNA mutations to be the cause of degenerative diseases. In 1997, Boyer received the Nobel Prize for his work on ATP synthase.

STRUCTURE

Mitochondria vary greatly in both size (0.5 micrometers-10 micrometers) and number (1 over 1000) per cell. However, regardless of their size, number per cell, plant or animal origin, they have similar structures.

Mitochondria (*singular*: mitochondrion) are bounded by two membranes, an inner and an outer. The two membranes create distinct compartments within the organelle, and are different in structure and function. The outer membrane is a relatively simple phospholipid bilayer, containing protein structures called porins, which make it permeable to molecules of about 10 kD or less (the size of the smallest proteins). Besides, ions, nutrient molecules, ATP, ADP, etc. can pass through the outer membrane easily.

The inner membrane has inward directed folds, or ridges, called cristae (*singlular*: crista). This inner membrane is permeable to oxygen, carbon dioxide, and water. The cristae greatly increase the total surface area of the inner membrane. The structure of the inner membrane is highly complex and is the seat of all of the complexes of the electron transport system, the ATP synthase complex, and transport proteins. Both the outer and inner membranes have a typical unit membrane structure, however, the protein and phospholipid composition of both mitochondrial membranes are very different.

The double membrane structure makes two compartments in mitochondria. The inter-membrane space or the peri-mitochondrial space is the region between the inner and outer membranes. This space is continuous into the core of the cristae. It has an important role in the primary function of mitochondria, which is oxidative phosphorylation that is to convert organic materials into energy in the form of adenosine triphosphate (ATP). The internal space is filled with fluid called the mitochondrial matrix. The matrix contains the enzymes that are responsible for the citric acid cycle reactions. The matrix also contains dissolved oxygen, water, carbon dioxide, and the recyclable intermediates that serve as energy shuttles. The side of the inner membrane facing the matrix is called the M-side, while the side facing the outer chamber is called the cytosol or C-side.

The existence of this double membrane has led many biologists to theorize that mitochondria are the descendants of some bacteria that were endocytosed by a larger cell, billions of years ago, but not digested. This fascinating theory of symbiosis, which might lend an explanation to the development of eukaryotic cells, has additional supporting evidence.

Mitochondria have their own DNA and ribosomes. Mitochondrial ribosomes are more similar to bacterial ribosomes than to eukaryotic ribosomes (Fig. 6.1).

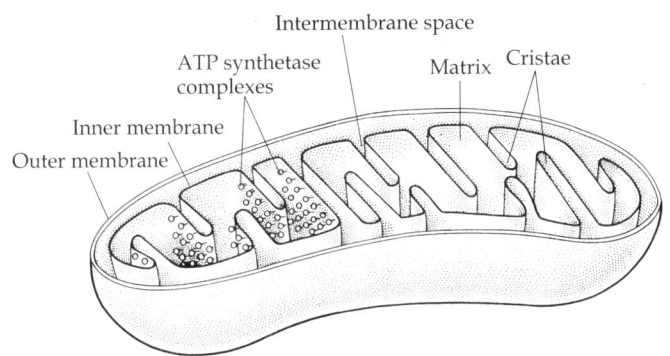

FIG. 6.1 Diagram of Mitochondria showing internal structure

Elementary Particles

Elementary particles or oxysomes or F_1 particles are the structures present on the mitochondrial inner membrane that consist of a spherical or polyhedral head piece which is 80-100 Å units in diameter, a cylindrical stalk that is ~50 Å long and 30-40 Å wide, and a base piece (40 × 110 Å).

The elementary or F_1 particle was first isolated by E. Racker in 1961. The stalk and base of the elementary particle includes a transmembrane domain, which carries protons across the membranes of the cristae (the F_0 portion of the molecule). The head comprises the F_1 ATPase, which synthesizes ATP when protons pass through F_0 down an electrochemical gradient. Both the F_0 and F_1 portions are composed of multiple protein subunits. The F_1 portion remains

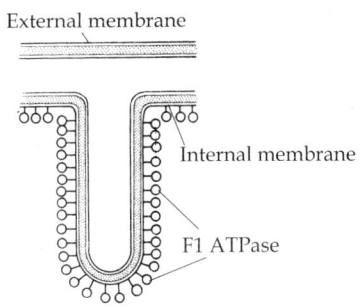

FIG. 6.2 Location of ATP synthetase on the inner membrane of mitochondria

above the membrane, the F_0 portion is embedded within the membrane. These particles are regularly spaced at intervals of 10 nm (Fig. 6.2).

Mitochondrial DNA

Mitochondria possess their own genetic material, and the machinery to manufacture their own RNA and protein. Mitochondrial DNA is a double stranded, circular molecule. It varies in length from about 5μm in most animal species to 30 μm in higher plants. It is localized in the matrix and may be attached to the inner membrane. The point of its attachment to the membrane determines the point from where the duplication begins. This non-chromosomal DNA encodes a small number of mitochondrial peptides (13 in humans) that are integrated into the inner mitochondrial membrane, along with polypeptides encoded by genes that reside in the host cell's nucleus. The mitochondrial DNA does not have many non-coding areas (junk DNA, or introns).

DISTRIBUTION

The cytoplasm of nearly all eukaryotic cells contains mitochondria, with one exception, the protist *Chaos (Pelomyxa) carolinensis*. They are especially abundant in cells and parts of cells that are associated with active processes. For example, in flagellated protozoa or in mammalian sperm, mitochondria are concentrated around the base of the flagellum or flagella. In cardiac muscle, mitochondria surround the contractile elements and reflect the great amount of work done by these cells. Hummingbird flight muscle is one of the richest sources of mitochondria known. Thus, from their distribution it is clear that they are involved in energy production.

FUNCTIONS

1. Cell Respiration/Cellular Respiration

The main function of the mitochondria is the production of energy, in the form of adenosine triphosphate (ATP). The cell uses this energy to perform the specific work necessary for cell survival and function.

The breaking down of complex food into simpler molecules such as carbohydrates, fats, and protein is called catabolism which is a form of metabolism, the other form being anabolism which is the building up process in which complex molecules are produced from simpler molecules.

As a result of digestion, the proteins are broken down into amino acids, carbohydrates into simple sugars (mainly glucose) and fats into fatty acids and glycerol. In the intestine the amino acids and sugars are absorbed into the blood vessels, the fatty acids and glycerol into lymphatic vessels and from there they are carried to the tissues. These molecules are then transferred into the mitochondria, where further processing occurs.

As stated above, mitochondria are the energy generators of the cell therefore the primary function of mitochondria is cell respiration which is divided into four phases: (1) glycolysis, (2) oxidation of pyruvic acid, (3) Kreb's citric acid cycle and (4) oxidative phophorylation.

Glycolysis

Oxidation of glucose is known as glycolysis. Glucose is oxidized to either lactate or pyruvate. The enzymes for glycolysis are found in the cytosol outside the mitochondrion. Under aerobic conditions, the dominant product in most tissues is pyruvate and the pathway is known as aerobic glycolysis. When oxygen is depleted, as for instance during prolonged vigorous exercise, the dominant glycolytic product in many tissues is lactate (lactic acid) and the process is known as anaerobic glycolysis. Other substances such as glycogen and fructose may also be the starting compounds in glycolysis.

Glycolysis is also called the Embden-Meyerhof-Parnas (EMP) pathway. Aerobic glycolysis of glucose yields two molecules of pyruvic acid. The process requires two molecules of ATP to activate the process, with the subsequent production of four molecules of ATP and two of NADH. Thus, conversion of one mole of glucose to two moles of pyruvic acid is accompanied by the net production of two moles each of ATP and NADH.

$$\text{Glucose} + 2ADP + 2NAD^+ + 2P_i \rightarrow 2\text{Pyruvate} + 2ATP + 2NADH + 2H^+$$

The pathway of glycolysis can be seen as consisting of two separate phases. The first is the preparatory phase that requires energy in the form of ATP, and the second is considered the energy-yielding phase. In the first phase, 2 molecules of ATP are used to convert glucose to fructose 1, 6-bisphosphate. In the second phase, fructose 1, 6-bisphosphate is degraded to pyruvic acid, with the production of 4 molecules of ATP and 2 equivalents of NADH.

Phosphorylation: This is the first reaction of glycolysis in which ATP-dependent phosphorylation of glucose to form glucose-6-phosphate is catalyzed by tissue-specific isoenzymes known as hexokinases.

Isomerization: The second reaction of glycolysis is isomerization, in which glucose-6-phosphate is converted to fructose-6-phosphate. The enzyme catalyzing this reaction is phosphohexose isomerase (also known as phosphoglucose isomerase).

Second Phosphorylation: The next reaction of glycolysis involves the utilization of a second ATP to convert fructose-6-phosphate to fructose-1, 6-bisphosphate. This reaction is catalyzed by 6-phosphofructo-1-kinase, also known as phosphofructokinase-1.

Cleavage: Aldolase catalyses the hydrolysis of fructose-1, 6-bisphosphate into two 3-carbon products: dihydroxyacetone phosphate and glyceraldehyde-3-phosphate. The aldolase reaction proceeds readily in the reverse direction, being utilized for both glycolysis and gluconeogenesis. The two products of the aldolase reaction equilibrate readily in a reaction catalyzed by triose phosphate isomerase. Succeeding reactions of glycolysis utilize glyceraldehyde-3-phosphate as a substrate.

Glyceraldehyde-3-Phosphate Dehydrogenase: The second phase of glucose catabolism features the energy-yielding glycolytic reactions that produce ATP and NADH. In the first of these reactions, glyceraldehyde-3-P-dehydrogenase catalyzes the NAD^+-dependent oxidation of glyceraldehyde-3-phosphate to 1, 3-bisphosphoglycerate and NADH.

Phosphoglycerate Kinase: The high-energy phosphate of 1, 3-bisphosphoglycerate is used to form ATP and 3-phosphoglycerate by the enzyme phosphoglycerate kinase.

Isomerization: The remaining reactions of glycolysis are aimed at converting the relatively low energy phosphoacyl-ester of 3-phosphoglycerate to a high-energy form and harvesting the phosphate as ATP. The 3-phosphoglycerate is first converted to 2-phosphoglycerate by phosphoglycerate mutase and the 2-phosphoglycerate conversion to phosphoenol pyruvate is catalyzed by enolase.

ATP Generation: The final reaction of aerobic glycolysis is catalyzed by the highly regulated enzyme pyruvate kinase. In this strongly exergonic reaction, the high-energy phosphate of phosphoenol pyruvate is conserved as ATP. The loss of phosphate by phosphoenol pyruvate leads to the production of pyruvate in an unstable enol form, which spontaneously tautomerizes to the more stable, keto form of pyruvate. This reaction contributes a large proportion of the free energy of hydrolysis of phosphoenol pyruvate.

It is to be noted that two molecules of pyruvate are formed per molecule of glucose metabolized. During the cleavage step fructose-1, 6-bisphosphate is cleaved into two triose phosphate molecules, each of which then passes through the reaction.

Lactic acid Metabolism: During anaerobic glycolysis, which occurs in anaerobic organisms or when glycolysis is proceeding at a high rate, the oxidation of NADH occurs through the reduction of an organic substrate. Lactic acid fermentation is the process by which pyruvic acid is converted to lactic acid. Lactic acid fermentation by microorganisms plays an essential role in the manufacture of food products such as yogurt and cheese. Also, certain animal cells (muscle cells) convert pyruvic acid to lactic acid during strenuous exercise when muscles run out of oxygen. This process provides the muscles with the energy they need during stress. The side effects of lactic acid fermentation is muscle fatigue, pain, cramps, soreness. Most lactic acid made in the muscles diffuse into the bloodstream, then to the liver, where it is converted back to pyruvic acid when oxygen becomes available.

Ethyl alcohol metabolism: Alcoholic fermentation converts pyruvic acid to carbon dioxide and ethanol. Animal cells (primarily hepatocytes) contain the cytosolic enzyme alcohol dehydrogenase which oxidizes ethanol to acetaldehyde. Acetaldehyde then enters the mitochondria where it is oxidized to acetate by acetaldehyde dehydrogenase. Bakers use alcoholic fermentation

Glycolysis: Embden-Meyerhof-Parnas pathway

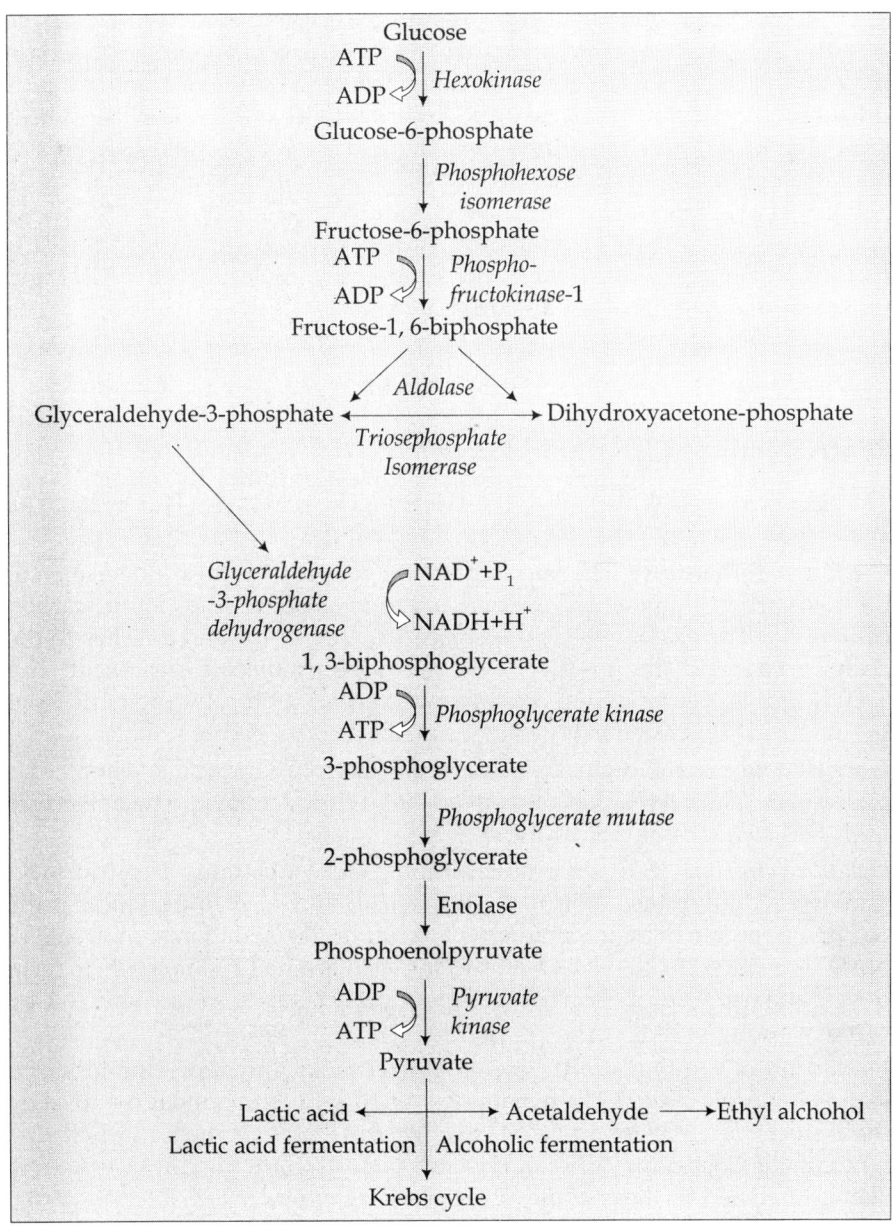

of yeast to make bread. As yeast ferments the carbohydrates in dough, CO_2 is produced and trapped in the dough, causing it to rise.

Pyruvic Acid Oxidation

The end product of glycolysis, i.e. pyruvic acid, does not enter the Krebs cycle directly. The 3-carbon compound pyruvic acid is first converted into 2-carbon molecule, acetic acid. This step is catalyzed by a multi-enzyme complex (including coenzyme A, pyruvate dehydrogenase and vitamin B_1) which removes the carboxyl group of pyruvic acid as a CO_2 molecule. This is accompanied by the reduction of a NAD^+ molecule to NADH. This leaves behind a two-carbon fragment (acetyl) attached to a molecule of CoA $\rightarrow$ Acetyl-CoA. Since there are two molecules of pyruvic acid formed for each molecule of glucose which enters the system, two molecules of NADH and two molecules of acetyl-CoA are formed.

2 pyruvic acid + 2NAD$^+$ + 2CoA $\rightarrow$ O$_2$ + 2NADH + H$^+$ + 2 acetyl-CoA

Acetyl-CoA is now ready to enter the Citric Acid Cycle.

Krebs Cycle or Citric Acid Cycle

The citric acid cycle is also called the Tricarboxylic acid cycle (TCA cycle). The Krebs cycle reactions take place in the matrix of the mitochondria.

Citric acid formation: The first step in the Krebs cycle is the condensation of the two-carbon fragment of Acetyl-CoA to oxaloacetic acid, a four-carbon compound. The oxaloacetic acid displaces the coenzyme and attaches to the acetyl group and the product is the six-carbon molecule citric acid. The acetyl-CoA is free to prime another two-carbon molecule fragment from pyruvic acid. This reaction is catalyzed by the enzyme citrate synthase.

Isocitric acid: A molecule of water is removed from citric acid and another is added back. The net result is the conversion of citric acid to its isomer isocitric acid. This step is catalyzed by the enzyme aconitase.

Dehydrogenation and decarboxylation: The substrate, isocitric acid undergoes dehydrogenation to form oxalosuccinic acid in presence of the enzyme isocitrate dehydrogenase. As a result of the oxidation, a molecule of NAD^+ is reduced to NADH. Oxalosuccinic acid loses a CO_2 molecule to form α-Ketoglutaric acid, a 5-carbon compound. This reaction is catalyzed by carboxylase.

Formation of Succinyl-CoA: α-Ketoglutaric acid undergoes the loss of a second molecule of CO_2. The remaining four-carbon compound is oxidized by the transfer of electrons to NAD^+ to form NADH. It is then attached to a molecule of Coenzyme A by an unstable bond to form Succinyl Co-A. This step is catalyzed by the enzyme Ketoglutaric dehydrogenase.

Phosphorylation of ADP: Succinic acid is formed from succinyl Co-A in the presence of the enzyme succinic thiokinase. Substrate-level phosphorylation occurs in this step. CoA is displaced by a phosphate group, which is then transferred to GDP to form guanine triphosphate (GTP). GTP then donates a phosphate group to ADP forming a molecule of ATP.

Fumaric acid formation: Succinic acid is oxidized transferring two hydrogens to FAD to form $FADH_2$. This coenzyme is similar to NADH but it stores less energy. The enzyme that catalyzes this step, succinic dehydrogenase, is the only enzyme of the cycle that is embedded in the mitochondrial membrane. All the other enzymes of the citric acid cycle are dissolved in the mitochondrial matrix.

Hydration: Fumaric acid is hydrated to form malic acid. The reaction is catalyzed by the enzyme fumarase.

Regeneration of oxaloacetic acid: The last oxidative step produces another molecule of NADH and regenerate oxaloacetic acid in the presence of malate dehydrogenase. This molecule will again accept a two-carbon fragment from acetyl CoA for another turn of the cycle.

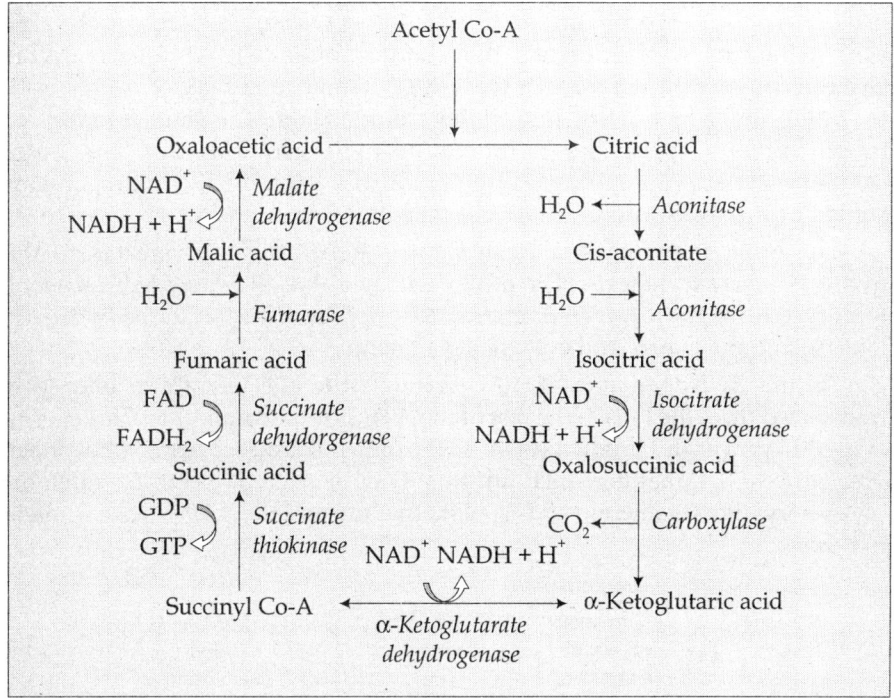

KREBS CYCLE OR CITRIC ACID CYCLE

Remember that each turn of Krebs cycle consumes only one of the two molecules of pyruvic acid formed via glycolysis therefore, each glucose molecule produces two turns of the cycle and twice the product.

$$2 \text{ acetyl-CoA} + 2\text{ADP} + 2\text{P}_i + 6\text{NAD} + 2\text{FAD} \rightarrow 2 \text{ CoA} + 2\text{ATP} + 6\text{NADH} + 2\text{FADH}_2 + 2\text{CO}_2$$

Oxidative phosphorylation and the electron transport system:

The electron transport system takes electrons that it receives from NADH and $FADH_2$ and transfers those electrons from one protein to another. The energy from those electrons is used to pump hydrogen ions into the intermembrane space of mitochondrion. The reason for doing this is that the hydrogen ions then diffuse back to the inner compartment through special channels in a protein structure called ATP synthase. The hydrogen ions transfer energy to the ATP synthase which uses the energy to make ATP from ADP and inorganic phosphate. In many cells electron transport phosphorylation is the main way in which ATP is synthesized.

NAD$^+$ and FAD are electron acceptors that carry electrons from the energy containing molecules that enter the Krebs cycle. These electron acceptors pick up two electrons along with hydrogen ions yielding NADH and FADH$_2$.

The Krebs cycle is the source of the hydrogen ions and the electrons needed to make ATP from the electron transport system. The Krebs cycle takes place in the inner compartment of mitochondria often called the matrix. As the electrons travel through the ETS the energy contained in the electrons is used to pump hydrogen ions into outer compartment of mitochondria or intermembrane space. The pumping of hydrogen ions into the outer compartment sets up a concentration gradient. Naturally, the hydrogen ions should diffuse back into the matrix of mitochondria.

The inner mitochondria membrane is generally impermeable to the hydrogen ions except for ATP synthase channels that allow the hydrogen ions to diffuse through. It is a combination channel for diffusion of the hydrogen ions back to the matrix and enzyme for making ATP.

As the hydrogen ions diffuse through the ATP synthase energy is transferred from the ions to the rest of the ATP synthase molecule. This energy is used to power the production of ATP from ADP and inorganic phosphate. The process of making ATP in this way is called electron transport phosphorylation or chemiosmosis. Note that most of the ATP we use is made in this way.

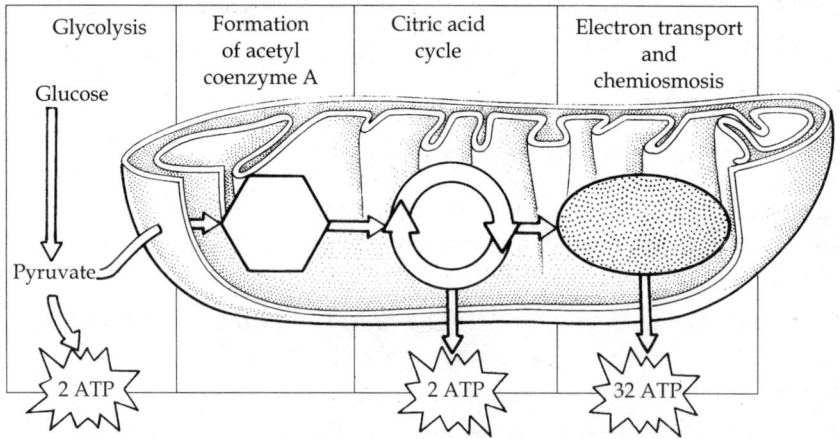

FIG. 6.3   The overall equation for Aerobic Respiration

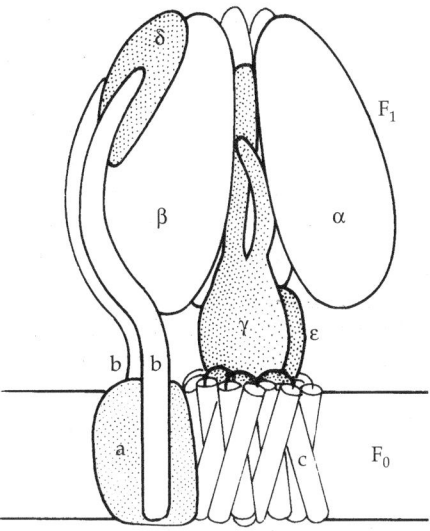

FIG. 6.4 ATP Synthetase

It is important to realize that ATP production by electron transport phosphorylation uses inorganic phosphate from the cytoplasm as the source of phosphate. In contrast substrate level phosphorylation uses phosphate transferred from other organic compounds to the ADP to make ATP (Fig. 6.3 and Fig. 6.4).

The role of oxygen: As the electrons in the ETS (electron transport chain) are used to do work, the electrons lose energy and reach a point at the end of the ETS where they have to be get rid of. The scheme the cell uses to do this is to combine the electrons with hydrogen ions and oxygen to produce water. Also, oxygen is an excellent electron acceptor.

In anaerobic respiration, as opposed to aerobic respiration, other electron acceptors are used to accept electrons from the electron transport systems.

Mitochondrial Redox Carriers

We have seen that mitochondria perform the process of catabolism for the break down of complex food materials. The end result of this catabolic process is the production of two energy-rich electron donors, NADH and $FADH_2$. Electrons from these donors are passed through an electron transport chain to oxygen, which is reduced to water. This is a multi-step redox process that occurs on the mitochondrial inner membrane. The enzymes that catalyze these reactions have the remarkable ability to simultaneously create a proton gradient across the membrane, producing a thermodynamically unlikely high-energy state with the potential to do work.

Four membrane-bound complexes have been identified in mitochondria. Each is an extremely complex transmembrane structure that is embedded in

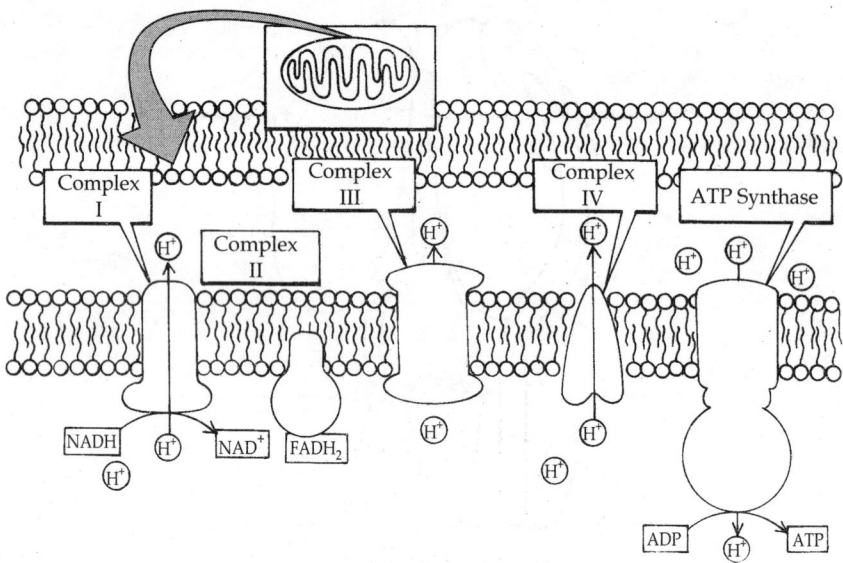

FIG. 6.5 The Electron Transport Chain

the inner membrane. Three of them are proton pumps. The structures are electrically connected by lipid-soluble electron carriers and water-soluble electron carriers. The overall electron transport chain is (Fig. 6.5):

NADH → *Complex I* → **Q** → *Complex III* → **cytochrome** *c* → *Complex IV* →
O_2 ↑
      Complex II

Complex I

The NADH dehydrogenase complex (also called NADH-Q reductase) is the largest of the respiratory enzyme complexes, with a mass of about 880 kD and more than 34 polypeptide chains. It accepts electrons from NADH and passes them through a flavin and at least five iron-sulfur centers to coenzyme-Q.

The initial step is the binding of NADH and the transfer of its two high-potential electrons to the flavin mononucleotide (FMN) prosthetic group of this complex to give the reduced form, $FMNH_2$. Electrons are then transferred from $FMNH_2$ to a series of iron-sulfur clusters (abbreviated as Fe-S), the second type of prosthetic group in NADH-dehydrogenase. Iron atoms in these Fe-S complexes cycle between Fe^{2+} (reduced) and Fe^{3+} (oxidized) states. Electrons in the iron-sulfur clusters of NADH-dehydrogenase are then shuttled to ubiquinone, also known as coenzyme-Q.

Ubiquinone is reduced to a free-radical semiquinone anion by the uptake of a single electron. Reduction of this enzyme-bound intermediate by a second electron yields ubiquinol (QH_2). The flow of two electrons from NADH to QH_2

through NADH-dehydrogenase leads to the pumping of four H^+ from the matrix to the cytosolic side of the inner mitochondrial membrane.

Complex II

$FADH_2$ is formed in the citric acid cycle in the oxidation of succinate to fumarate by succinate dehydrogenase. This enzyme is part of the succinate-Q reductase complex (Complex II). The succinate-Q reductase complex is an integral membrane protein of the inner mitochondrial membrane. $FADH_2$ does not leave the complex, rather its electrons are transferred to Fe-S centers and then to ubiquinone for entry into the electron transport chain. The succinate-Q reductase complex and other enzymes that transfer electrons from $FADH_2$ to ubiquinone, in contrast with NADH-dehydrogenase, are not proton pumps because the free-energy change of the catalyzed reaction is too small. Consequently, less ATP is formed from the oxidation of $FADH_2$ than from NADH.

Complex III

Cytochrome c co-enzyme Q reductase (also called cytochrome b-c_1 complex) contains at least 8 different polypeptide chains and is thought to function as a dimer of about 500 kd. Each monomer contains three hemes bound to cytochromes and an iron-sulfur protein. The complex accepts electrons from co-enzyme Q and passes them on to cytochrome c.

A cytochrome is an electron-transferring protein that contains a heme prosthetic group. Their iron atoms alternate between a reduced ferrous (Fe^{+2}) state and an oxidized ferric (Fe^{+3}) state during electron transport. The function of cytochrome reductase is to catalyze the transfer of electrons from QH_2 to cytochrome c, a water-soluble protein, and concomitantly pump protons across the inner mitochondrial membrane. The flow of a pair of electrons through this complex leads to the effective net transport of $2H^+$ to the cytosolic side, half the yield obtained with NADH-dehydrogenase because of a smaller thermodynamic driving force. Cytochrome reductase itself contains two types of cytochromes, namely b and c_1. The reductase also contains an Fe-S protein and several other polypepetide chains. The prosthetic group of cytochrome b, c_1, and c is iron-protoporphyrin IX, the same heme as in myoglobin and hemoglobin.

Complex IV

The cytochrome oxidase complex (cytochrome a-a_3) is isolated as a dimer of about 300 kD; each monomer contains at least 9 different polypeptide chains, including two heme A, one of them is called heme a and the other heme a_3. The two heme A molecules, though chemically identical, have different properties because they are located in different parts of the cytochrome oxidase.

The complex accepts electrons from cytochrome c and passes them to oxygen. Four electrons are funneled into O_2 to completely reduce it to H_2O and concomitantly pump protons from the matrix to the cytosolic side of the inner mitochondrial membrane.

The free energy available as a consequence of transferring 2 electrons from NADH or succinate to molecular oxygen is -57 and -36 kcal/mol, respectively. Oxidative phosphorylation traps this energy as the high-energy phosphate of ATP. In order for oxidative phosphorylation to proceed, two principal conditions must be met. First, the inner mitochondrial membrane must be physically intact so that protons can only re-enter the mitochondrion by a process coupled to ATP synthesis. Second, a high concentration of protons must be developed on the outside of the inner membrane.

Complex V

Electrons return to the mitochondrion through the integral membrane protein known as ATP synthase (or Complex V). ATP synthase is a multiple subunit complex that binds ADP and inorganic phosphate at its catalytic site inside the mitochondrion, and requires a proton gradient for activity in the forward direction.

Summary

The electron transport chain takes place in the inner mitochondrion membrane. It follows the citric acid cycle, where NADH and $FADH_2$ are reduced. These coenzymes then enter the electron transport chain. The first step is the transfer of high-energy electrons from NADH $+ H^+$ to FMN, the first carrier in the chain. From each molecule of glucose, two NADH $+ 2H^+$ are generated from glycolysis, two from the formation of acetyl-CoA, and six from the citric acid cycle. In this transfer, a hydride ion (H^- ion) passes to FMN, which then picks up an additional H^+ from the surrounding aqueous medium. As a result, NADH $+ H^+$ is oxidized to NAD^+, and FMN is reduced to $FMNH_2$. In the second step of electron transport chain, $FMNH_2$ passes electrons to several iron-sulfur centers and then to coenzyme Q, which picks up an additional H^+ from the surrounding aqueous medium. As a result, $FMNH_2$ is oxidized to FMN.

The next sequence in the transport chain involves cytochromes, iron-sulfur clusters, and copper atoms located between coenzyme Q and molecular oxygen. Electrons are passed successively from coenzyme Q to cytochrome b, to Fe-S, to cytochrome c_1, to cytochrome c, to Cu, to cytochrome a, and finally to cytochrome a_3. Each carrier in the chain is reduced as it picks up electrons and is oxidized as it gives up electrons. The last cytochrome, cytochrome a_3, passes its electrons to one-half of a molecule of oxygen, which becomes negatively charged and then picks up $2H^+$ from the surrounding medium to form H_2O. This is the only point in aerobic cellular respiration where O_2 is consumed.

Note that $FADH_2$, derived from the citric acid cycle, is another source of electrons. However, $FADH_2$ adds its electrons to the electron transport chain at a lower energy level than does NADH $+ H^+$. Because of this, the electron transport chain produces about one-third less energy for ATP generation when $FADH_2$ donates electrons as compared with NADH $+ H^+$ (Fig. 6.6).

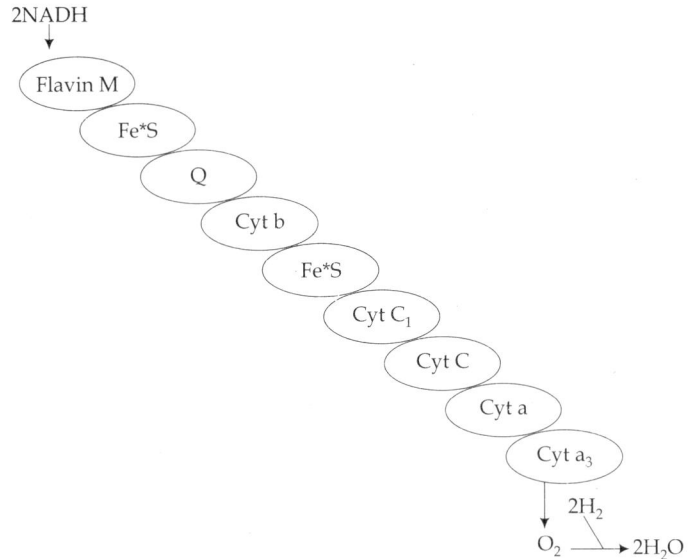

FIG. 6.6 The Components of respiratory chain

The various electron transfers in the electron transport chain generate 32 to 34 ATP molecules from each molecule of glucose that is oxidized: 28 or 30 from the 10 molecules of NADH + H^+ and 2 from each of the 2 molecules of $FADH_2$ (4 total). Thus, during aerobic respiration, 36 or 38 ATPs can be generated from one molecule of glucose. Note that two of those ATPs come from substrate-level phosphorylation in glycolysis and two come from substrate-level phosphorylation in the citric acid cycle.

The overall reaction for aerobic respiration is:

$$C_6H_{12}O_6 + 6O_2 + 36 \text{ or } 38 \text{ ADPs} + 36 \text{ or } 38 \text{ P}_i \rightarrow 6CO_2 + 6H_2O + 36 \text{ or } 38 \text{ ATPs}$$

2. Transport

Mitochondrial membranes contain numerous transport systems for the import of metabolites and high energy intermediates, export of ATP which is utilized in the cytosol, and inorganic phosphate, which is returned to the matrix via phosphate-proton symport that is driven by the chemiosmotic gradient. Thus some of the gradient energy is always used for purposes other than synthesis of ATP.

3. Calcium Storage

Mitochondria are exceedingly important as storage tanks for calcium ions. Calcium ion concentration is an important second messenger in cells. It must be precisely controlled in various intracellular compartments, or cellular function is compromised. Indeed, calcium itself is a mediator of many toxins. Mitochondria may act as sinks to buffer the effects of calcium overload.

4. Reproduction

Among other processes, mitochondria contain their own independent machinery for protein synthesis, including DNA, messenger RNA, transfer RNA and ribosomes. They reproduce by fission in a manner similar to that of bacterial cells. In fact, the seeming independence of mitochondria from genetic machinery of eukaryotic cell, as well as the resemblance of mitochondria-associated macromolecules to those of bacteria provides strong evidence for an endosymbiotic origin for the organelles.

When the energy needs of a cell are high, mitochondria grow and divide. When the energy use is low, mitochondria are destroyed or become inactive. At cell division, mitochondria are distributed to the daughter cells more or less randomly during the division of the cytoplasm.

Gene Inheritance in Mitochondria

Mitochondrial genes are not inherited by the same mechanism as nuclear genes. At fertilization of an egg by a sperm, the egg nucleus and sperm nucleus each contribute equally to the genetic makeup of the zygote nucleus. In contrast, the mitochondria, and therefore the mitochondrial DNA, usually comes from the egg only. At fertilization of an egg, a single sperm enters the egg along with the mitochondria that it uses to provide the energy needed for its swimming behavior. However, the mitochondria provided by the sperm are targeted for destruction very soon after entry into the egg. The egg itself contains relatively few mitochondria, but it is these mitochondria that survive and divide to populate the cells of the adult organism. This means that mitochondria are, in most cases, inherited down the female line.

OXIDATION OF PROTEINS

A major component of food is protein. The proteins ingested as part of our diet are not the same proteins required by the body, nor can large molecules be absorbed from the gut. Therefore, these proteins are digested and their component amino acids absorbed into the blood stream.

Uses of amino acids: Amino acids are used in three ways in the body:

1) Protein synthesis: The synthesis of new proteins is very important during growth. In adults new protein synthesis is directed towards replacement of proteins as they are constantly turned over.

2) Synthesis of a variety of other compounds: Examples of compounds synthesized from amino acids include purines and pyrimidines (components of nucleotides), catecholamines (adrenalin and noradrenalin), neurotransmitters (serotonin), histamine, porphyrins (the central oxygen binding component of haemoglobin).

3) As a biological fuel: About 10% of energy production in humans is from amino acids. The percentage is much higher in carnivores, whose diet is almost entirely protein.

Amino Acid Catabolism

The other biological fuels discussed (carbohydrates and fats) contain only the elements carbon, hydrogen and oxygen. Amino acids contain nitrogen as well. The first step in amino acid catabolism is the removal of the nitrogen (the amino group).

$$\boxed{\text{Transamination}}$$

COO		COO		
C = O	COO⁻	H − C − NH₃⁺	COO⁻	
CH₂ +	H − C − NH₃⁺ →	CH₂ +	C = O	
CH₂	R	CH₂	R	
α-Ketoglutrate	Amino acid	Glutamate	α-Keto acid	

$$\text{C} = \text{O}, \quad \text{COO}^-, \quad \text{H}-\text{C}-\text{NH}_3^+ \longrightarrow \text{CH}_2 \;+\; \text{C}=\text{O}$$

Deamination: The removal of the amino groups of all twenty amino acids begins with the transfer of amino groups to just one amino acid — **glutamic acid** (or glutamate ion). This is catalyzed by transaminase enzymes which transfer the amino group from amino acids to a compound called **alpha-ketoglutarate**. The product is an alpha-keto acid formed from the amino acid and glutamate (formed from the addition of the amino group to alpha-ketoglutarate).

Once the amino groups have all been collected in the form of the one amino acid, glutamate, and this amino acid has its amino group removed (a process termed 'oxidative deamination'). This reaction reforms alpha-ketoglutarate with the other product being **ammonia** (NH_4^+).

$$\text{Glutamate} + \text{NAD}^+ + \text{H}_2\text{O} \xrightleftharpoons[]{\substack{\text{Glutamate} \\ \text{dehydrogenase}}} \text{NH}_4^+ + \alpha\text{-ketoglutarate} + \text{H}^+$$

Ammonia and Urea

Ammonia is toxic to the nervous system and its accumulation rapidly causes death. Therefore it must be detoxified to a form which can be readily removed from the body. Ammonia is converted to urea, which is water soluble and is readily excreted via the kidneys in urine.

The remainder of the amino acid is referred to as the "carbon skeleton". Depending on the particular amino acid being catabolized, its carbon skeleton will be converted to:

(1) Acetyl CoA: Those carbon skeletons which end up as acetyl CoA are committed to energy production. They will either be immediately oxidized via the citric acid cycle or they may be converted to ketone bodies. Because the amino acids whose carbon skeletons yield acetyl

CoA are *potentially* a source of ketone bodies they are referred to as ketogenic amino acids.

(2) or pyruvate

(3) or a citric acid cycle intermediate

The carbon skeletons which end up as either pyruvate or a citric acid cycle intermediate may be used for energy production or they may be used to synthesize glucose by the pathway known as gluconeogenesis. Because the amino acids whose carbon skeletons yield pyruvate or a citric acid cycle intermediate are *potentially* a source of glucose they are referred to as glucogenic amino acids.

OXIDATION OF FATS

Fatty acids must be activated in the cytoplasm before being oxidized in the mitochondria. Activation is catalyzed by fatty acyl-CoA ligase (also called acyl-CoA synthase or thiokinase). The net result of this activation process is the consumption of 2 molar equivalents of ATP.

$$\textbf{Fatty acid + ATP + CoA} \rightarrow \textbf{Acyl-CoA + PP}_i \textbf{ + AMP}$$

Oxidation of fatty acids occurs in the mitochondria. The transport of fatty acyl-CoA into the mitochondria is accomplished via an acyl-carnitine intermediate, which itself is generated by the action of carnitine acyltransferase I, an enzyme that resides in the outer mitochondrial membrane. The acylcarnitine molecule then is transported into the mitochondria where carnitine acyltransferase II catalyzes the regeneration of the fatty acyl-CoA molecule.

The process of fatty acid oxidation is termed β-oxidation since it occurs through the sequential removal of 2-carbon units by oxidation at the β-carbon position of the fatty acyl-CoA molecule.

Each round of β-oxidation produces one mole of NADH, one mole of $FADH_2$ and one mole of acetyl-CoA. The acetyl-CoA, the end product of each round of β-oxidation, then enters the TCA cycle, where it is further oxidized to CO_2 with the concomitant generation of three moles of NADH, one mole of $FADH_2$ and one mole of ATP. The NADH and $FADH_2$ generated during the fat oxidation and acetyl-CoA oxidation in the TCA cycle then can enter the respiratory pathway for the production of ATP.

The oxidation of fatty acids yields significantly more energy per carbon atom than does the oxidation of carbohydrates. The net result of the oxidation of one mole of oleic acid (an 18-carbon fatty acid) will be 146 moles of ATP (2 mole equivalents are used during the activation of the fatty acid), as compared with 114 moles from an equivalent number of glucose carbon atoms.

OBJECTIVE TYPE QUESTIONS

1. Mitochondria are called the powerhouses of the cell because they make energy available for cellular metabolism. Which of the following statements is most convincing in supporting this concept of mitochondrial function?
   A. ATP occurs in the mitochondria
   B. Mitochondria have a double membrane
   C. The enzymes of the Krebs cycle, and molecules required for terminal respiration, are found in mitochondria
   D. Mitochondria are found in almost all kinds of plant and animal cells

2. When more carbohydrate is consumed than can be used for present energy needs and glycogen storage, the carbohydrate is converted to fatty acid in the liver. All of the following statements about the first part of this pathway are true EXCEPT
   A. Carbohydrate is converted to pyruvate using glycolysis
   B. Activated by dephosphorylation of phosphofructokinase-2/fructose-2, 6-bisphosphatase
   C. The allosteric activator fructose-2, 6-bisphosphatase is required
   D. The inhibition of protein phosphatases by insulin is required

3. What happens when carbohydrate is converted to fatty acid in the liver? All of the following statements are true EXCEPT
   A. Some pyruvate is converted to acetyl CoA by pyruvate dehydrogenase
   B. Some pyruvate is converted to oxaloacetate by pyruvate carboxylase in the presence of high concentrations of acetyl CoA
   C. Acetyl CoA and oxaloacetate form citrate and the citrate leaves the mitochondria
   D. In the cytosol, citrate synthase produces ATP, oxaloacetate and acetyl CoA from citrate

4. The control enzyme for fatty acid synthesis
   A. Is activated by glucagon, epinephrine, and AMP kinase
   B. Is inhibited by insulin and protein phosphatase
   C. Produces malonyl CoA and ATP
   D. Requires biotin as a cofactor

5. During fasting, the blood glucose is maintained by the liver. Because of low insulin to glucagon ratios, all of the following contribute to this process EXCEPT
   A. The cAMP cascade is active so phosphorylase b is converted to phosphorylase a
   B. Glycerol is available for gluconeogenesis

    C. Phosphofructokinase-1 and phosphofructokinase-2 are both active

    D. Beta-oxidation and the TCA cycle are providing energy for gluconeogenesis

6. If living cells similar to those found on earth were found on another planet where there was no molecular oxygen, which cell part would most likely be absent?

    A. Cell membrane   B. Nucleus

    C. Mitochondria   D. Ribosome

7. Mitochondria are concerned with

    A. Glucose formation   B. ATP formation

    C. ADP formation   D. None of the above

8. Ultimately in glycolysis

    A. Glucose is converted into sucrose

    B. Glucose is converted into pyruvic acid

    C. Protein is converted into glucose

    D. Glucose is converted into glycogen

9. In respiration, pyruvic acid is

    A. Formed only when oxygen is available

    B. One of the products of the Kreb's cycle

    C. A result of protein breakdown

    D. Broken down into a 2-carbon fragment

10. What are cristae?

    A. Cell organelles

    B. Folds of inner wall of mitochondria

    C. Folds of outer wall of mitochondria

    D. Outer membrane of mitochondria

Answers

1. C	2. D	3. D	4. D	5. C
6. C	7. B	8. B	9. D	10. B

SHORT ANSWER TYPE QUESTIONS

1. Write a short note on oxidation of proteins.
2. List the steps in glycolysis which are not reversible.
3. Draw up a balance sheet for ATP production during glycolysis.
4. Calculate the energy yield from a representative fatty acid, such as palmitic acid.
5. Where does fatty acid synthesis take place?

LONG ANSWER TYPE QUESTIONS

1. Write a note on the electron transport chain. Where does it take place?
2. How is the free energy released by electron transport used to form ATP? Explain.
3. Write a note on the citric acid cycle. Why does the citric acid cycle stop under anaerobic conditions?
4. Outline the four steps in the β-oxidation of fatty acids.
5. Discuss the similarities and differences between the process of fatty acid synthesis and breakdown.

Plastids

7

INTRODUCTION

Plastids are spherical or disc shaped organelles that only exist in plant cells and photosynthetic protists. They are found in the cytoplasm of the cell and are bounded by a double membrane. The number of plastids in a cell varies. This number depends on the changing environmental conditions that the plant encounters and how the plant adjusts to these changes. The number of plastids in a cell also depends on the type of species. Plastids are responsible for photosynthesis, storage of products like starch and for the synthesis of many classes of molecules such as fatty acids and terpenes which are needed as cellular building blocks and for the function of the plant.

HISTORY

The term 'Plastid' was coined by A. F.W. Schimper in 1885, while detailed cytological studies on plastids were done by A. Meyer (1883) and F. Schmitz (1884) and Schimper (1885). However, the chloroplasts in plant cells were identified by early microscopists, including Antony von Leeuwenhoek, in the 17th century and Nehemiah Grew, who in 1682, described some green precipitates in leaves; this event is considered the first report of the existence of chloroplasts.

Although some of the steps in photosynthesis are still not completely understood, the overall photosynthetic equation has been known since the 1800s. It was Jan van Helmont who began the research of the process in the mid-1600s, when he carefully measured the mass of the soil used by a plant and the mass of the plant as it grew. He came to the idea that the plant's biomass comes from the inputs of photosynthesis and not from the soil as previously thought.

The light-dependent production of oxygen by plants was discovered by Joseph Priestley in 1771. Jean Senebier in 1796 showed that CO_2 was taken up by plants in photosynthesis.

In 1931, Cornelis Van Niel made key discoveries explaining the chemistry of photosynthesis. By studying purple sulfur bacteria and green bacteria, he was the first scientist to demonstrate that photosynthesis is a light-dependent redox reaction, in which hydrogen reduces carbon dioxide.

Further experiments to prove that the oxygen developed during the photosynthesis of green plants came from water were performed by Robert Hill in 1937 and 1939. He showed that isolated chloroplasts give off oxygen in the presence of unnatural reducing agents like iron oxalate, ferricyanide or benzoquinone after exposure to light. The Hill reaction is as follows:

$$2 H_2O + 2 A + (light, chloroplasts) \rightarrow 2 AH_2 + O_2$$

where A is the electron acceptor. Therefore, in light, the electron acceptor is reduced and oxygen is evolved.

Samuel Ruben and Martin Kamen in 1941 used radioactive isotopes to determine that the oxygen liberated in photosynthesis came from the water.

Melvin Calvin and Andrew Benson, along with James Bassham in 1948, elucidated the path of carbon assimilation (the photosynthetic carbon reduction cycle) in plants. The carbon reduction cycle is known as the Calvin cycle.

In 1960, Hill and Bendall proposed Z-scheme for electron transport from water to NADPH during photosynthesis.

In 1966, Hatch and Slack proposed the C_4 cycle.

TYPES OF PLASTIDS

In plants, plastids may be differentiated into several forms, depending upon the function they play in the cell. Undifferentiated plastids (proplastids) may develop into any of the following plastids:

Chromoplasts: These are the plastids meant for pigment synthesis and storage. Chromoplasts are red, orange or yellow plastids. The colour is usually the result of yellow xanthophyll and red carotenoids. They are found in coloured organs of plants such as fruit and floral petals, to which they give their distinctive colors. The probable main evolutionary role of chromoplasts is to act as an attractant for pollinating animals (for example, insects) or for seed dispersal via the eating of coloured fruits. They allow the accumulation of large quantities of water-insoluble compounds in otherwise watery parts of plants.

Leucoplasts: These are meant for monoterpene synthesis. They are non-pigmented and are located in roots and non-photosynthetic tissues of plants. They are specialized for bulk storage of starch, lipid or proteins and are further differentiated as **(a) Amyloplasts:** for starch storage, **(b) Elaioplasts:** for storing fat, **(c) Proteinoplasts:** for storing proteins. However, in many cell types, leucoplasts do not have a major storage function and are present to provide a wide range of essential biosynthetic functions, including the synthesis of fatty acids, many amino acids, and tetrapyrrole compounds such as haem.

Chloroplasts: Chloroplasts serve as the site of photosynthesis. They are flat, disc shaped structures present in the leaves of the plant. They are usually 2-10 µm in diameter and 1µm thick. Leaves have about 500,000 chloroplasts per square millimeter of the surface. They are found primarily in the mesophyll cells. Each mesophyll cell contains 30-40 chloroplasts. They contain chlorophyll, the green pigment that absorbs energy from sunlight for use in photosynthesis, which means putting together with light. Photosynthesis is the process by which plants make food. In green plants, sunlight captured by chlorophyll enables carbon dioxide from the air to unite with water and minerals from the soil and synthesize food. This process also releases oxygen into the air, which is utilized by humans and animals to breathe. Energy from the sun splits water molecules into hydrogen and oxygen. The hydrogen joins with the carbon from carbon dioxide to produce sugar. The sugar—together with nitrogen, sulfur, and phosphorus from the soil—helps a plant make the fat, protein, starch, vitamins, and other materials that it needs to survive.

Structure: The chloroplast has a double membrane envelope termed the inner and outer membrane, respectively. Each membrane is a phospholipid bilayer, between 6 and 8 nm thick, and the two are separated by a gap of 10-20nm, called the intermembrane space. The fluid within the chloroplast is called the stroma. It contains tiny circular DNA and ribosomes, though most of their proteins are encoded by genes contained in the cell nucleus, and the protein products are carried to the chloroplast. Within the stroma are stacks of thylakoids which contain chlorophyll, the sub-organelles where photosynthesis actually takes place. A stack of thylakoids is called a granum (*plural*: grana). A thylakoid looks like a flattened disc, and inside is an empty area called the thylakoid space or lumen. The photosynthesis reaction takes place on the membrane of the thylakoid. These are the sites of light absorption and ATP synthesis, and contain many proteins, including those involved in the electron transport chain. Photosynthetic pigments such as chlorophyll a and b, and some others, for example, xanthophylls and carotenoids are also located within this space (Fig. 7.1).

Function of thylakoids: The membranes of the thylakoids contain photosystems I and II which harvest solar energy in order to excite electrons which travel down the electron transport chain. This exergonic fall in potential energy along the way is used to pump H^+ ions from the stroma into the

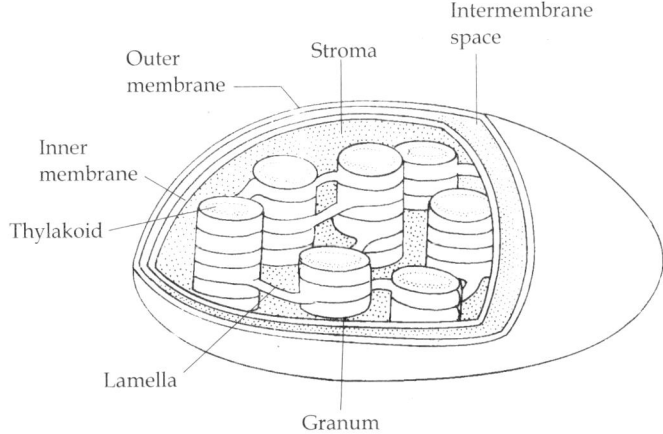

FIG. 7.1 Ultrastructure of a Chloroplast

thylakoid space. A concentration gradient is formed, which allows chemiosmosis to occur, where the protein ATP synthase harvests the potential energy of the hydrogen ions and uses it to combine ADP and a phosphate group to form ATP.

Etioplasts: These are the chloroplasts that have not been exposed to light.

FUNCTION

Photosynthesis

Photosynthesis is the process of converting light energy to chemical energy and storing it in the bonds of sugar. This process occurs in plants and some algae. Plants need CO_2, light energy and H_2O to make sugar. The process of photosynthesis takes place in the chloroplasts, specifically using chlorophyll, the green pigment.

Photosynthesis takes place primarily in plant leaves. The anatomy of a typical leaf includes the upper and lower epidermis, the mesophyll, the vascular bundle(s) (veins), and the stomata. The upper and lower epidermal cells do not have chloroplasts, thus photosynthesis does not occur there. They serve primarily as protection for the rest of the leaf. The stomata are the holes, present in the epidermis meant for air exchange, that is, they allow the exchange of CO_2 and O_2. The vascular bundles or veins in a leaf are part of the plant's transportation system, moving water and nutrients around the plant as needed. The mesophyll cells have chloroplasts and this is where photosynthesis occurs.

The overall chemical reaction involved in photosynthesis is:

$$6CO_2 + 12H_2O + \text{light energy} \xrightarrow{\text{chlorophyll}} C_6H_{12}O_6 + 6O_2 + 6H_2O$$

There are two phases to photosynthesis (Fig. 7.2):

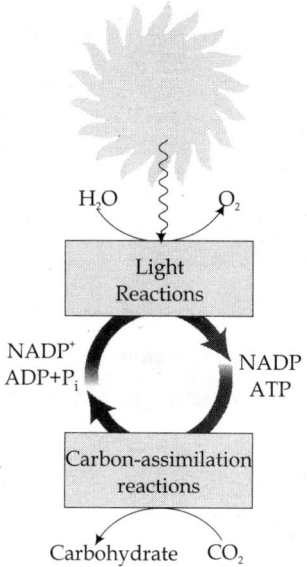

FIG. 7.2 The process of Photosynthesis

Light Reaction/Hill Reaction (Fig. 7.5): The light reaction happens in the thylakoid membrane and converts light energy to chemical energy. This chemical reaction must, therefore, take place in the light. Chlorophyll and several other pigments such as β-carotene are organized in clusters in the thylakoid membrane and are involved in the light reaction. Each of these differently coloured pigments can absorb a slightly different color of light and pass its energy to the central chlorophyll molecule to do photosynthesis. The central part of the chemical structure of a chlorophyll molecule is a Porphyrin ring, which consists of several fused rings of carbon and nitrogen with a magnesium ion in the center.

Chlorophyll is a complex molecule. Several modifications of chlorophyll occur among plants and other photosynthetic organisms. All photosynthetic organisms (plants, certain protistans, prochlorobacteria, and cyanobacteria) have chlorophyll-a (Fig. 7.3). Accessory pigments absorb energy that chlorophyll-a does not absorb. Accessory pigments include chlorophyll-b (Fig. 7.3) (also c, d, and e in algae and protistans), xanthophylls, and carotenoids (such as β-carotene). Chlorophyll-a absorbs its energy from the violet-blue and reddish orange-red wavelengths, and little from the intermediate (green-yellow-orange) wavelengths.

The energy harvested via the light reaction is stored by forming a chemical called adenosine triphosphate (ATP), a compound used by cells for energy storage. ATP is made up of the nucleotide adenine bonded to a ribose sugar, and that is bonded to three phosphate groups. This molecule is very similar to the building blocks for our DNA.

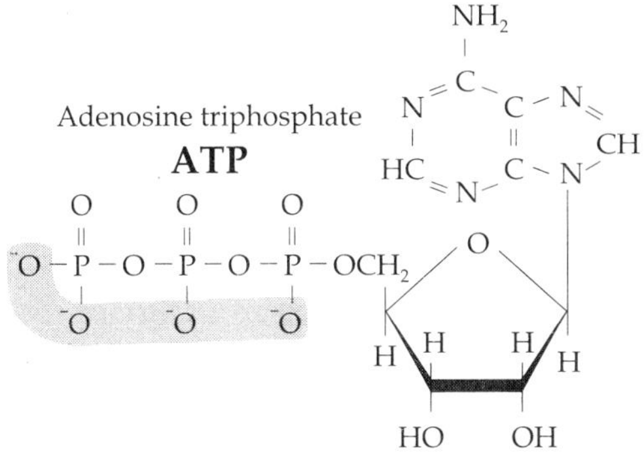

Adenosine triphosphate

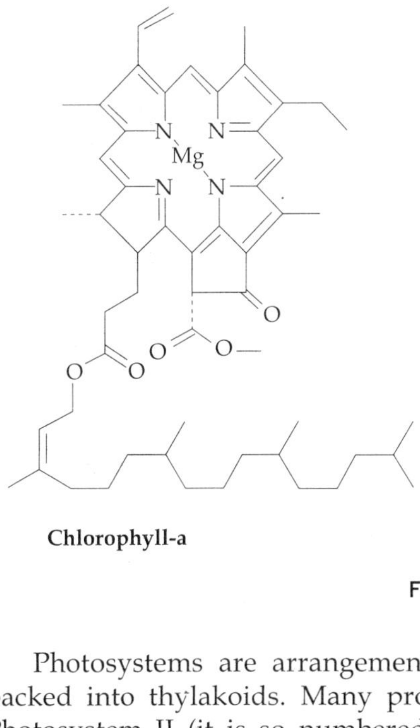

Chlorophyll-a 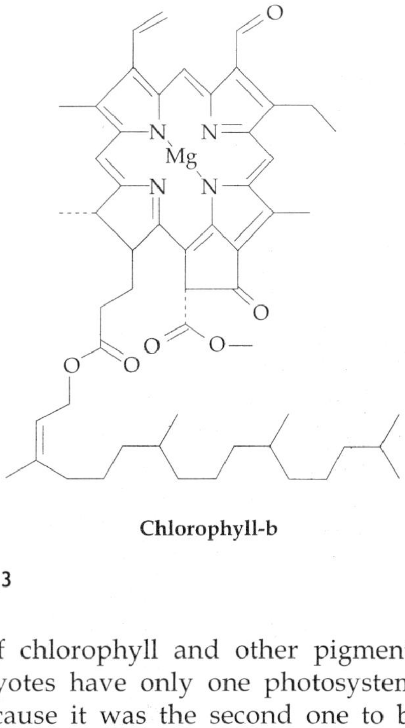   Chlorophyll-b

FIG. 7.3

Photosystems are arrangements of chlorophyll and other pigments packed into thylakoids. Many prokaryotes have only one photosystem, Photosystem II (it is so numbered because it was the second one to be discovered, though it is the first to evolve). Eukaryotes have Photosystem II plus Photosystem I. Photosystem I uses chlorophyll-a, in the form referred to as P_{700}. Photosystem II uses a form of chlorophyll a known as P_{680}. Both 'active' forms of chlorophyll-a function in photosynthesis due to their association with proteins in the thylakoid membrane.

Photophosphorylation is the process of converting energy from a light-excited electron into the pyrophosphate bond of an ADP molecule. This occurs when the electrons from water are excited by the light in the presence of P_{680}. The energy transfer is similar to the chemiosmotic electron transport occurring in the mitochondria. Light energy causes the removal of an electron from a molecule of P_{680} that is part of Photosystem II. The P_{680} requires an electron, which is taken from a water molecule, breaking the water into H^+ ions and O^{-2} ions. These O^{-2} ions combine to form the diatomic O_2 that is released. The electron is boosted to a higher energy state and attached to a primary electron acceptor, which begins a series of redox reactions, passing the electron through a series of electron carriers, eventually attaching it to a molecule in Photosystem I. Light acts on a molecule of P_{700} in Photosystem I, causing an electron to be boosted to a still higher potential. The electron is attached to a different primary electron acceptor (i.e., a different molecule from the one associated with Photosystem II). The electron is passed again through a series of redox reactions, eventually being attached to $NADP^+$ and H^+ to form NADPH, an energy carrier needed in the Light Independent Reaction. The electron from Photosystem II replaces the excited electron in the P_{700} molecule. There is thus a continuous flow of electrons from water to NADPH. This energy is used in Carbon Fixation. Cyclic Electron Flow takes place in some eukaryotes and primitive photosynthetic bacteria wherein no NADPH is produced but only ATP is synthesized. This occurs when cells may require additional ATP, or when there is no $NADP^+$ to reduce to NADPH. In Photosystem II, the pumping to H^+ ions into the thylakoid and the conversion of $ADP + P_i$ into ATP is driven by electron gradients established in the thylakoid membrane.

Dark Reaction/Calvin Reaction (Fig. 7.5): The dark reaction takes place in the stroma within the chloroplast, and converts CO_2 to sugar. This reaction doesn't directly need light in order to occur, but it does need the products of the light reaction (ATP and another chemical called NADPH). The dark reaction involves a cycle called the Calvin cycle (Fig. 7.4) in which CO_2 and

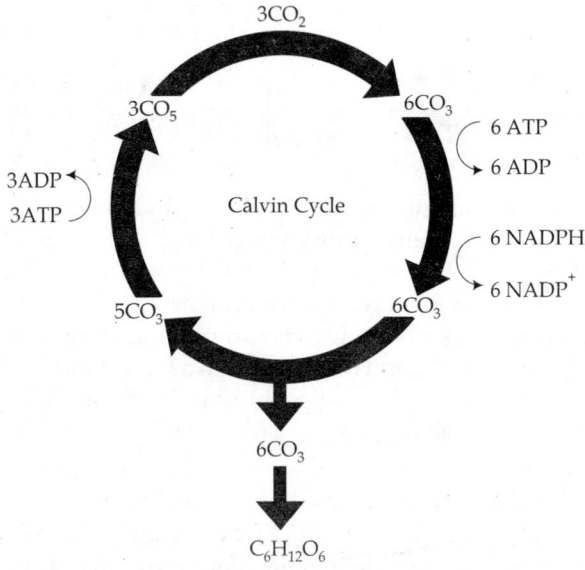

FIG. 7.4 Calvin Cycle

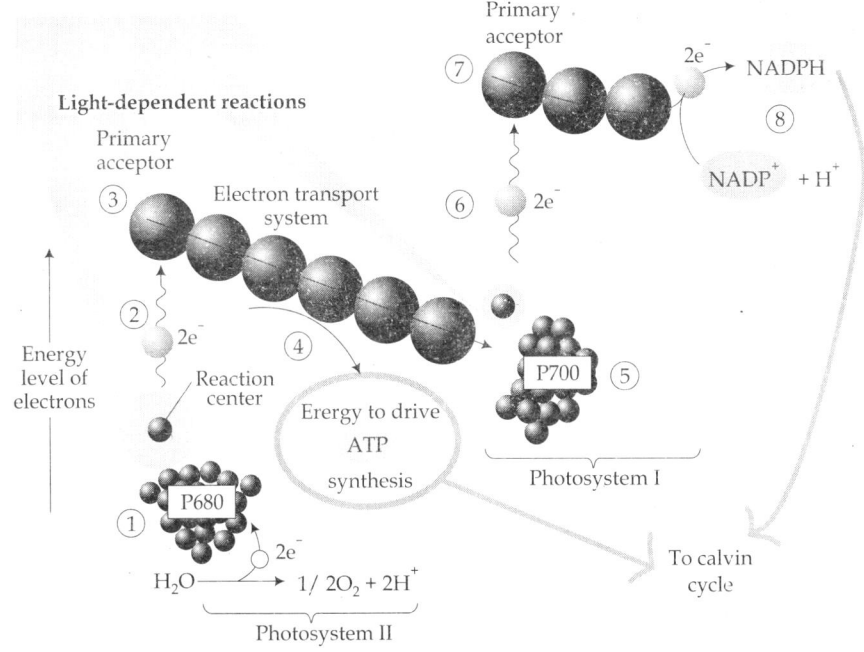

Light-dependent reactions

Primary
acceptor

Energy
level of
electrons

Electron transport
system

Reaction
center

Energy to drive
ATP
synthesis

P680

P700

Photosystem I

Primary
acceptor

$2e^-$ → NADPH

$NADP^+ + H^+$

$2e^-$

H_2O → $1/2O_2 + 2H^+$

$2e^-$

Photosystem II

To calvin
cycle

FIG. 7.5 An overview of Light and Dark Reactions of Photosynthesis

energy from ATP are used to form sugar. Actually, notice that the first product of photosynthesis is a three-carbon compound called glyceraldehyde-3-phosphate. Almost immediately, two of these join to form a glucose molecule.

Most plants put CO_2 directly into the Calvin cycle. Thus the first stable organic compound formed is the glyceraldehyde-3-phosphate formed from CO_2 and H_2O. Since that molecule contains three carbon atoms, these plants are called **C_3 plants**. For all plants, hot summer weather increases the amount of water that evaporates from the plant. Plants reduce the rate of water evaporation by keeping their stomata closed during hot, dry weather. Unfortunately, this means that once the CO_2 in their leaves reaches a low level, they must stop doing photosynthesis. Even if there is a tiny bit of CO_2 left, the enzymes used to grab it and put it into the Calvin cycle. Typically the grass in our yards just turns brown and goes dormant. Some plants like crabgrass, corn, and sugarcane have a special modification to conserve water. These plants capture CO_2 in a different way and have a special enzyme that can work better, even at very low CO_2 levels, to grab CO_2 and turn it first into oxaloacetate, which contains four carbons. Thus, these plants are called **C_4 plants**. The CO_2 is then released from the oxaloacetate and enters into the Calvin cycle. This is why crabgrass can stay green and keep growing when all the rest of your grass is dried up and brown.

There is yet another strategy to cope with very hot, dry, desert weather and conserve water. Some plants (e.g., cacti and pineapple) that live in extremely hot, dry areas like deserts, can only safely open their stomata at night when the weather is cool. Thus, there is no chance for them to get the CO_2 needed for the dark reaction during the daytime. At night when they can open their stomata and take in CO_2, these plants incorporate the CO_2 into various organic compounds to store it. In the daytime, when the light reaction is occurring and ATP is available (but the stomata must remain closed), they take the CO_2 from these organic compounds and put it into the Calvin cycle. These plants are called **CAM plants**, which stands for Crassulacean Acid Metabolism after the plant family, Crassulaceae (which includes the garden plant *Sedum*) where this process was first discovered.

OBJECTIVE TYPE QUESTIONS

1. The Calvin Cycle
   A. Is part of the light reactions of photosynthesis
   B. Is part of respiration
   C. Occurs in mitochondria
   D. Takes place in the thylakoids of the chloroplast
2. The Krebs Cycle
   A. Is part of the light reactions of photosynthesis
   B. Is part of respiration
   C. Takes place in the thylakoids of the chloroplast
   D. Produces no ATP
3. The light reactions of photosynthesis occur in the _____, and the dark reactions occur in the _____ of the chloroplast.
   A. Thylakoid membrane, Stroma
   B. Stroma, Thylakoid membrane
   C. Intermembrane space, Stroma
   D. Thylakoid membrane, Intermembrane space
4. The overall source of energy for photosynthesis is
   A. Energy of electron transport in thylakoid membrane
   B. Energy from hydrolysis of ATP
   C. Light energy from the sun
   D. Energy released when water is oxidized and oxygen is produced
5. An overall result of photosynthesis in plants is the use of electrons from water to reduce
   A. Glucose   B. Carbon dioxide
   C. Oxygen   D. Chlorophyll
6. Which of the following is NOT true of photosystem II?
   A. It is located in thylakoid membranes
   B. It is involved in the oxidation of water

C. It has a special oxidizable chlorophyll, P680

D. It is required for cyclic photophosphorylation

7. Plants need _____ and _____ to carry on photosynthesis.

A. Oxygen; Water

B. Carbon dioxide; Water

C. Carbon dioxide; Oxygen

D. Glucose; Water

8. Which of these could best be described as the membrane that contains chlorophyll molecules in or on its outer surface?

A. Thylakoid

B. Stroma

C. Lumen

D. Cell

9. What are the best terms to describe the photosynthetic reactions?

A. Light and dark reactions

B. Glycolysis and Kreb's cycle

C. Electron transport chains

D. Light dependent and light independent reactions

10. Where would you expect to see chlorophyll molecules in a green plant?

A. In the grana

B. In the stroma

C. In the outer membrane of the chloroplast

D. Dissolved in the cell cytoplasm

Answers

1. A	2. B	3. A	4. C	5. B
6. D	7. B	8. A	9. D	10. A

SHORT ANSWER TYPE QUESTIONS

1. Write a short note on different types of plastids.
2. Describe the structure and compartments of the chloroplast.
3. Discuss the structure and function of the photosystems.
4. Discuss the carbon fixation reactions of photosynthesis.
5. Explain how C-4 photosynthesis provides an advantage for plants in certain environments.

LONG ANSWER TYPE QUESTIONS

1. List the two major processes of photosynthesis and state what occurs in those sets of reactions.
2. Describe the Calvin-Benson cycle in terms of its reactants and products.
3. Explain the role of the two energy-carrying molecules produced in the light-dependent reactions (ATP and NADPH) in the light-independent reactions.

4. Describe how the pigments found on thylakoid membranes are organized into photosystems and how they relate to photon light energy.
5. Describe the function of electron transport systems in the thylakoid membrane.

Endoplasmic Reticulum

8

INTRODUCTION

The endoplasmic reticulum (endoplasmic means 'within the cytoplasm' and reticulum means 'little net') is a network of flattened sacs and branching tubules that extends throughout the cytoplasm in plant and animal cells. It is continuous with the nuclear envelope and the plasma membrane. These sacs and tubules are all interconnected by a single continuous membrane so that the organelle has only one large, highly convoluted and complexly arranged lumen (internal space). The endoplasmic reticulum cisternal space, the lumen of the organelle often takes up more than 10% of the total volume of a cell. The function of the endoplasmic reticulum membrane is that it modifies proteins, makes macromolecules and allows molecules to be selectively transferred between the lumen and the cytoplasm, and since it is connected to the double-layered nuclear envelope, it further provides a pipeline between the nucleus and the cytoplasm.

HISTORY

Endoplasmic reticulum was discovered by Garmier in 1897. Its electron microscopic structure was revealed by Porter, Claude and Fullam in 1945 as a network of delicate strands and vesicles in the cytoplasm. The term 'endoplasmic reticulum' was first used by Porter and Kallman in 1952.

STRUCTURE

Morphologically, endoplasmic reticulum exhibits great variation in form, viz., Vesicular form, Cisternae and Tubular form (Fig. 8.1).

Vesicles: These are rounded, spherical or ovoid spaces measuring about 25-500 μm in diameter. These occur abundantly in cells involved in protein synthesis.

Cisternae: These are elongated, flattened and usually unbranched tubular vesicles arranged in parallel rows. These form successive layers around the nucleolus. These are about 40-50 μm thick.

Tubules: The tubules are small, smooth walled branched tubular spaces about 50-200 μm in diameter. These occur in the cells involved in the synthesis of steroids like cholesterol, glycerides and hormones.

These three forms may appear in the cell at the same time but their distribution varies in the cells of different tissues and different species.

There are two basic kinds of endoplasmic reticulum: rough and smooth.

Rough Endoplasmic Reticulum (RER): The rough endoplasmic reticulum is so called because its surface is covered with ribosomes, giving it a rough appearance when viewed through the microscope (Fig. 8.2). This type of endoplasmic reticulum is involved mainly with the production and processing of proteins that will be exported, or secreted, from the cell. The ribosomes assemble amino acids into protein units, which are transported into the rough endoplasmic reticulum for further processing. These proteins may be either transmembrane proteins, which become embedded in the membrane of the endoplasmic reticulum, or water-soluble proteins, which are able to

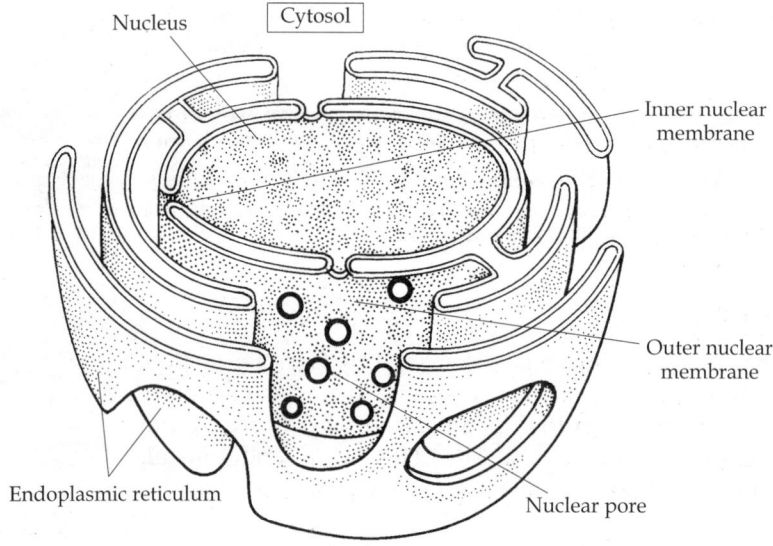

FIG. 8.1 Endoplasmic Reticulum surrounding the Nucleus

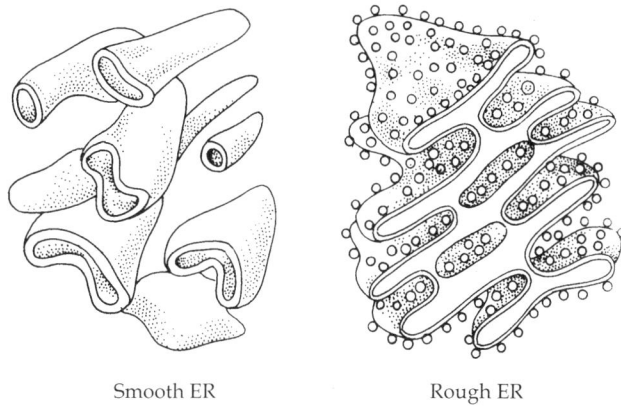

Smooth ER   Rough ER

FIG. 8.2 Structure of Endoplasmic Reticulum

pass completely through the membrane into the lumen. Those proteins that reach inside of the endoplasmic reticulum are folded into the correct three-dimensional conformation, as a flattened cardboard box might be opened up and folded into its proper shape in order to become a useful container. Chemicals, such as carbohydrates or sugars, are added, then the endoplasmic reticulum either transports the completed proteins to areas of the cell where they are needed, or they are sent to the Golgi apparatus for further processing and modification.

Smooth Endoplasmic Reticulum (SER): Most proteins exported from the endoplasmic reticulum exit the organelle in vesicles budded from the smooth portion, which has a more even appearance than rough endoplasmic reticulum when viewed through the electron microscope because of the lack of ribosomes (Fig. 8.2). The smooth endoplasmic reticulum in most cells is much less extensive than the rough endoplasmic reticulum and is sometimes alternatively termed transitional. Smooth endoplasmic reticulum is chiefly involved with the production of lipids, building blocks for carbohydrate metabolism, and the detoxification of drugs and poisons. Therefore, in some specialized cells, such as those that are occupied chiefly in lipid and carbohydrate metabolism (brain and muscle) or detoxification (liver), the smooth endoplasmic reticulum is much more extensive and is crucial to cellular function. Smooth endoplasmic reticulum also plays a role in various cellular activities through its storage of calcium and involvement in calcium metabolism. In muscle cells, smooth endoplasmic reticulum releases calcium to trigger muscle contractions.

SARCOPLASMIC RETICULUM

Sarcoplasmic Reticulum is the form of endoplasmic network found in the striated muscle cells. It is physically separate from the sarcolemma and surrounds each myofibril. Sarcoplasmic reticulum membrane contains high levels of calcium ATPase. It functions to uptake calcium from the sarcoplasm and to release calcium into the sarcoplasm to initiate contraction and sequester it during relaxation.

FUNCTIONS

The endoplasmic reticulum acts as a secretory, storage, circulatory and nervous system for the cell. It performs following important fuctions:

1. **Protein synthesis:** It is the rough ER that plays a role in protein synthesis, especially proteins that are to be secreted to outside the cell (e.g., hormones). It provides the surface for the attachment of polyribosomes and facilitates the activity of mRNA. Proteins enter the lumen of the endoplasmic reticulum while still being synthesized. In addition to protein synthesis, the rough endoplasmic reticulum also functions in the modification of newly formed proteins. For example, some enzymes may add carbohydrate chains forming glycoproteins. Other enzymes function to fold the newly-synthesized proteins into their proper shape.

2. **Synthesis of lipoproteins:** Smooth endoplasmic reticulum generally functions to produce lipid compounds such as phospholipids, steroids, and fatty acids.

3. **Detoxification:** The smooth endoplasmic reticulum brings about the detoxification of drugs and other poisons especially in liver cells, thus, making it easier to flush the toxics out of the body.

4. **ATP synthesis:** The membranes of endoplasmic reticulum are the sites of ATP synthesis in the cell. This ATP is used as a source of energy for all the metabolic processes and transport of materials.

5. **Glycogen metabolism:** The enzyme glucose-6-phosphatase has been found to be present in the endoplasmic reticulum. It is associated with the breakdown of glycogen, a process called glycogenolysis. It has also been suggested that SER membranes play a role in the synthesis of glycogen by a process termed glycogenesis.

6. **Mechanical support:** The network of the endomembrane system provides support to the colloidal matrix of cytoplasm.

7. **Exchange of materials:** The membranes of ER maintain osmotic pressure within the cells, isolate materials synthesized or delivered and regulate their exchange between the inner and outer compartments of the cell.

8. **Enzyme activities and cellular metabolism:** A number of enzymes are associated with the smooth endoplasmic reticulum which provides an increased surface area for various metabolic reactions.

9. **Formation of cytomembranes:** Other membranous structures of the cell like nuclear membrane and Golgi complex differentiate from ER.

OBJECTIVE TYPE QUESTIONS

1. Which of the following matches is not correct?
   A. Ribosomes—Rough ER
   B. Protein synthesis—Smooth ER
   C. Rough ER—Export of proteins out of cell
   D. Smooth ER—Cells in intestine
2. Which are structurally most similar in eukaryotes and prokaryotes?
   A. Rough endoplasmic reticulum   B. Plasma membrane
   C. Flagella   D. Nucleus
3. Part of the endoplasmic reticulum is labeled rough because it is studded with
   A. Chloroplasts   B. Ribosomes
   C. Vesicles   D. Chromosomes
4. Within eukaryotic cells, an extensive system of membranes called _____ separates various regions of the cytoplasm from each other.
   A. Tubules   B. Golgi apparatus
   C. Endoplasmic reticulum   D. None of the above
5. Proteins manufactured in the endoplasmic reticulum are secreted from the cell bywhich of the following processes?
   A. Diffusion   B. Osmosis
   C. Pinocytosis   D. Exocytosis

Answer

1. B   2. B   3. B   4. C   5. D

SHORT ANSWER TYPE QUESTIONS

1. Describe the structure of endoplasmic reticulum.
2. Explain the functions of smooth endoplasmic reticulum.
3. What is the difference between the rough endoplasmic reticulum and the smooth endoplasmic reticulum?

LONG ANSWER TYPE QUESTIONS

1. What is endoplasmic reticulum? What are its distinguishing characteristics?
2. Write a note on the Endoplasmic Reticulum as a gateway of the secretory pathway.
3. What are the functions of endoplasmic reticulum?

Golgi Complex

9

INTRODUCTION

The Golgi apparatus or complex is a membranous system of cisternae and vesicles located in the cytoplasm of most eukaryotic cells, including plants, animals and fungi. It is involved in the intracellular transport and modification of secretory proteins, membrane proteins, and proteins that remain membrane-bounded within the cell, in distinction to proteins such as hemoglobin and keratin that lie free in the cytoplasm.

Most of the transport vesicles that leave the endoplasmic reticulum, specifically smooth ER, are transported to the Golgi complex, where they are modified, sorted, and shipped towards their final destination. Thus, Golgi complex tends to be more prominent in the regions where substances, such as proteins, are being secreted.

HISTORY

The existence of Golgi apparatus or Golgi complex was first reported by Camillo Golgi in 1898, when he described in nerve cells of barn owl an *'apparata reticolare interno'* (internal reticular apparatus) impregnated by a variant of his chromoargentic staining. It soon became clear that the newly-identified cytoplasmic structure occurred in a variety of cell types.

Holmgren, in 1900, described a system of clear canals which he called trophospongium. This structure was earlier described as being homologous to internal reticular

FIG. 9.1 Camillo Golgi (1843-1926)

apparatus but later it was found to be more like endoplasmic reticulum. In 1910, Perroncito described the division of the Golgi complex and called it 'dictyokinesis'. R.Y. Cajal, in 1914, perfomed experiments in order to study the functional aspects of Golgi complex. Gatenby, in 1917, pointed out similarities between Golgi complex and dictyosomes of invertebrates. Nassonov, in 1924, and Bowen, in 1929, described the relationship between Golgi complex and the process of secretion. Baker, in 1951, coined the term 'Lipochondrion' for Golgi complex. Dalton, in 1952, and Dalton and Felix, in 1954, studied the detailed ultrasrtucture of the complex. In 1958, Perner coined the term 'dictyosomes' for the Golgi bodies in plants.

STRUCTURE

The Golgi complex is an organelle found in eukaryotic cells. It functions as the cell's packaging and shipping department. It has three distinct components (Fig. 9.2):

(a) **Flattened sacs or Cisternae:** These are elongated, flattened and slightly curved fluid filled sacs having swollen ends stacked parallel on one another. Each cisterna is a double membrane structure enclosing a thin space of about 20 Å while the inter-cisternal space is about 200-300 Å. The Golgi complex consists of 3 to 15 large flat sacs or cisternae apposed to one another. They are relatively compressed at their centers and somewhat dilated peripherally. The sacs tend to be bowed, presenting a convex proximal face (toward the nucleus) and concave distal face (away from the nucleus).

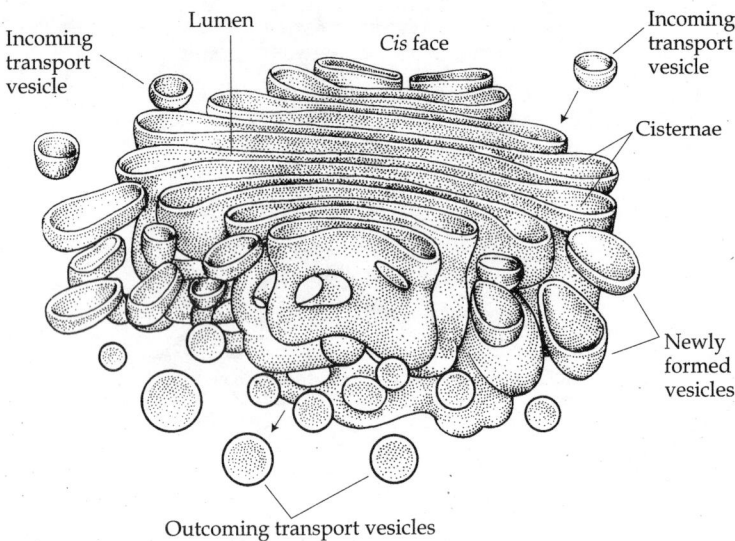

FIG. 9.2 The Golgi apparatus

(b) **Vesicles:** These are small drop like structures having a diameter of about 40-100 Å. They are of three types:

(i) **Transition vesicles:** These are found associated with the convex surface of Golgi and lie between smooth ER and Golgi cisternae. These are known to detach from the cisternae of endoplasmic reticulum.

(ii) **Secretory vesicles:** These are known to detach from the cisternae of Golgi apparatus and are present towards the concave face of Golgi. They contain secretory products of Golgi and are finally converted into zymogen granules or Lysosomes.

(iii) **Clathrin-coated vesicles:** These are found at the periphery of the organelle. They are known to play a role in intracellular traffic of membranes and of secretory products between ER and Golgi and between GERL region and the endosomal and lysosomal compartments.

(c) **Vacuoles:** These are spherical structures having a diameter of abut 600 Å and are present towards the concave side of the Golgi apparatus. These are formed either by the expanded cisternae or by the fusion of the secretory vesicles. They are filled by some amorphous or granular substance.

Polarity in Golgi Cisternae

The stacked cisternae of the Golgi complex form a bowl-shaped structure and the Golgi complex as a whole looks like a stack of shallow bowls with the concavity directed away from the nucleus. The cisternae may communicate with one another by slender channels at places along their contiguous surfaces. The proximal membranes (those near the nucleus) are thinner than the distal membranes (facing out toward the bulk of the cytoplasm), which are more like those of plasmalemma. At the edge of the flattened sacs, near their expanded peripheries, the vesicles are typically present. Similar vesicles may also be abundant at the distal face. The vesicles vary in size and probably fuse to form larger vesicles. Like lateral vesicles, they may contain a dense material. The distal concave face, which is typically engaged in granule formation, has been termed the maturation face (also known as the *trans* face). The proximal convex face, relatively free of vesicles, has been termed the forming face (also known as the *cis* face) (Fig. 9.3).

The GERL Region

The GERL (Golgi + smooth ER + Lysosome) is the region in the cell between Golgi and the plasma membrane, where secretory vesicles are converted into zymogen granules or lysosomes by the concentration of secretory products of ER. It is rich in acid phosphates.

The Endomembrane Concept

The concept of an endomembrane system (Morré and Mollenhauer, 1974) was introduced to indicate the possibility that various membranous compartments of the eukaryotic cells were interrelated and interconnected. Included within the endomembrane system were the nuclear envelope, rough and smooth endoplasmic reticulum, Golgi apparatus, and various cytoplasmic vesicles. Plasma membranes, vacuole membranes, and/or lysosomes were regarded as

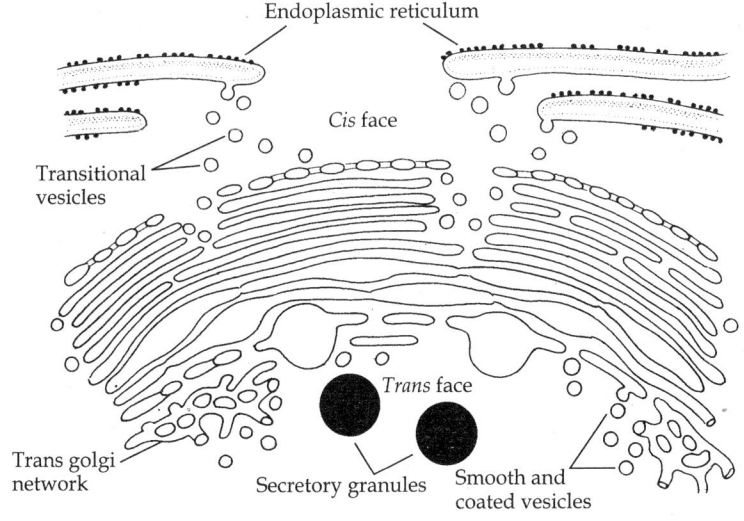

FIG. 9.3 Diagram showing an actual interface between the ER and the Golgi complex

end products of the system. Organelles such as mitochondria, chloroplasts, and peroxisomes were not included as part of the endomembrane system even though their outer membranes may contact closely or even connect directly with endoplasmic reticulum.

Secretory vesicles that move from the Golgi apparatus to the cell surface provide continuity with the plasma membrane. At the cell surface the membranes of these vesicles fuse with the plasma membrane. The vesicle membrane is incorporated, at least transiently, into the plasma membrane and the vesicle contents are delivered to the cell surface.

The vesicular traffic into and out of the Golgi apparatus involves structures other than the secretory vesicles and the transition vesicles that bleb off the rough endoplasmic reticulum and presumably join to form new Golgi apparatus cisternae. Also involved are clathrin- (spiny-) coated vesicles at the mature or *trans* face of Golgi apparatus, condensing vacuoles that give rise to secretory granules, fusiform vesicles, and cisternal remnants as well as various structures apparently derived from the plasma membrane through endocytosis and/or membrane recycling or belonging to the endosome/ lysosome/vacuole system (Fig. 9.4).

ORIGIN

Different views regarding the origin of Golgi complex have been presented from time to time. In 1910, Perroncito suggested that the division of Golgi body by a process called dictyokinesis led to the formation of new Golgi complexes. Later, in 1965, Bloch proposed that Golgi complex originated from

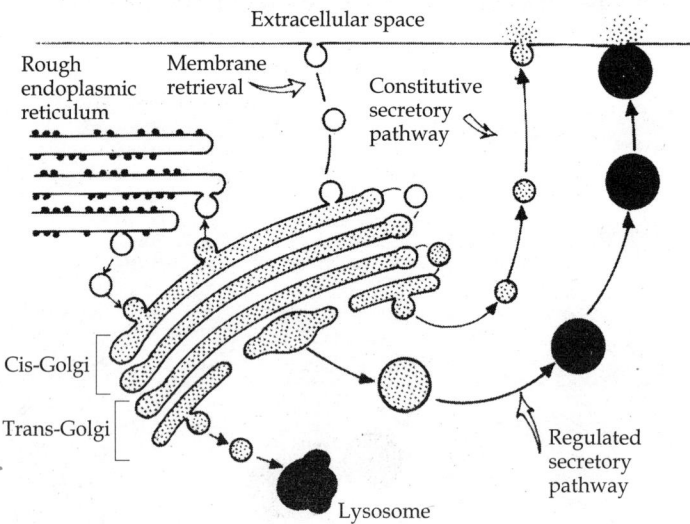

FIG. 9.4 The regulated secretory pathyway and constitutive pathyway that controls trafficking of different types of proteins

the outer membrane of the nuclear envelope. Daniels, in 1964, studied the Golgi apparatus in the amoeba, *Palomyx illinoisensis* and suggested its origin from the plasmalemma. However, the findings of Essner and Novikoff (1962) and Beams and Kessel (1968) led to the most accepted view that Golgi body originates from endoplasmic reticulum. This theory is supported by the fact that the cisternae of Golgi body are continuous with smooth ER at certain places.

FUNCTIONS

1. Membrane Flow (Vesicular transport):

The Golgi apparatus is considered more or less the 'Post office' of the cell. It handles all the incoming lipids, proteins, etc., and controls their export, as well.

The transport vesicles from the endoplasmic reticulum (ER) fuse with the convex face of the Golgi apparatus (to the cisternae) and empty their protein content into the inner space of Golgi. The proteins are then transported through the medial region toward the concave or maturation face and are modified on their way. Possible modifications include the processes like glycosylation and phosphorylation. The proteins are also labeled with a sequence of molecules according to their final destination (Fig. 9.4).

The mechanism of transport, though not very clear is assumed to happen by cisternae progression (the movement of the apparatus itself, building new cisternae at the proximal face and destroying them at the maturation face) or by vesicular transport (small vesicles transport the proteins from one cisterna to the next, while the cisternae remain unchanged). Lately, it is also proposed that the cisternae are interconnected and the transport of molecules within the Golgi occurs by diffusion.

Once the proteins reach the *trans* face, they are embedded into coated transport vesicles and brought to their final destinations. The form of the vesicle is determined by the type of protein and the label it acquired (Fig. 9.5).

An example of the Golgi complex's functioning is the modification of glycoproteins (used in cell membranes). Vesicles from the ER contain simplified glycosylated proteins. In the Golgi apparatus, carbohydrates are attached and removed from these glycoproteins, creating a diversity of carbohydrate structures on the proteins. After they have been secreted into the cell the vesicles fuse to the cell membrane and release their contents.

Along with protein modification, Golgi apparatus is involved in the transport of lipids around the cell as well creating lysosomes — organelles involved in digestion.

2. Formation of acrosome

Burgos and Fawcet (1966) discovered that the Golgi complex forms the acrosome during sperm maturation (spermeogenesis). Acrosome is a vesicle at

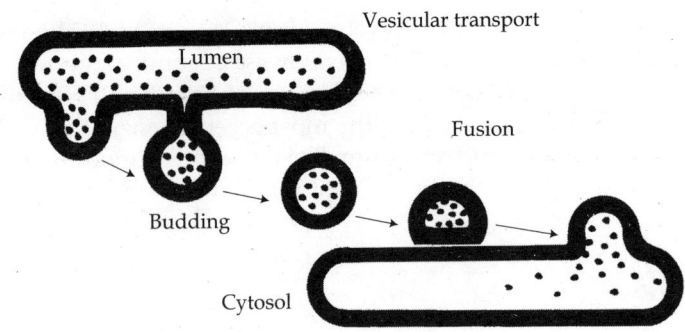

FIG. 9.5 Transport of material in and out of the Golgi complex involving budding and fusion of vesicles

the tip of the nucleus of sperm. It contains hydrolytic enzymes which aid in the breaking down of the protective covering of egg during fertilization.

In early stages, Golgi body is a spherical structure comprising of numerous vesicles and surrounded by several rows of cisternae. The vesicles, later, fuse to form one or two large vacuoles during spermiogenesis. Each vacuole then develops a small dense body called the proacrosomal granule, the vacuoles later fuse together forming a single vacuole. The vacuole enlarges greatly in size and migrates towards the anterior pole of nucleus and attaches to its tip. The proacrosomal granule develops into acrosomal granule and forms the major portion while vacuolar membrane spreads over the acrosome and the anterior portion of the nucleus.

3.   Formation of Cell Plate and Cell Wall

In plant cells, the cell plate and cell wall are formed during the telophase of the nuclear division by the deposition of pectin and hemicellulose. It has been observed that during anaphase, vesicles released from Golgi migrate to the centre of the spindle and aggregate above the spindle fibres. These Golgi vesicles provide carbohydrates like pectin, hemicellulose, etc. that form the cell plate and cell wall eventually.

OBJECTIVE TYPE QUESTIONS

1. The Golgi apparatus is a cell structure mainly devoted to processing the proteins   synthesized in the _____.

    A. Nucleus   B. Endoplasmic reticulum

    C. Nucleolus   D. None of the above

2. _____ receives materials through the endoplasmic reticulum and sends them to other parts of the cell.

    A. Nucleus   B. Lysosomes

    C. Golgi apparatus   D. Nucleolus

3. In invertebrates and plants the Golgi complex is known usually as
   _____ .

   A. Sacked cisternae   B. *Apparata reticolare*

   C. Lipochondrion   D. Dictyosome

4. The Golgi apparatus processes _____ produced on the ribosomes of
   the rough endoplasmic reticulum.

   A. Proteins   B. Carbohydrates

   C. Lipids   D. All of the above

5. The processing of proteins includes modification of the core _____
   of glycoproteins.

   A. Amino acids   B. Oligosaccharides

   C. Sialic acid   D. Glycosylated proteins

Answers

1. B   2. C   3. D   4. A   5. B

SHORT ANSWER TYPE QUESTIONS

1. Enlist three major differences between Golgi complex and endoplasmic
   reticulum.

2. Write a short note on polarity of Golgi complex.

3. Describe the structure of Golgi apparatus.

LONG ANSWER TYPE QUESTIONS

1. What is the Golgi apparatus or dictyosome?

2. Describe the functions of Golgi complex.

3. Explain how Golgi complex helps in the formation of acrosome.

Lysosomes

10

INTRODUCTION

Lysosomes (Gr., *lyso* = digestive; *soma* = body) are small membrane bound organelles that contain digestive enzymes (acid hydrolases) to digest macromolecules. They are found in both plant and animal cells, and are built in the Golgi apparatus by budding. The interior of the lysosomes is more acidic having a pH around 4.8 than the cytosol whose pH is maintained at 7. The single membrane of lysosome stabilizes the low pH by pumping in protons (H^+) from the cytosol, and also protects the cytosol, and therefore the rest of the cell, from the degradative enzymes within the lysosome. All these enzymes are produced in the endoplasmic reticulum, and transported and processed through the Golgi apparatus. Each acid hydrolase is then targeted to a lysosome by phosphorylation. The lysosome itself is likely safe from enzymatic action due to having proteins in the inner membrane which has a three-dimensional molecular structure that protects vulnerable bonds from enzymatic attack. The main function of lysosomes is digestion of food materials (intracellular digestion), degradation of worn out organelles and autophagic cell death (Fig. 10.1).

HISTORY

Christian de Duve was a Belgian researcher who discovered lysosomes in 1949 (Fig. 10.2). He discovered them when he homogenized some animal cells into various components by running them through an ultracentrifuge. After a few days the

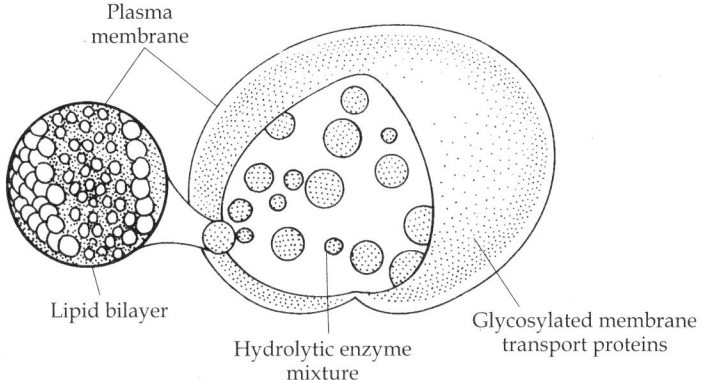

Plasma membrane

Lipid bilayer

Hydrolytic enzyme mixture

Glycosylated membrane transport proteins

FIG. 10.1 Anatomy of the lysosome

FIG. 10.2 Christian De Duve

level of a few enzymes rose significantly implying that they were kept separated in the cell and had not attacked any part of the cell before they were broken down. De Duve reasoned that these hydrolytic enzymes were enclosed within the cell in some kind of membranous envelope that formed a self-contained organelle. He calculated the probable size of this organelle and called it the lysosome. For this work he got the Nobel Prize in 1974 which he shared with Albert Claude and George Palade.

TYPES OF LYSOSOMES

Lysosomes exhibit polymorphism in their morphology. Following different types of lysosomes have been recognized in different types of cells or in the same cell at different times.

Primary lysosomes: These are newly formed organelles bounded by a single membrane and contain hydrolytic enzymes necessary for intracellular digestion.

Heterophagosomes: The term hetero (different) phagy (to eat) refers to a process whereby lysosomes aid in the intracellular digestion of material gathered from outside of the cell by some kind of endocytotic mechanism. The endocytotically ingested material is sequestered in a membrane bounded vesicle. The heterophagosomes are formed by the fusion of primary lysosomes with cytoplasmic vacuoles containing these extracellular substances.

Autophagosomes: The term auto (self) phage (to eat) literally refers to a process whereby the cell digests its own contents. Many important cellular constituents, such as proteins, organelles and membranes, are in a constant state of flux, that is, they are constantly being synthesized and degraded. The primary lysosomes that aid in such type of process are called autophagosomes.

Residual bodies: They are formed if the digestion inside the vacuole remains incomplete due to the absence of some lysosomal enzymes.

LYSOSOMAL ENZYMES

According to estimate lysosomes contain over 60 enzymes called hydrolases that degrade nucleotides, proteins, lipids, phospholipids, and also remove carbohydrate, sulfate, or phosphate groups from molecules. The hydrolases are active at an acid pH which is fortunate because if they leak out of the lysosome, they are not likely to do damage (at pH 7.2) unless the cell has become acidic. A Hydrogen ion ATPase is found in the membrane of lysosomes to acidify the environment. These enzymes include:

ENZYMES	SUBSTRATES
Proteases and peptidases	**Proteins**
1. Cathepsins (A, B_1, B_2, C, D and E)	Proteins
2. Collagenase	Collagen
Nucleases	**Polynucleotides**
1. DNases	DNA
2. RNases	RNA
Lipases	**Lipids**
1. Esterases	Fatty acid esters
2. Phospholipases	Phospholipids
Phosphatases	**Phosphates**
1. Acid phosphatase	Phosphomonoesters

2. Acidphosphodiesterase	Oligonucleotides and Phosphodiesters
3. Acid pyrophosphatases	ATP
Glycosidases	**Carbohydrates**
1. β-galactosidase	Lactose
2. α-mannosidase	β-mannosides
3. β-glucuronidase	Polysaccharides and mucopolysaccharides
4. α-glucosidase	Maltose
5. Hyaluronidase	Hyaluronic acid
6. Sucrase	Sucrose
Sulphatases	**Sulphates**

LYSOSOMAL MEMBRANE

The lysosome membrane has three special properties:

1. It has an ATP driven proton [H^+] pump which maintains a low pH (4.5-5.5) in the lysosomal compartment. The proton pump has a structure similar to the F_1 and F_0 of mitochondria which consists of a head containing 6 polypeptide units and a stalk that contains 5 polypeptide units. It is inhibited by n-ethylmaleimide.

2. The membrane has a glycoprotein coat, rich in carbohydrates, on its inner surface to protect it against hydrolysis by its own enzymes.

3. The transporter channels transport the breakdown products such as amino acids, glucose, nucleotides and other small molecules out of the lysosome. These molecules may move out by facilitated diffusion, by active transport or by co-transport using the proton (H^+) gradient to provide the energy for transport.

FUNCTIONS

Lysosomes may be involved in the following pathways:

1. Phagocytosis

Phagocytosis is the process by which the cell engulfs particulate matter (>0.5 mm) from the extracellular space, and digests it by lysosomal action. The lysosomes infuse with vesicles of engulfed material and release the digestive enzymes to break up the material. The large molecules of food are broken down into smaller particles. The products diffuse through the lysosomal membrane and are distributed throughout the cell. The products serve as the building blocks of new materials (Fig. 10.3). Three main cell types are capable of performing this function:

Macrophages: They are widely distributed in the connective tissue around the body. Their function is to phagocytose and digest all particulate matter in the extracellular space, including microorganisms, foreign particles and damaged cells. They are derived from monocytes, circulating in the peripheral

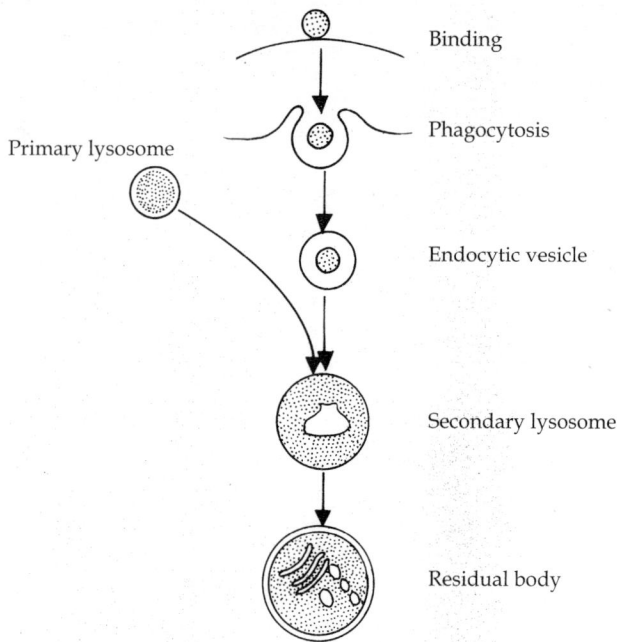

Binding

Phagocytosis

Primary lysosome

Endocytic vesicle

Secondary lysosome

Residual body

FIG. 10.3 The process of phagocytosis

blood. The macrophages may have different names in different tissues such as Kupffer cells in the liver, osteoclasts in bone, and microglia in the central nervous system.

Neutrophils: They are present in the peripheral blood and phagocytose and digest microorganisms.

Eosinophils: These are also present in the peripheral blood and phagocytose and digest antigen-antibody complexes.

The macrophages have a gylcocalyx coat, rich in glycosaminoglycans, on their outer surface, which causes particulate matter to adhere to them. Neutrophils and eosinophils have specific receptors to recognize specific particles to be engulfed. The specific molecules that bind to the receptors are called ligands.

The process of phagocytosis involves the adhesion of the particle to the glycocalyx or specific receptor on the plasma membrane. This is followed by the extrusion of pseudopodia to surround the particle, a process mediated by actin filaments. This causes the formation of a phagocytic vacuole containing the engulfed particle which then fuses with a primary lysosome.

2. Autophagocytosis

This is the process in which old or damaged organelles are broken down. It occurs in practically all cells as a recycling system. It is most marked in cells that are not replaced, such as neurons. The process involves formation of a membrane around the organelle by vesicles. This structure is called an autophagic vacuole. The autophagic vacuole fuses with one or more primary lysosomes to form secondary lysosomes and the structures with partially undigested material at the end are called residual bodies (Fig. 10.4).

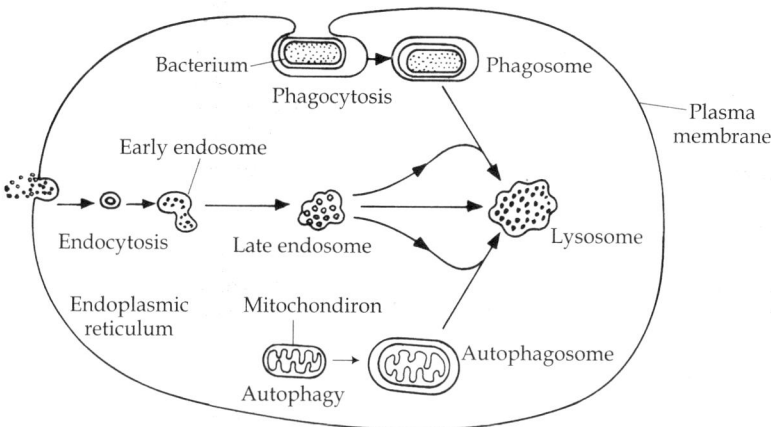

FIG. 10.4 Three routes to degradation in Lysosomes

3. Endocytosis and Exocytosis

Endocytosis is the uptake of extracellular fluid by infolding of the plasma membrane, and formation of a vesicle containing the extracellular material. This process was formerly called pinocytosis.

Exocytosis is the reverse of pinocytosis, i.e. the extrusion of fluid contained in vesicles into the extracellular space.

Endocytosis and exocytosis are important mostly for membrane flow. For example, exocytosis replaces the plasma membrane removed by phagocytosis. Similarly, exocytosis of secretion vesicles must be balanced by endocytosis.

4. Receptor-mediated Endocytosis

Receptor-mediated endocytosis is the process whereby cells that have a specific receptor take up specific macromolecules. This process is used for the uptake of hormones, growth factors, antibodies, lipoproteins, etc. The receptors are integral membrane proteins and the molecule that binds to the receptor is termed the ligand.

The process of receptor-mediated endocytosis involves the following steps:

I. Binding of the ligand to the receptor

II. Lateral diffusion of the ligand-receptor complex

III. Accumulation of clathrin, adaptor protein and dynamin on the cytoplasmic surface of the plasma membrane at a particular site

IV. Formation of a pit, and accumulation of the ligand receptor complex at the site of clathrin acculmulation

V. Deepening of the pit and formation of clathrin-coated vesicles containing the ligand-receptor complex

VI. The vesicles lose their clathrin coat

VII. The vesicles fuse with a primary lysosome (early endosome) and the ligand is cleaved from the receptor

LYSOSOMAL STORAGE DISEASES

There are several lysosomal storage disorders. All are associated with a deficiency of a particular lysosomal enzyme, resulting in accumulation of an undigested substrate within the lysosomes. The following are a few examples of lysosomal storage disease with the associated substrate:

Disease	Substrate
Pompe's disease	Glycogen
Hunter disease	Heparan and dermatan sulphates
Morquio's disease	Keratan and chondroitin sulphate
Tay-Sachs disease	GM2 ganglioside
Niemann-Pick	Sphingomyelin
Farber disease	Ceremide
Gangliosidosis	Galactose

Familial hypercholesterolaemia is a condition in which there is defective binding of low density lipoprotein (LDL) to its receptor. Receptor-mediated endocytosis of the LDL does not occur, and it accumulates as high levels in the blood. The process of receptor-mediated endocytosis was first described through the study of LDL uptake by cultured fibroblasts.

Three molecules are involved in forming the coat on the cytoplasmic face of coated vesicles:

1. **Clathrin**
   - A molecule that has a triskelion (three-pronged) structure.
   - Under appropriate conditions it forms a hexagonal (geodesic) lattice work on the cytoplasmic surface of the plasma membrane.
   - It causes the membrane to invaginate and form a coated pit and vesicle.

2. **Adaptor protein**
   - It binds to the cytoplasmic end of the transmembrane receptor.
   - Mediates the attachment of clathrin to the plasma membrane.

- Regulates clathrin assembly and disassembly.
- Is itself regulated by phosphorylation and dephosphorylation.

3. Dynamin
- Incorporates a GTPase that provides the energy for the formation of coated pit and vesicle.
- Undergoes a configuration change that brings about closure of the vesicle.

OBJECTIVE TYPE QUESTIONS

1. The lysosome is a membrane bound structure that contains _____ for digesting macromolecules.
   A. Peptidases　　　　　　　　B. Villi
   C. Hydrolytic enzymes　　　　D. Acids

2. The enzymes, called hydrolases, are made in the _____ and transported to the lysosome by the _____, using a vesicle.
   A. Endoplasmic reticulum, Golgi apparatus
   B. Nucleus, Endoplasmic reticulum
   C. Golgi apparatus, Endoplasmic reticulum
   D. Golgi apparatus, Plasma membrane

3. The hydrolases are active at an _____ pH.
   A. Basic　　　　　　　　　　B. Acid
   C. Neutral　　　　　　　　　D. All of the above

4. Lysosomes are manufactured by _____ .
   A. Nucleus　　　　　　　　　B. Endoplasmic reticulum
   C. Cytoplasm　　　　　　　　D. Golgi apparatus

5. Which one of the following is lysosomal storage disease?
   A. Huntington chorea　　　　B. Klinefelter syndrome
   C. Tay-Sachs disease　　　　　D. Phenylketonuria

Answers

1. C　　　2. A　　　3. B　　　4. D　　　5. C

SHORT ANSWER TYPE QUESTIONS

1. Write a short note on different types of lysosomes.
2. Enlist the different lysosomal enzymes.
3. How are lysosomes produced?

LONG ANSWER TYPE QUESTIONS

1. What are lysosomes?
2. Explain any two lysosomal storage disorders.
3. Write an account on the functions of lysosomes.

Microbodies: Peroxisomes and Glyoxysomes

11

INTRODUCTION

Microbodies are roughly spherical in shape, bound by a single membrane, and are usually 0.5 to 1 μm in diameter. They occur in many types of eukaryotic cells, including those of animals, higher plants, and protozoa and contain oxidative enzymes. In vertebrates, they are particularly numerous in liver and kidney cells. They are believed to originate in the endoplasmic reticulum. Two principal types are peroxisomes, found in vertebrates, and glyoxysomes, found in plants and microorganisms.

HISTORY

Peroxisomes were discovered by Rhodin in 1954 in liver and kidney cells and their name was coined by De Duve 1965 for their role in production and disposing of hydrogen peroxide.

Glyoxysomes were discovered by Beevers in 1961, they were so named because of their involvement in the formation of carbohydrates from fats by glyoxylate cycle.

PEROXISOMES

Peroxisomes are small, spherical membrane bound organelles of the cytoplasm of animal cells having a diameter of 0.2-2.0 μm and pH optimum of 7.5. They contain enzymes involved in the degradation of fatty acids and amino acids. The three oxidation enzymes contained in the peroxisome are D-amino acid oxidase, urate oxidase, and catalase. The

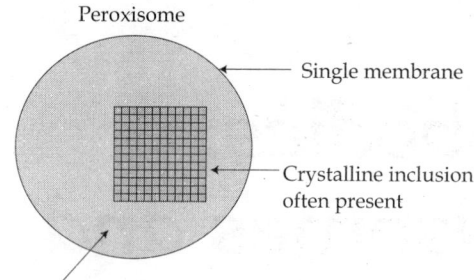

Peroxisome

Single membrane

Crystalline inclusion
often present

Enzymes, some of which produce H_2O_2 and
always including Catalase, that degrades H_2O_2

FIG. 11.1 Diagrammatic representation of a peroxisome

enzymes in the peroxisomes are synthesized on the RER and are sorted into
the peroxisomes in the Golgi apparatus. Like mitochondria, peroxisomes are
self-replicating and the process of reproduction is called peroxisomal
biogenesis (Fig. 11.1).

Peroxisomes function to rid the body of toxic substances like hydrogen
peroxide, or other metabolites. They are the major site of oxygen utilization
and are numerous in the liver where toxic by-products accumulate.

The enzymes of peroxisomes remove hydrogen atoms from small
molecules and join them to oxygen creating hydrogen peroxide. They
consume up to 20% of the oxygen in the liver cell. The peroxisomal enzyme
catalase uses this oxygen to convert hydrogen peroxide to water and oxygen
neutralizing it. In the liver this method is used to break down molecules of
alcohol into substances that can be eliminated from the body.

Functions

Peroxisomes contain oxidative enzymes such as D-amino acid oxidase, urate
oxidase, and catalase. By using molecular oxygen, hydrogen atoms are
removed from specific organic substrates (labeled as R), in an oxidative
reaction, producing hydrogen peroxide (H_2O_2, a toxic by- product of cellular
metabolism):

$$RH_2 + O_2 \rightarrow R + H_2O_2$$

Catalase uses H_2O_2 generated by other enzymes in the peroxisome to
oxidize other substrates, including phenols, formic acid, formaldehyde and
alcohol, by means of the peroxidation reaction:

$$H_2O_2 + R' \, H_2 \rightarrow R' + 2H_2O$$

This reaction is important in liver and kidney cells where the peroxisomes
detoxify various toxic substances that enter the blood. About 25% of the
ethanol that we drink is oxidized to acetaldehyde in this way. In addition,
when excess H_2O_2 accumulates in the cell, catalase converts it to H_2O through
this reaction:

$$2H_2O_2 \rightarrow 2H_2O + O_2$$

A major function of the peroxisome is the breakdown of fatty acid molecules, in a process called β-oxidation. In this process, the fatty acids are broken down and reduced by two carbons at a time, converted to acetyl-CoA, which is then transported back to the cytosol for further use. In animal cells, β-oxidation can also occur in the mitochondria. In yeast and plant cells this process is exclusive for the peroxisome.

The first reactions in the formation of plasmalogen in animal cells also occurs in peroxisomes. Plasmalogen is the most abundant phospholipid in myelin. Deficiency of plasmalogen causes profound abnormalities in the myelination of nerve cells, which is one of the reasons that many peroxisomal disorders lead to neurological disease.

Peroxisomes also play a role in the production of bile acids.

Deficiencies

The lack of specific peroxisome enzymes may cause following diseases:

Adrenoleukodystrophy: It is a progressive neurological disorder caused by a defect in β-oxidation of long chain fatty acid. It results in accumulation of long chain fatty acids in neurons and other cells.

Gout: It occurs due to accumulation of uric acid. It is the failure of conversion of uric acid to allantoin. It leads to the deposition of uric acid in joints, and consequent painful arthritis.

Zellweger syndrome: It is caused by the absence of peroxisomes. It causes severe neurological disorder, metabolic defects, and early death.

GLYOXYSOMES

Glyoxysomes are membrane bound organelles of about 0.5-1.0μm diameter. They are found in yeast, *Neurospora* and particularly in the fat storage tissues of germinating seeds of higher plants. Glyoxysomes contain enzymes that initiate the breakdown and conversion of fatty acids to sugars, which the emerging seedling uses as an energy and carbon source until it is able to produce its own sugar by photosynthesis. In this pathway, fatty acids are hydroylzed to acetyl-CoA for the glyoxylate bypass.

Functions

Glyoxysomes are the site of the glyoxylate cycle that is tightly linked to the breakdown of fatty acids. During plant germination glyoxysomes are in a key-position. They control and catalyze the degradation of storage fat and channel the degradation products towards the synthesis of numerous carbon compounds (mainly carbohydrates). The overall reaction proceeds as:

$$2 \text{ Acetyl CoA} \rightarrow \text{Succinate} + 2H^+ + 2 \text{ CoA}$$

The succinic acid formed may diffuse from glyoxysomes into the mitochondria in order to enter the Krebs cycle, or may serve as a raw material for the synthesis of glucose, a process called gluconeogenesis.

OBJECTIVE TYPE QUESTIONS

1. _____ are the sites of photorespiration, while _____ are in charge of the mobilization of storage compounds (fats).
   A. Lysosomes; glyoxysomes
   B. Peroxisomes; glyoxysomes
   C. Peroxisomes; lysosomes
   D. Glyoxysomes; peroxisomes

2. _____ is an enzyme specific for peroxisomes.
   A. Hyaluronidase
   B. Sucrase
   C. Catalase
   D. Peroxidase

3. Glyoxysomes are the site of the _____ that is tightly linked to the breakdown of fatty acids.
   A. Gluconeogenesis
   B. Krebs cycle
   C. Glyoxylate cycle
   D. None of the above

4. Peroxisomes contain high concentrations of _____ .
   A. Sucrose
   B. Hydrogen peroxide
   C. Succinic acid
   D. Hydrolytic enzymes

5. Peroxisomes were discovered by _____ .
   A. De Duve
   B. Palade
   C. Claude
   D. Rhodin

Answers

1. B   2. C   3. C   4. B   5. D

SHORT ANSWER TYPE QUESTIONS

1. Write a short note on glyoxysomes.
2. Write a note on a few inherited diseases that are caused by peroxisome malfunction.
3. Write a brief comparative account on peroxisomes and glyoxysomes.

LONG ANSWER TYPE QUESTIONS

1. Describe in detail the structure and function of peroxisomes.
2. Explain how the presence of working peroxisomes is essential for the normal functioning of the cell.
3. Explain how peroxisomes are helpful in fatty acid breakdown.

Ribosomes 12

INTRODUCTION

Ribosomes are small, dense, roughly spherical particles with a diameter of about 20 nm and can be seen only with the electron microscope. They are composed of ribosomal RNA and ribosomal proteins (known as a ribonucleoprotein or RNP). The free ribosomes occur in the cytoplasm of all cells and in mitochondria and chloroplast of eukaryotic cells while the ribosomes in bound form remain attached with the membranes of endoplasmic reticulum and nucleus.

HISTORY

Ribosomes were first isolated from the cell cytoplasm by centrifugation by Albert Claude in 1943 and he called them 'microsomes'. Later, they were reported in the plant cells by Robinson and Brown in 1953 and in animal cells by G.E. Palade in 1955. Palade called them 'ribosomes' due to the presence of high amounts of RNA.

STRUCTURE

The ribosome is a large ribonucleoprotein (RNA-protein) complex. It is formed from two unequally sized subunits, referred to as the small subunit and the large subunit. The ribosomal subunits of prokaryotes and eukaryotes are quite similar. However, prokaryotes have 70S ribosomes, each consisting of a small 30S subunit and a large 50S subunit (Fig. 12.1), whereas eukaryotes have 80S ribosomes,

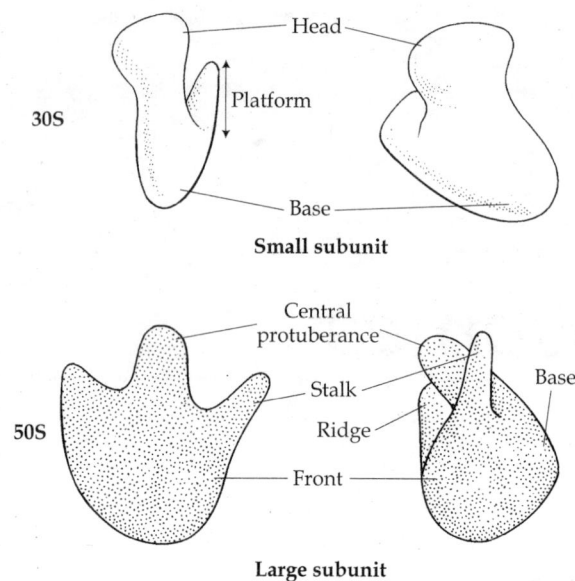

30S

Head

Platform

Base

Small subunit

50S

Central protuberance

Stalk

Ridge

Front

Base

Large subunit

FIG. 12.1 Prokaryotic Ribosome 70S

each consisting of a small 40S subunit and a large 60S subunit. However, the ribosomes found in chloroplasts and mitochondria of eukaryotes are 70S, which is one of the observations supporting the endosymbiotic theory. The unit 'S' means Svedberg unit, a measure of the rate of sedimentation of a particle in a centrifuge, where the sedimentation rate is associated with the size of the particle. It is important to note that Svedberg units are not additive—two subunits together can have Svedberg values that do not add up to that of the entire ribosome. This is resulting from the loss of surface area when the two subunits are bound. The two ribosomal subunits remain united with each other due to high concentration of Mg^{++} ions. When the concentration of Mg^{++} ions reduces in the matrix, the two ribosomal subunits get separated. At a high concentration, however, the two subunits get associated forming a dimer.

The RNA and proteins occur in approximately equal proportions in smaller as well as larger subunit. In Eukaryotes, the co-efficient of ribosomes are 80s, of which is divided into 60s for the large, and 40s for the small subunit. The 60s subunit contains 28s rRNA, with a small fragment that is attached non-covalently and can be released upon heating; a 5.8s, and a very small 5sRNA. Whereas, the 40s subunit has only a single 18s rRNA. In prokaryotes, however, the large and small subunits are split into 50s and 30s, making a total of 70s. The 50s subunit has two types of rRNA 23s and a 5s. It also has 32 different proteins. On the other hand, the 30s subunit contains a single 16s rRNA plus 21 different types of proteins.

Thus, in prokaryotes, the small subunit contains one RNA molecule and about twenty different proteins, while the large subunit contains two different RNAs and about thirty different proteins. Eukaryotic ribosomes are even more complex: the small subunit contains one RNA and over thirty proteins, while the large subunit is formed from three RNAs and about fifty proteins. Mitochondrial and chloroplast ribosomes are similar to prokaryotic ribosomes. In addition, there are certain low molecular weight components in the form of divalent metallic ions such as Mg^{++}, Ca^{++} and Mn^{++}.

In spite of its complex composition, the architecture of the ribosome is very precise. Even more remarkable, ribosomes from all organisms, ranging from bacteria to humans, are very similar in their form and function. Recent breakthroughs in studies of ribosome structure, using techniques such as scanning, cryo-electron microscopy, and X-ray crystallography, have provided scientists with highly refined structures of this complex organelle. One particularly exciting conclusion from studies of the large subunit is that it is ribosomal RNA (rRNA), and not protein, that provides the catalytic activity for peptide bond formation. That is, it forms the chemical linkage between the amino acids of the growing protein molecule.

Characteristics of Different Ribosome Types

Ribosome Types	Occurrence	Large subunit Size	Large subunit RNAs	Small subunit Size	Small subunit RNAs
80S	Eukaryotes (animals)	60S	28-29S+5S+5.8S	40S	18S
80S	Eukaryotes (Plants)	60S	25S+5S+5.8S	40S	16S-18S
70S	Prokaryotes	50S	23S+ 5S	30S	16S
55S	Mitochondria (Vertebrates)	40S	16-17S + 5S	30S	12-13S

ULTRASTRUCTURE

The structure of 70S ribosomes (prokaryotic) has been studied in detail as compared to the 80S ribosomes of eukaryotes. Recently, following two models have been suggested to explain the three-dimensional structure of 70S ribosomes:

Stoffler and Wittman's model: This model was suggested by Stoffler and Wittman in 1977. According to this model, the 30S subunit has an elongated, slightly bent prolate shape. It is a bipartite structure. A transverse cleft divides the 30S subunit into two parts, a smaller head and a larger body. The two lobes protrude unequally and to different extents. In a frontal view the 50S subunit appears bilaterally symmetrical, and shows three protuberances arising from a rounded base (maple leaf structure).

Of these the central protuberance is the most prominent. The 50S subunit has been compared to an armchair, with the rounded base forming a vaulted seat, the central protuberance the back, and the lateral protuberances the arms. The 30S and the 50S subunits are associated to form the 70S ribosome. The frontal face of the 30S subunit with its hollow faces the vaulted seat of the 50S subunit.

The long axis of the 30S subunit is oriented transversely to the central protuberance of the 50S subunit. A tunnel is formed between the hollow of the small subunit and the vaulted seat of the large subunit.

Lake's Asymmetrical Model of Ribosome Structure: This model was suggested by James A. Lake in 1981. In Lake's model the 30S subunit is considered to be completely asymmetrical. An indentation divides the subunit into two unequal parts, referred to as the upper one-third (head) and the lower two-third (body) and extending from the lower two-third is a region called-the platform. There is a cleft between the platform and the upper one third. This cleft is an important functional region. It has been proposed as the site of the codon-anticodon interaction and as a part of the binding site for initiation factors IF-I, IF-2 and IF-3. The 30S subunit in Lake's model differs from that of Stoffler and Wittmann's model in that it is completely asymmetrical, and does not contain a mirror plane.

BIOGENESIS OF RIBOSOMES

In eukaryotes most of the ribosomal RNA (rRNA) is synthesized in the nucleolus. The nucleolar organizer contains many copies of ribosomal DNA (repetitive DNA). There are several distinct types of rRNA. Of these only four classes, namely 28S, 5.8S and 5S have been found in the ribosomes. The other types are intermediate stages in the formation of ribosomal RNA. The scheme for the biosynthesis of ribosomes is as follows:

(1) The RNA cistron of nucleolar DNA transcribes 45S precursor RNA in the presence of the enzyme RNA polymerase.

(2) The ribose sugar of certain regions of 45S RNA undergoes methylation (addition of methyl groups). It is the methylated regions which gives rise to 28S and 18S RNA of the ribosomes. The non-methylated regions have a higher content of guanine and cytosine than the methylated regions.

(3) Cleavage at site 1 removes the transcribed spacer sequence from the 5'P end of 45S RNA. Cleavage at site 2 separates 18S rRNA. Cleavage at site 3 results in a large segment containing 28S RNA and 5.8S RNA along with spacer segments. Cleavage 4 results in the final trimming of this segment.

(4) The 5S RNA is synthesized outside the nucleolus.

(5) Ribosomal proteins are synthesized in the cytoplasm and translocated to the nucleus where they become associated with RNA. Structural core

proteins first associate with 45S RNA to form 80S ribonucleoprotein particles. Other proteins are probably bound later. Ultimately the 40S and 60S subunits of the ribosome are formed, containing 18S and 28S+5.8S RNA, respectively. 5S RNA, which is synthesized outside the nucleolus, also becomes associated with the large 60S subunit. The two types of subunits pass through pores in the nuclear membrane to the cytoplasm.

(6) The 40S subunit binds to mRNA in the cytoplasm to form a 40S — mRNA complex. The 60S subunit now becomes associated with the 40S — mRNA complex to form the 80S ribosome with bound mRNA.

(7) In prokaryotic cells the subunit RNA precursors are trimmed to form 16S and 23S RNA. These become associated with protein to form the 30S and 50S subunits, respectively, of the 70S ribosome.

The RNA genes are present in only a few copies closely grouped together, i.e., the ribosomal genes occur in a single operon.

OVERVIEW

The 70S ribosomes are vulnerable to some antibiotic that the 80S ribosomes are not. This helps create drugs that can destroy a bacterial infection without harming the animal/human host cells. Even though human mitochondria possess 70S ribosomes, mitochondria are not affected by these antibiotics because the mitochondria is covered by a double membrane that does not admit these antibiotics into the organelle. Many of the antibiotics used in humans and other animals to treat bacterial infections specifically inhibit ribosome activity in the disease-causing bacteria, without affecting ribosome function in the animal's host cells. These antibiotics work by binding to a protein or RNA target in the bacterial ribosome and inhibiting translation. In recent years, the misuse of antibiotics has resulted in the natural selection of bacteria that are resistant to many of these antibiotics, either because they have mutations in the antibiotic's target in the ribosome or because they have acquired a mechanism for excluding or inactivating the antibiotic.

FUNCTION

Ribosomes are the sites where the cell assembles proteins according to genetic instructions. A bacterial cell may have a few thousand ribosomes, although a human cell has a few million. Cells with high rates of protein synthesis have a particularly great number of ribosomes. These cells, which are active in protein synthesis, also have prominent nucleoli, that make ribosomes.

Ribosomes function in two cytoplasmic areas. Free ribosomes are spread throughout the cytosol, while bound ribosomes are attached to the outside of a membranous network, endoplasmic reticulum. Most of the proteins that are made by free ribosomes will function inside the cytosol. The proteins produced by bound ribosomes usually exported from the cell. Each ribosome is built from two subunits, each having its own mix of ribosomal RNA and proteins. Ribosomes are built with RNA from the nucleolus and are made in the nucleolus itself. These subunits join together to form a functional ribosome only when they attach to a messenger RNA molecule. The ribosomes present in eukaryotic cells are slightly larger than those found in prokaryotic cells. Two or more ribosomes simultaneously engaged in protein synthesis on the same mRNA strand form polyribosomes (Fig. 12.2).

The ribosome functions as a platform, bringing together different components involved in the synthesis of proteins. Interaction of the tRNA-amino acid complex with mRNA, which brings about translation of the genetic code, is coordinated by the ribosome (Fig. 12.3).

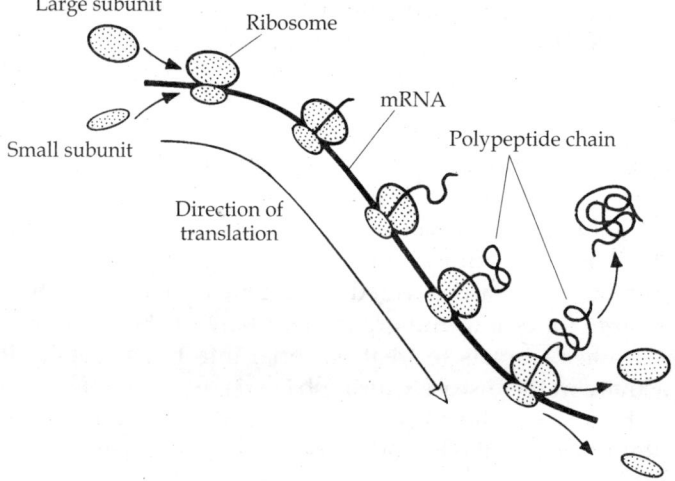

FIG. 12.2 Polyribosomes in action

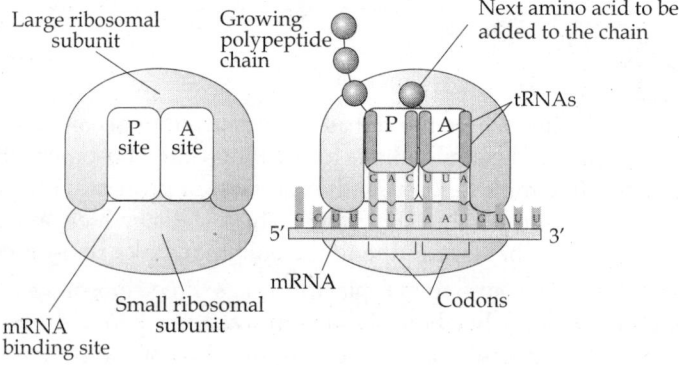

FIG. 12.3 Ribosomes performing translation

OBJECTIVE TYPE QUESTIONS

1. Ribosomes are responsible for assembling the _____ of the cell.
   - A. Carbohydrates
   - B. Proteins
   - C. Lipids
   - D. Nucleic acids
2. Ribosomal subunits are synthesized by the _____.
   - A. Nucleus
   - B. Nucleolus
   - C. Golgi apparatus
   - D. Endoplasmic reticulum
3. There are two places that ribosomes usually exist in the cell: suspended in the _____ and bound to the _____.
   - A. Cytosol, endoplasmic reticulum
   - B. Nucleus, endoplasmic reticulum
   - C. Nucleoplasm, endoplasmic reticulum
   - D. All of the above
4. The ribosomes found in chloroplasts and mitochondria of eukaryotes are _____ type.
   - A. 70S
   - B. 60S
   - C. 55S
   - D. 80S
5. Ribosomes are composed of _____ and _____.
   - A. DNA, RNA
   - B. DNA, protein
   - C. RNA, protein
   - D. Varies according to the type

Answers

1. B   2. B   3. A   4. C   5. C

SHORT ANSWER TYPE QUESTIONS

1. Write a note on Lake's Asymmetrical Model of Ribosome Structure.
2. Explain the functions of ribosomes.
3. Describe in general terms the structure of the large ribosome subunit. What makes up the core of the particle?

LONG ANSWER TYPE QUESTIONS

1. Write a note on the structure of 70 S and 80S ribosomes.
2. Explain the Stoffler and Wittman's model of ribosome.
3. Write a note on biogenesis of ribosomes.

Cytoskeleton: Microtubules, Microfilaments and Microvilli

13

INTRODUCTION

The cytoskeleton is unique to eukaryotic cells. It is a dynamic three-dimensional structure that fills the cytoplasm and helps the cells to have definite shape and carry out coordinated and direct movements. This structure acts as both muscle and skeleton, for movement and stability. The long fibers of the cytoskeleton are polymers of subunits. The primary types of fibers comprising the cytoskeleton are microfilaments, microtubules, microvilli.

MICROTUBULES

Microtubules are filamentous intracellular structures that are responsible for various kinds of movements in all eukaryotic cells. They are made up of proteins and have a diameter of about 24 nm and varying length. Microtubules are involved in nuclear and cell division, organization of intracellular structure, and intracellular transport, as well as ciliary and flagellar motility.

History

Microtubules were first observed by Robertis and Franchi, in 1953, in the axoplasm of nerve fibres and they called them 'neurotubules'. Later, in 1963, Sabatini, Bensch and Barnett studied their exact structure in the electron microscope. Ledbetter and Porter, in 1963, described the structure of microtubules in plant cells.

Structure

Microtubules are polymers of α- and β-tubulin heterodimers (Fig. 13.1). The heterodimer does not come apart once it is formed. The α-tublin has a bound molecule of GTP, on the other hand, the β-tubulin may have bound GTP or GDP. The tubulin dimers polymerize end-to-end in protofilaments. The protofilaments then bundle in hollow cylindrical filaments. Typically, the protofilaments arrange themselves in an imperfect helix with one turn of the helix containing 13 tubulin dimers each from a different protofilament (Fig. 13.2).

Another important feature of microtubule structure is polarity. Tubulin polymerizes end-to-end with the α-subunit of one tubulin dimer contacting the β-subunit of the next. Therefore, in a protofilament, one end will have the α-subunit exposed while the other end will have the β-subunit exposed. These ends are designated (–) and (+) respectively. The protofilaments bundle parallel to one another, so in a microtubule, there is one end, the (+) end, with only β-subunits exposed while the other end, the (–) end, only has α-subunits exposed (Fig. 13.3).

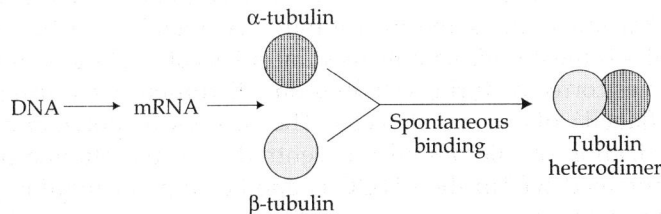

FIG. 13.1 Formation of a tubulin heterodimer

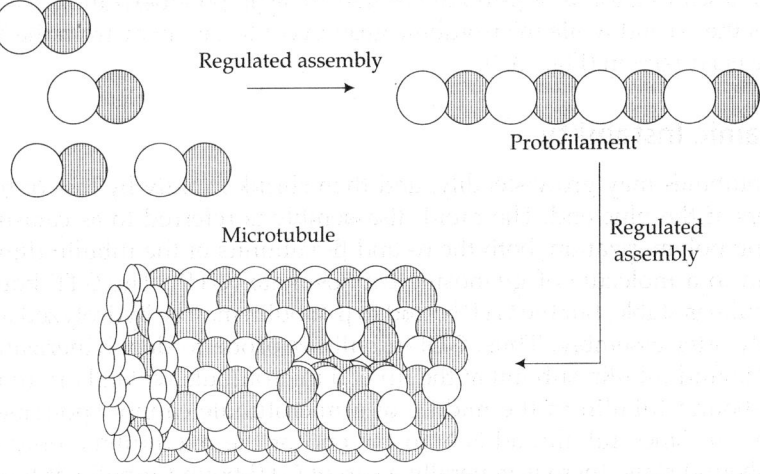

FIG. 13.2 Formation of a microtubule by tubulin molecules

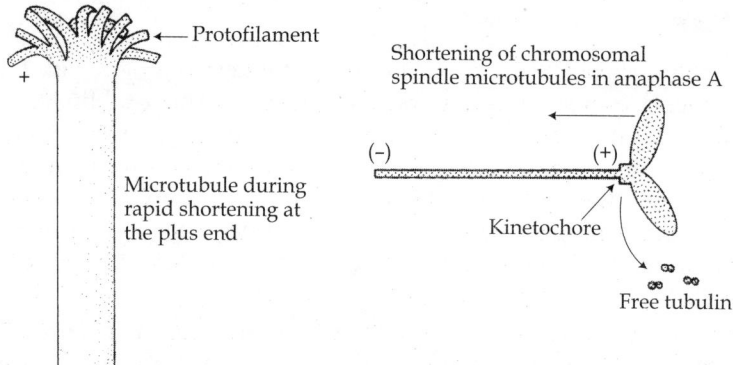

FIG. 13.3 The Organization of Microtubules

Organization

Microtubules are nucleated and organized by the microtubule organizing centers (MTOCs), such as centrosomes and basal bodies. They are capable of growing and shrinking in order to generate force, and there are also motor proteins that move along the microtubule. A notable structure involving microtubules is the mitotic spindle used by eukaryotic cells to segregate their chromosomes correctly during cell division. Microtubules are also part of the cilia and flagella of eukaryotic cells. The process of polymerization and depolymerization of microtubules is centred in a microtubule organizing center. Contained within the MTOC is another type of tubulin, γ-tubulin, which is distinct from the α- and β-subunits which compose the microtubules themselves. The γ-tubulin combines with several other associated proteins to form a circular structure known as the 'γ-tubulin ring complex'. This complex acts as a scaffold for α- or β-tubulin dimers to begin polymerization; it acts as a cap of the (–) end while mictrotubule growth continues away from the MTOC in the (+) direction (Fig. 13.3).

Dynamic Instability

Microtubules may grow steadily, and then shrink rapidly by loss of tubulin dimers at the plus end. The rapid disassembly is referred to as catastrophe. During polymerization, both the α- and β- subunits of the tubulin dimer are bound to a molecule of guanosine triphosphate (GTP). The GTP bound to α-tubulin is stable, but the GTP bound to β-tubulin may be hydrolyzed to GDP shortly after assembly. Thus, GDP-tubulin is prone to depolymerization. A GDP-bound tubulin subunit at the tip of a microtubule will fall off, though a GDP-bound tubulin in the middle of a microtubule cannot spontaneously come out. Since tubulin adds onto the end of the microtubule only in the GTP-bound state, there is generally a cap of GTP-bound tubulin at the tip of the microtubule, protecting it from disassembly. When hydrolysis catches up to the tip of the microtubule, it begins a rapid depolymerization and shrinkage. GTP-bound tubulin can begin adding to the tip of the microtubule again, providing a new cap and protecting the microtubule from shrinking. This is referred to as rescue.

Dynamic instability of microtubules *in vivo* is regulated by interaction with other proteins. For example, during prophase of mitosis, microtubules grow out from the centrosome. If the plus end of a microtubule makes contact with a chromosome, it becomes stabilized. Otherwise rapid disassembly at the plus end begins, and the tubulin dimers are available for growth of another microtubule.

Microtubule-associated Proteins (MAPs)

MAPs are a diverse class of proteins that bind to microtubules. Binding of MAPs may be regulated by phosphorylation, which causes some MAPs to detach from microtubules.

Plus-end tracking proteins: These are proteins that associate with the plus ends of microtubules, many of them are motor proteins. The example of such proteins that stabilize or promote growth of microtubules includes members of the XMAP215 family of proteins and CLASPs.

Catastrophe-promoting proteins (catastrophins): These are the proteins that bind to plus ends of microtubules and promote dissociation of tubulin dimers. They may activate GTP hydrolysis or induce a curved protofilament conformation. An example is MCAK, a member of the kinesin family of proteins found in kinetochores.

Other proteins that promote microtubule disassembly include the following:

Stathmin: It is a microtubule destabilizing protein that increases in abundance in some cancer cells.

Katanin: It severs microtubules, generating new plus ends that lack a stabilizing GTP cap, and minus ends that are not stabilized by being capped by γ-ring complexes of the centrosome.

KLP10A and related members of the Kin I subfamily of kinesins associate with uncapped minus ends of microtubules at the spindle poles during mitosis.

Some MAPs cross-link adjacent microtubules or link microtubules to membranes or to intermediate filaments. The length of intervening segments between microtubule-binding domains in particular MAPs may determine the spacing of microtubules in parallel arrays. Some examples are:

Type I MAPs: These are found in axons and dendrites of nerve cells and in some non-neural cells, have several repeats of the sequence KKEX (Lys-Lys-Glu-X) that binds to negatively charged tubulin domains.

Type II MAPs: These are found in axons, dendrites and non-neural cells. They have 3-4 repeats of an 18-residue sequence that binds tubulin (e.g., MAP4 and Tau).

Some toxins and drugs (all of which inhibit mitosis) affect polymerization or depolymerization of tubulin.

Taxol: This is an anti-cancer drug which is known to stabilize microtubules.

Colchicine: It binds tubulin and blocks polymerization. Microtubules are depolymerized at high colchicine concentration.

Vinblastine: It causes depolymerization and formation of vinblastine-tubulin paracrystals.

Nocodazole: It causes depolymerization of microtubules.

Functions

Microtubules perform several functions in eukaryotic cells. Some of them are enlisted below:

1. **Maintenance of cell shape:** The shape of the cell, for example, axon of neurons, cilia, flagella etc., has been correlated to the orientation and distribution of microtubules.

2. **Locomotion:** The microtubules in cilia and flagella help these structures to perform locomotion.

3. **Morphogenesis:** The cells change their shape during cell differentiation. Microtubules play an important role in the elongation of cells in the lens of eye and elongation of the nucleus of spermatid during spermiogenesis.

4. **Contraction:** Microtubules cause the contraction of the spindle and thus chromosome movement during cell division.

5. **Intracellular transport:** Microtubules perform the movement of pigment vesicles for protective coloration and help in the discharge of vesicle content for water regulation in protozoa.

MICROFILAMENTS

Microfilaments are fine, thread-like protein fibers, 3-6 nm in diameter. They are composed predominantly of a contractile protein called actin, which is the most abundant cellular protein. Microfilaments and their association with the protein myosin is responsible for muscle contraction. Microfilaments can also carry out cellular movements including gliding, contraction, and cytokinesis.

History

Microfilaments were discovered by Paleviz and his coworkers in 1974.

Structure

Microfilaments are solid rods made up of a protein known as **actin**. Individual units of actin appear in globular form and are called **G-actin**. In microfilaments, however, which are also often referred to as actin filaments,

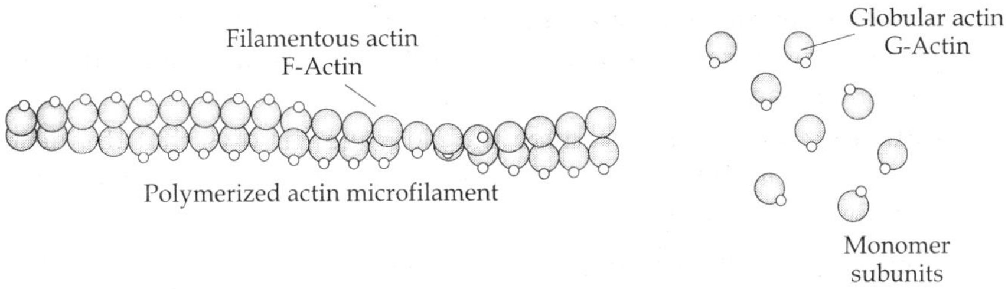

FIG. 13.4 Microfilament structure and assembly

long polymerized chains of the G-actin molecules are intertwined in a helix, creating a filamentous form of the protein (**F-actin**) (Fig. 13.4). All of the subunits that compose a microfilament are connected in such a way that they have the same orientation. Due to this fact, each microfilament exhibits **polarity**, the two ends of the filament being distinctly different. This polarity affects the growth rate of microfilaments, one end (termed the plus end) typically assembling and disassembling faster than the other (the minus end). The filaments elongate approximately 10 times faster at the (+) end than at the (–) end. This phenomenon is known as the treadmill effect. The process of polymerization of actin begins with the association of three G-actin monomers, thus, forming a trimer. After this ATP-actin binds to the (+) end and subsequently ATP is hydrolyzed thus reducing the binding strength between neighboring units. ADP-actin dissociates from the (–) end. This rapid turnover is important for the movement of the cell. End-capping proteins such as Cap Z prevent the addition or loss of monomers at the filament end.

Unlike microtubules, which typically extend out from the centrosome of a cell, microfilaments are typically centered at the plasma membrane. Therefore, the periphery (edges) of a cell generally contains the highest concentration of microfilaments. A number of external factors and a group of special proteins influence microfilament characteristics, however, and enable them to make rapid changes if needed, even if the filaments must be completely disassembled in one region of the cell and reassembled somewhere else. When found directly beneath the plasma membrane, microfilaments are considered part of the cell cortex, which regulates the shape and movement of the cell's surface. Consequently, microfilaments play a key role in the development of various cell surface projections including filopodia, lamellipodia, and stereocilia.

Functions

Common to all eukaryotic cells, microfilaments are primarily structural in function and are an important component of the cytoskeleton. They perform a variety of functions:

1. **Cytoskeleton:** Microfilaments make up a major portion of all cell cytoskeletons and are best known for their role in the contractile fibrils of muscle cells.

2. **Movements and locomotion:** They can form connections with the plasma membrane and thereby influence locomotion, amoeboid movement, and cytoplasmic streaming.

3. **Cell division:** They produce the cleavage furrows that divide the cytoplasm of cells after chromosomes have been separated by the spindle fibers during mitosis.

4. **Maintenance of cell shape:** Microfilaments also help develop and maintain cell shape by acting as tensile elements in the cellular structure, pulling the plasma membrane and all of the cell's internal constituents toward the nucleus at the core.

5. **Mechanical signals:** Microfilaments also conduct mechanical signals throughout the cell at propagation speeds of 100-1200 m/sec, allowing 0.01-2 Hz signals to cross a cell in 2-20 nsec.

MICROVILLI

The microvilli are tiny, hair like structures about 0.08μm in diameter, present on the surface of epithelial cells involved in absorption and secretion, and help in increasing the surface area of the cell. They also occur in sensory cells like those of the inner ear, the taste buds, and the olfactory receptor cells (Fig. 13.5). This is because they are so tiny and numerous they can attach to remarkably small amounts of substances (in taste and smell functions). In the ear, hearing loss due to loud noise is actually caused by the destruction of the microvilli through sonic energy. The primary component of microvilli is actin filaments. These filaments are cross-linked by fimbrin and villin.

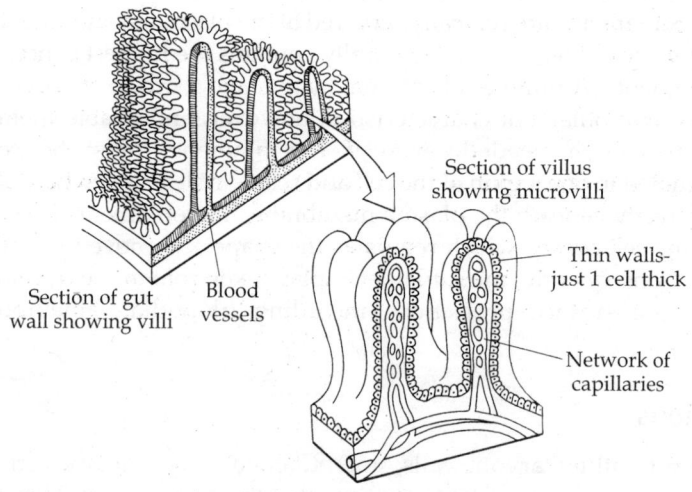

Section of villus showing microvilli

Section of gut wall showing villi

Blood vessels

Thin walls- just 1 cell thick

Network of capillaries

FIG. 13.5 Structure of Microvilli

Structure

Microvilli are covered in plasma membrane, which encloses cytoplasm and microfilaments. Though these are cellular extensions, there are little or no cellular organelles present in the microvilli. Each microvillus has a dense bundle of cross-linked actin filaments, which serves as its structural core. 20 to 30 tightly bundled actin filaments are cross-linked by bundling proteins fimbrin and villin to form the core of the microvilli. The actin filaments render the microvilli capable of contracting motion, though the motion is limited, similar to flexing of fingers. The structural core is attached to the plasma membrane along its length by lateral arms made of myosin I and Ca^{2+} binding protein calmodulin. Myosin I functions through a binding site for filamentous actin on one end and a lipid binding domain on the other. The plus ends of the actin filaments are collected in the tip of the microvillus, while the minus ends bind to a 'terminal web' composed of thin filaments, linked together by a complicated set of proteins including spectrin and myosin II.

Functions

Microvilli perform several important functions in the cell such as:

1. **Determining shape and movement:** Microvilli are formed as cell extensions from the plasma membrane surface. Actin filaments, present in the cytosol, are most abundant near the cell surface. These filaments are thought to determine the shape and movement of the plasma membrane. The nucleation of actin fibers occurs as a response to stimuli, thus allowing the cells surface to alter its shape and stiffness.

2. **Increasing surface area:** These structures increase the surface area of cells by approximately 600 fold in humans, thus facilitating absorption and secretion. There are several thousand microvilli present on the apical surface of a single cell in human small intestinal cells.

3. **Increasing the area for enzymatic action:** Microvilli have enzymes associated with them that aid their function. For example, lactase and other enzymes that can help hydrolyze carbohydrates are present on microvilli in intestinal epithelial cells. These enzymes are localized in the amorphous dark staining tip of the microvilli. Thus, they are not only increasing the area for absorption, they are also increasing the area for enzymes involved in digestion to anchor on the cell surface and perform final stages of extracellular digestion, breaking down small peptides and disaccharides for transport across the membrane.

4. Microvilli are also of importance on the cell surface of white blood cells, as they aid in the migration of white blood cells.

Certain diseases can lead to the destruction of microvilli and alter the rearrangement of cytoskeleton in host cells. This can lead to malabsorption of nutrients and persistent osmotic diarrhoea, often accompanied by fever. This is seen in infections caused by EPEC subgroup *Escherichia coli*, in celiac disease, and Microvillus Inclusion Disease, which is an inherited disease characterized by defective microvilli and presence of cytoplasmic inclusions of the cell membrane other than the apical surface. However, the destruction of microvilli can be beneficial sometimes, as in the case of elimination of microvilli on white blood cells which can be used to combat autoimmune diseases.

OBJECTIVE TYPE QUESTIONS

1. The cytoskeleton of a cell is made of
   A. Microtubules, lignin, and microfilaments
   B. Microtubules, intermediate filaments, and microfilaments
   C. Microtubules, flagellin, and microfilaments
   D. Pectins, intermediate filaments, and microfilaments
2. Microtubules are components of all _____ cells.
   A. Plant   B. Prokaryotic
   C. Eukaryotic   D. Human
3. Microtubules are built by the assembly of dimers of _____ and
   _____ .
   A. α-tubulin, β-tubulin   B. γ-tubulin, β-tubulin
   C. α-tubulin, γ-tubulin   D. Two γ-tubulins
4. Microfilaments are composed of a contractile protein called _____ .
   A. Tubulin   B. Actin
   C. Myosin   D. Katanin
5. In animal cells, the microtubules originate at the _____ .
   A. Nucleus   B. Animal pole
   C. Centrosome   D. There is no fixed position

Answers

1. B   2. C   3. A   4. B   5. C

SHORT ANSWER TYPE QUESTIONS

1. Write a short note on microfilaments.
2. How is the cytoskeleton polarized in many cells resulting in structures such as the apical brush border microvilli of epithelial cells?
3. How cells change shape or form and how that is precisely regulated using the cytoskeleton?

LONG ANSWER TYPE QUESTIONS

1. Explain the structure of cytoskeleton.
2. Discuss the various roles played by microtubules. How do their functions differ in plant and animal cells?
3. List the functions of microtubule associated proteins (MAPS)? Specifically mention the functions of kinesin and dynein.

Cilia and Flagella

14

INTRODUCTION

Cilia (L., *cili* = eyelash) and flagella (L., little whip) are projections from the cell. They are made up of microtubules and it is the specific arrangement of microtubules within cilia and flagella that contributes to their movement. They are motile and designed either to move the cell itself or to move substances over or around the cell. Cilia and flagella have the same internal structure. Nine pairs of microtubules are arranged at the periphery around the cilium or flagellum, and two microtubules are centrally positioned. The major difference is in their length, cilia are more in number and generally shorter (about 0.2 micrometers) and more abundant whereas flagella are less in number and are longer (up to 1000 micrometers).

HISTORY

Flagellum was discovered by Engelmann in 1868 and its structure was first studied by Jansen in 1887. The protein responsible for movement of cilia and flagella was first discovered and named dynein in 1963.

STRUCTURE

Cilia and flagella have similar organization in all eukaryotic cells; they possess a central bundle of microtubules, called the axoneme in which nine outer doublet microtubules surround a central pair of singlet microtubules. This characteristic 9 + 2 arrangement of microtubules is seen when the axoneme is viewed in cross section with the electron microscope. Each doublet microtubule on the outer side consists of A and B tubules, or subfibers: the A tubule is a complete microtubule with 13 protofilaments, while the B tubule contains 10 protofilaments. The bundle of microtubules comprising the axoneme is surrounded by the plasma membrane. Regardless of the organism or cell type, the axoneme is about 0.25 µm in diameter, but it varies greatly in length, from a few microns to more than 2 mm (Fig. 14.1).

At its point of attachment to the cell, the axoneme connects with the basal body. Basal bodies are cylindrical structures, about 0.4 µm long and 0.2 µm wide, which contain nine triplet microtubules. Each triplet contains one complete 13-protofilament microtubule, the A tubule which is fused to the incomplete B tubule, which in turn is fused to the incomplete C tubule. The A and B tubules of basal bodies continue into the axonemal shaft, whereas the C tubule terminates within the transition zone between the basal body and the shaft. The two central tubules in a flagellum or a cilium also end in the transition zone, above the basal body. The basal body plays an important role in initiating the growth of the axoneme.

Within the axoneme, the two central singlet and nine outer doublet microtubules are continuous for the entire length of the structure. Doublet microtubules, which represent a specialized polymer of tubulin, are found only in the axoneme. Permanently attached to the A tubule of each doublet microtubule is an inner and an outer row of dynein arms. These dyneins reach out to the B tubule of the neighboring doublet. The junction between A and B tubules of one doublet is probably strengthened by the protein tektin, a highly α-helical protein that is similar in structure to intermediate-filament proteins. Each tektin filament, which is 2 nm in diameter and approximately 48 nm long, runs longitudinally along the wall of the outer doublet where the A tubule is joined to the B tubule.

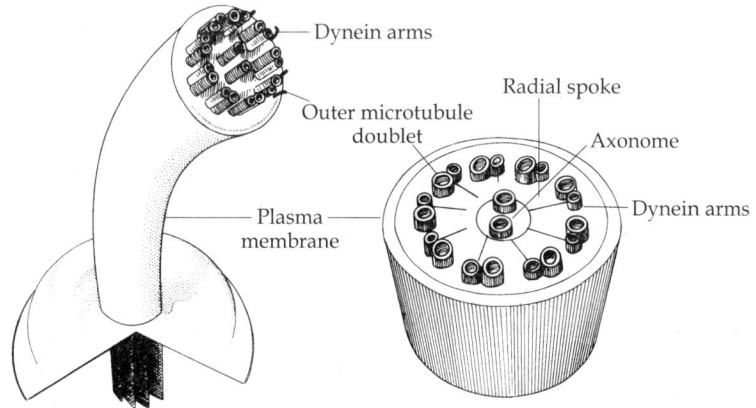

FIG. 14.1 Structure of cilia and flagella

The axoneme is held together by three sets of protein cross-links. The central pair of singlet microtubules is connected by periodic bridges, like rungs on a ladder, and is surrounded by a fibrous structure termed the inner sheath. A second set of linkers, composed of the protein nexin, joins adjacent outer doublet microtubules. Spaced every 86 nm along the axoneme, nexin is proposed to be part of a dynein regulatory complex. Radial spokes, which radiate from the central singlets to each A tubule of the outer doublets, form the third linkage system.

Although the 9 + 2 pattern is the fundamental pattern of virtually all cilia and flagella, the axonemes in certain protozoans and sperms of a few insects show some interesting variations. The simplest axoneme, containing three doublet microtubules and no central singlets (3 + 0) is found in *Daplius*, a parasitic protozoan. Its flagellum beats slowly (1.5 beats/s) in a helical pattern. Other axonemes consist of 6 + 0 or 9 + 0 arrangements of microtubules. These atypical cilia and flagella, which are all motile, show that the central pair of singlet microtubules is not necessary for axonemal beating and that fewer than nine outer doublets can sustain motility, but at a lower frequency.

THE SLIDING-FILAMENT MODEL OF BENDING

The structure of cilia and flagella has a circle of nine doublets, each of which have one complete (A Tubule) and one incomplete (B Tubule) microtubules, while both the core doublets are complete. There are sets of arms composed of the protein 'dynein' that extend from the doublets and join neighboring doublets and are spaced at 24 nm intervals. There are nexin links that are spaced along the microtubules to hold them together. Projecting inward are radial spokes that connect with a sheath enclosing the doublets.

The dynein arms have ATPase activity. In the presence of ATP, they can move from one tubulin to another. In this way, they enable the tubules to slide along one another so the cilium can bend. The dynein bridges are regulated so that sliding leads to synchronized bending. The doublets are held in place due to the nexin and radial spokes and therefore, the sliding is limited lengthwise. If nexin and the radial spokes are subjected to enzyme digestion, and exposed to ATP, the doublets will continue to slide and telescope up to 9× their length (Fig. 14.2).

The bending of cilia (and flagella) has many parallels to the contraction of skeletal muscle fibers. Thus, in the case of cilia and flagella, dynein powers the sliding of the microtubules against one another — first on one side, then on the other (Fig. 14.3).

AXONEMAL DYNEINS

Axonemal dyneins are complex multimers of heavy chains, intermediate chains, and light chains. Isolated axonemal dyneins, when slightly denatured and spread out on an electron microscope grid are seen as a bouquet of two or

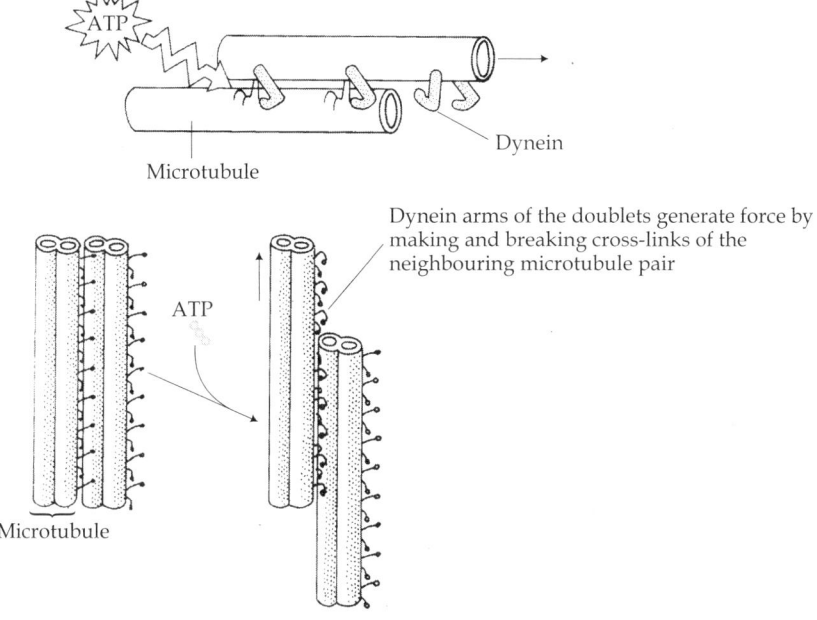

FIG. 14.2 Dynein arms have ATPase activity when the arms use ATP, microtubule doublets slide past each other

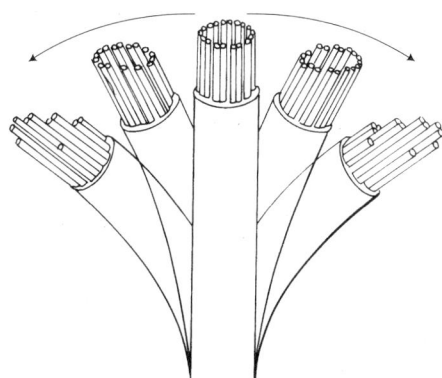

FIG. 14.3 When microtubules slide past each other the flagella changes shape

three 'blossoms'. Each blossom consists of a large globular domain attached to a small globular domain (the head) through a short stalk; another stalk connects one or more blossoms to a common base. The base is thought to be the site where the dynein arm attaches to the A tubule, while the small globular heads bind to the adjacent B tubule.

Each globular head and its stalk is formed from a single dynein heavy chain. The dynein heavy chain is enormous, approximately 4,500 amino acids in length with a molecular weight exceeding 540,000. Each heavy chain is capable of hydrolyzing ATP, and on the basis of sequences commonly found at the ATP-binding sites in other proteins, the ATP-binding domain of axonemal dynein is predicted to lie in the globular head portion of the heavy chain. The intermediate and light chains, thought to form the base of the dynein arm, help mediate attachment of the dynein arm to the A tubule and may also participate in regulating dynein activity. These base proteins thus are analogous to the MBP complexes associated with cytosolic dynein.

Axonemes contain at least eight or nine different dynein heavy chains. All inner dynein arms are two-headed structures, containing two heavy chains. The outer dynein arms have two heavy chains (e.g., in a sea urchin sperm flagellum) or three heavy chains (e.g., in *Chlamydomonas* flagella).

FUNCTIONS

Cilia and flagella are only present in a few types of specialized cells such as the epithelial cells lining the respiratory tract, the epithelial cells lining the gastrointestinal tract, and in motile cells such as spermatozoa. In the respiratory tract there are about 200 cilia per epithelial cell, giving a density of 109 cilia per cm^2. The beating action of these cilia moves substances such as mucus away from the lungs. In the gastrointestinal tract, the beating cilia aid digestion by mixing the ingested food with the digestive secretions. Flagella are even less common than cilia. They exist in motile cells such as the male gamete spermatozoa. Rotation of the flagellum propels the sperm forward in a swimming motion that enables the sperm to swim up the female reproductive tract (Fig. 14.4).

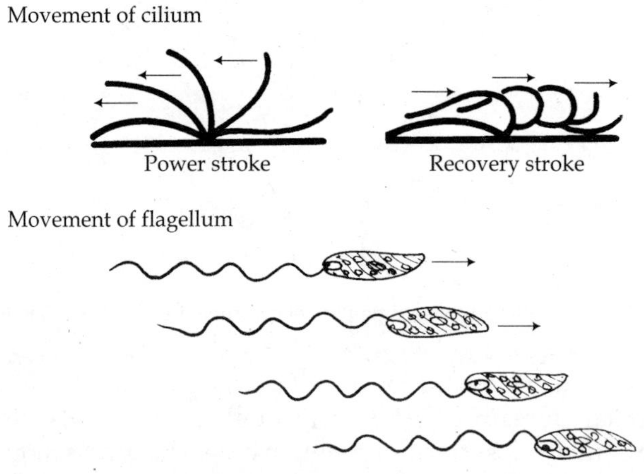

Movement of cilium

Power stroke   Recovery stroke

Movement of flagellum

FIG. 14.4 Movement of cilia and flagella

OBJECTIVE TYPE QUESTIONS

1. The respiratory tract in humans is lined with _____ that keep inhaled dust, smog, and potentially harmful microorganisms from entering the lungs.
   A. Flagella
   B. Villi
   C. Cilia
   D. Pili

2. The core of both cilia and flagella is termed the _____ .
   A. Hub
   B. Axoneme
   C. Shaft
   D. Basal body

3. A hereditary condition known as _____ is caused by problems with the dynein arms of cilia.
   A. Kartagener's syndrome
   B. Huntington chorea
   C. Klinefelter syndrome
   D. Tay-Sachs disease

4. Prokaryotic flagella are built from the protein _____ .
   A. Tubulin
   B. Flagellin
   C. Dyenin
   D. All of the above

5. Each cilium or flagellum grows out from, and remains attached to, a _____ embedded in the cytoplasm.
   A. Plasma membrane
   B. Nuclear membrane
   C. Basal body
   D. Centrosome

Answers

1. C   2. B   3. A   4. B   5. C

SHORT ANSWER TYPE QUESTIONS

1. Draw a neat and labeled diagram of the structure of cilia.
2. Compare and contrast the functions of cilia and flagella.
3. What mechanism is responsible for movements of cilia and flagella?

LONG ANSWER TYPE QUESTIONS

1. Describe the structure of cilia and flagella.
2. Write a note on ciliary and flagellar motion.
3. Explain the sliding-filament model of bending.

Centrioles and Basal Bodies

15

INTRODUCTION

Centrioles are barrel shaped microtubule structures found only in animal cells, these paired organelles are typically located together near the nucleus in the centrosome, a granular mass that serves as an organizing center for microtubules. Within the centrosome, the centrioles are positioned so that they are at right angles to each other. Each centriole is made up of nine bundles of microtubules (three per bundle) arranged in a ring.

A basal body is an organelle formed from a centriole. It is found at the base of a cilium or a flagellum and serves as a nucleation site for the growth of the axoneme microtubules. Basal bodies are structurally identical to centrioles.

HISTORY

Centrioles were first discovered by Van Benden as early as in 1887. Later, T. Boveri in 1888 studied its structure in detail. The works of Henneguy and Lenhossek (1897) have suggested that the structure of basal bodies is identical to centrioles.

STRUCTURE

Centrioles and basal bodies are cylindrical structures, usually in pairs oriented at right angles to one another (Fig. 15.1). They are about 0.15-0.25 μm in diameter and 0.3-0.7 μm in length. The wall of each centriole cylinder is made up of nine

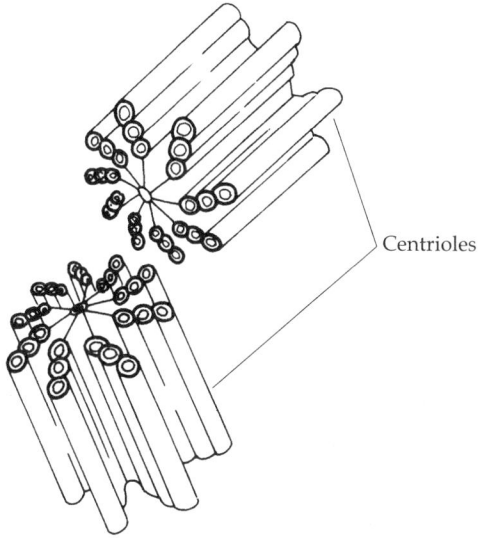

FIG. 15.1 A diplosome

interconnected triplet microtubules, arranged as a pinwheel. The interior of each centriole appears empty, except for a 'cartwheel' structure which is formed as a result of 9 protein spokes that radiate out from a central core region towards each triplet. The triplet consists of three subunit microtubules designated as A, B and C; the microtubule A is the innermost; it is round and complete and has 13 globular subunits while B and C are incomplete with 10 subunits each. The microtubule B shares 3 subunits with A while C shares 3 subunits with B. The A tubule of each triplet is linked with C tubule of adjacent triplet by protein linkers (Fig. 15.2).

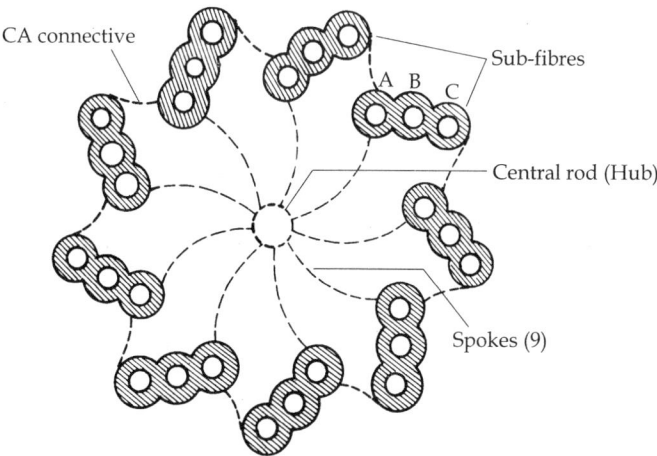

FIG. 15.2 Cartwheel structure of Centriole

Electron micrographs show appendages that protrude from the outer surface at one end of a mature centriole, and fibrous structures connecting the two centriole cylinders.

REPLICATION OF CENTRIOLES

The centrioles are known to replicate autonomously like mitochondria and peroxisomes. They begin from centers which contain proteins, like tubulin, needed for their formation. This leads to the formation of procentrioles. Each procentriole grows out a single microtubule from which the triplet can form. Once a centriole is made, daughter centrioles can grow out from the tubules at right angles. These then add to the daughter cell (in a dividing cell), or they move to the periphery and form the basal body for the cilium.

FUNCTIONS

1. Centrioles play a notable role in cell division. During interphase of an animal cell, the centrioles and other components of the centrosome are duplicated. At first the two pairs of centrioles remain in close proximity to each other, but as mitosis initiates, one set of centrioles is located in each of the new microtubule-organizing centers. These new centers radiate microtubules in star-shaped clusters known as asters. As the asters move to opposing poles of the cells, the microtubules, with the help of the centrioles, become organized into a spindle-shaped formation that spans the cell. These spindle fibers act as guides for the alignment of the chromosomes as they separate later during the process of cell division (Fig. 15.3).

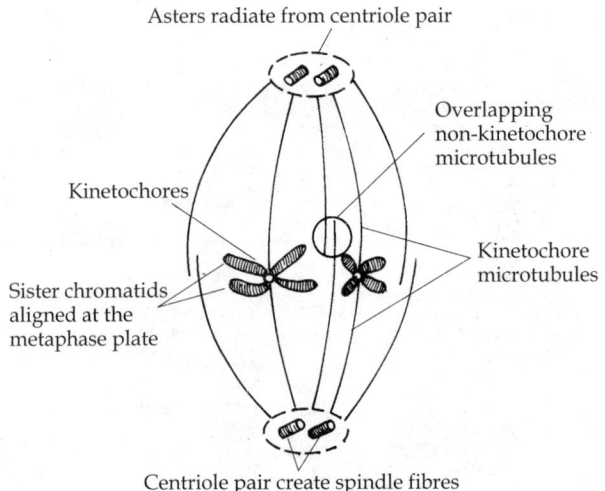

FIG. 15.3 Centrioles functioning during cell division

2. In cells that feature cilia or flagella, the basal bodies, which exhibit the same structural form as centrioles, are present. These assemblies are located, however, near the cell surface at the base of each cilium or flagellum, rather than in the centrosome near the nucleus. Basal bodies are anchored in their cytoplasmic locations by a rootlet system in the cell. In some organisms, such as the unicellular *Chlamydomonas*, basal bodies change their location and are functionally converted to centrioles before the mitotic process.

THE TRANSFORMATION OF THE CENTRIOLE TO BASAL BODY

As a new cell emerges from mitosis it inherits paired-centrioles (diplosome). This diplosome consists of a structurally mature mother centriole and an immature daughter centriole. The mother centriole functions as a microtubule organization center for the cell and later transforms into the basal body giving rise to the cilium. This is one reason why each set of newly replicated centrioles receives one centriole from the original cell and one newly formed centriole, so that the cell will know which one to allow to form flagella or cilia.

The transformation of the centriole into a basal body includes: (i) daughter centriole first matures to a mother centriole, (ii) produces basal body accessory structures, (iii) migrates to the plasma membrane, and (iv) docks to the cell membrane.

OBJECTIVE TYPE QUESTIONS

1. Centrioles are paired organelles typically located together near the nucleus in the _____.

   A. Centomere
   B. Kinetochore
   C. Centrosome
   D. Basal body

2. Each centriole is made of _____ bundles of microtubules arranged in a ring.

   A. Nine
   B. Eleven
   C. Thirteen
   D. Nineteen

3. Centrioles play a notable role in _____ .

   A. Heredity
   B. Cell division
   C. Locomotion
   D. Reproduction

4. During mitosis, a double-stranded chromosome is attached to a spindle fiber at the

   A. Centriole
   B. Centromere
   C. Centrosome
   D. Cell plate

5. Centriole replication is

   A. Semiconservative
   B. Conservative
   C. Dispersive
   D. All of the above

Answers

1. C   2. A   3. B   4. B   5. A

SHORT ANSWER TYPE QUESTIONS

1. Identify the protein components of the centriole.
2. Determine centriole function in chromosome segregation and cell division.
3. Why do plants not use centrioles as spindle poles during mitosis?

LONG ANSWER TYPE QUESTIONS

1. Describe the structure of the centriole.
2. Write a note on the duplication of centriole.
3. Explain the functions of centrosome.

The Nucleus

16

INTRODUCTION

The nucleus (L., *nucleus* = kernal) is a highly specialized organelle that serves as the information processing and administrative center of the cell. This organelle has two major functions: it stores the cell's hereditary material, DNA, and it coordinates the cell's activities, which include growth, intermediary metabolism, protein synthesis, and cell reproduction (cell division). Usually the nucleus is round and is the largest organelle in the cell. It is surrounded by a membrane, which is similar to the cell membrane that encloses the entire cell. The envelope is perforated. These perforations allow specific materials to pass in and out of the nucleus, just like proteins in the cell membrane regulate the movement of molecules in and out of the cell itself. Attached to the nuclear envelope is the endoplasmic reticulum.

Only the cells of advanced organisms, known as eukaryotes, have a nucleus. Simpler one-celled organisms (prokaryotes) like bacteria and cyanobacteria do not have a nucleus. In these organisms, all of the cell's information and administrative functions are dispersed throughout the cytoplasm.

HISTORY

The nucleus was discovered around 1833 by Robert Brown, a Scotch botanist, in orchid cells.

STRUCTURE

The spherical nucleus typically occupies about 10% of the volume of eukaryotic cell, making it one of the cell's most prominent features. The nucleus varies from 11-25 μm in diameter. It is enclosed by a double membrane called the nuclear envelope. The inner and outer nuclear membranes fuse at regular intervals, forming perforations or holes called nuclear pores. The nuclear envelope regulates and facilitates transport between the nucleus and the cytoplasm, while separating the chemical reactions taking place in cytoplasm from reactions happening within the nucleus. The outer membrane is continuous with the rough endoplasmic reticulum (RER) and may be studded with ribosomes. The space between the two membranes is called the perinuclear space. It is continuous with the lumen of the RER. The nuclear face of the nuclear envelope is surrounded by a scaffold of filaments called the nuclear lamina (Fig. 16.1).

Within the nucleus, the semifluid matrix called nucleoplasm is present. The nucleoplasm is a liquid with a gel-like consistency similar in this respect to the cytoplasm. The nucleoplasm contains many substances like nucleotide triphosphates, enzymes, proteins, and transcription factors. There also exists a network of fibers in the nucleoplasm known as the nuclear matrix. Most of the nuclear material that consists of chromatin exists in the nucleoplasm in the less condensed form of the cell's DNA that later organizes to form chromosomes during mitosis or cell division. There are two types of chromatin: euchromatin and heterochromatin. Euchromatin is the least compact form of DNA, and the regions of DNA which constitute euchromatin contain genes which are frequently expressed by the cell. In heterochromatin, DNA is more tightly compacted. Regions of DNA which constitute heterochromatin generally

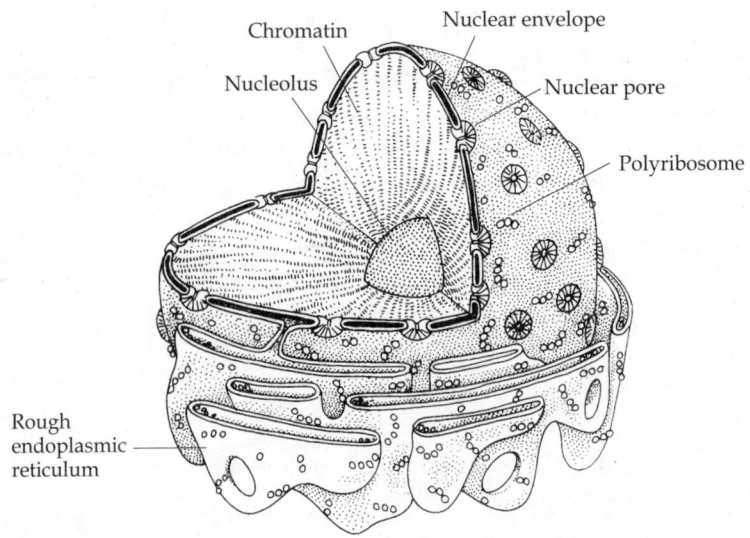

FIG. 16.1 Ultrastructure of the Nucleus

contain genes which are not expressed by the cell (this type of heterochromatin is known as facultative heterochromatin) or are regions which make up the telomeres and centromeres of the chromosomes (this type of heterochromatin is known as constitutive heterochromatin). In multicellular organisms, cells are highly specialized to perform particular functions, hence different sets of genes are required and expressed. Therefore, the regions of DNA that constitute heterochromatin vary between cell types. The nucleus also contains one or more nucleoli, the organelles that synthesize protein-producing macromolecular assemblies called ribosomes, and a variety of other smaller components, such as Cajal bodies.

Thus, we have seen that the nucleus is composed of (1) the nuclear membrane or the nuclear envelope, (2) the nucleoplam, (3) the chromatin fibres, and (4) the nucleolus. These structures will be explained in detail in the following sections.

THE NUCLEAR ENVELOPE

The nuclear envelope is a thin, elastic, semi-permeable bilayered outermost envelope of the interphase nucleus. It regulates the nucleo-cytoplasmic interactions and exchange of materials between the nucleus and the cytoplasm.

History

The ultrastructure of the nuclear envelope, nuclear pore complex and nuclear lamina was studied by Kirschner and his coworkers in 1977.

Structure

The nuclear envelope is a double-layered membrane, each 5-10 nm thick, that encloses the contents of the nucleus during most of the cell's life cycle. The space between the two layers is called the perinuclear space which is 10-50 nm wide; this space appears to be connected with the rough endoplasmic reticulum. The envelope is perforated with tiny holes called nuclear pores. These pores regulate the passage of molecules between the nucleus and cytoplasm, permitting some to pass through the membrane, but not others (Fig. 16.2).

Nuclear Lamina

The inner surface of the nuclear envelope has a protein lining called the nuclear lamina, which binds to chromatin and other nuclear components. During mitosis, or cell division, the nuclear envelope disintegrates, but reforms as the two cells complete their formation and the chromatin begins to loosen and disperse. The inner membrane of the nuclear envelope lies next to a layer of thin filaments which surrounds the nucleus except at the nuclear pores. These

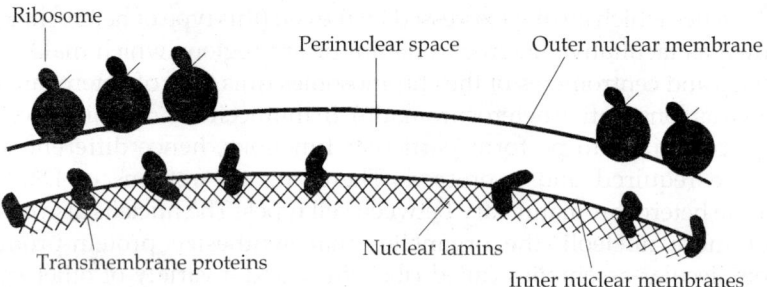

FIG. 16.2 The nuclear envelope

may also serve as stabilizing filaments. This structure is called the nuclear lamina. It consists of intermediate filaments which are 30-100 nm thick. These intermediate filaments are polymers of lamin, ranging from 60-75 kD, they are of three types, viz., A-type lamins that lie next to nucleoplasm, B-type lamins, near the nuclear membrane, and C-type lamins. They may bind to integral proteins inside the membrane. The lamins may be involved in the functional organization of the nucleus and may play a role in assembly and disassembly before and after mitosis. After they are phosphorylated, this triggers the disassembly of the lamina and causes the nuclear envelope to break up into vesicles. Dephosphorylation reverses this and allows the nucleus to reform.

Nuclear Pore Complex

Nuclear pores are formed at sites where the inner and outer membranes of the nuclear envelope are joined, leaving a space filled with filamentous material. Sometimes a thin diaphragm may be seen running horizontally through the pore. Also, the chromatin which carries the genetic material is organized so that a space is created to the nuclear pore. There are about 3,000-4,000 nuclear pore complexes in the nuclear envelope of an animal cell. The whole pore complex has a diameter of about 150 nm, and the diameter of the opening is about 50 nm wide (Fig. 16.3).

The nuclear pores allow the transport of water-soluble molecules across the nuclear envelope. This transport includes RNA and ribosomes moving from nucleus to the cytoplasm and proteins (such as DNA polymerase and lamins), carbohydrates, signal molecules and lipids moving into the nucleus. Although smaller molecules simply diffuse through the pores, larger molecules may be recognized by specific signal sequences and then be diffused with the help of nucleoporins into or out of the nucleus. This is known as the RAN cycle. There are a total of eight protein subunits surrounding the actual pore (the outer ring). Each of these subunits project a spoke-shaped protein into the pore channel. The center of the pore often appears to contain a plug-like structure.

Nuclear pore complexes control the flow into and out of the nucleus and check the credentials of all large molecules attempting to pass through.

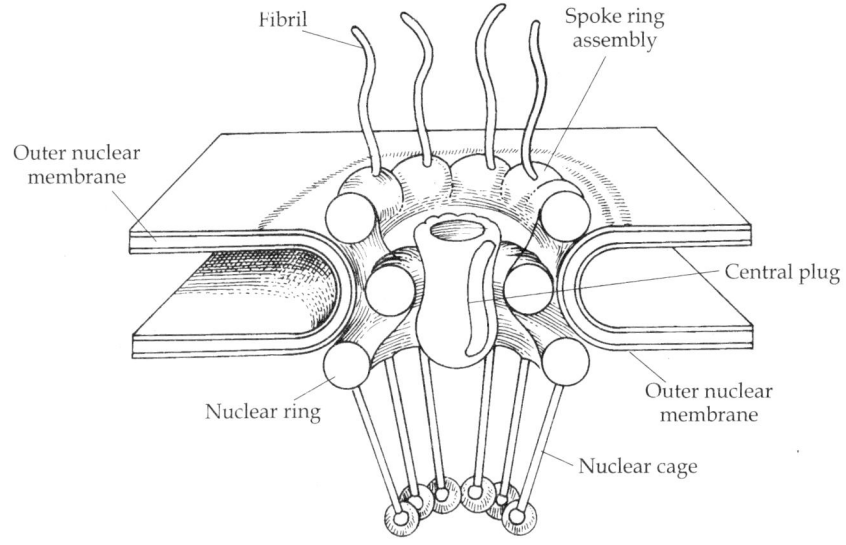

FIG. 16.3 The nuclear pore complex

NUCLEOPLASM

The nucleoplasm is a highly viscous liquid that surrounds the chromosomes and nucleoli. It may be called 'karyoplasm' when it is gel-like and 'karyolymph' when it is a colloidal fluid, but generally the two terms are synonymous. Many substsances such as nucleotides, proteins, enzymes and minerals are dissolved in the nucleoplasm. A network of fibers known as the nuclear matrix can also be found in the nucleoplasm.

History

The term nucleoplasm was coined by Strasburger in 1882.

Chemical Composition

Nucleoplasm contains nucleotides that are the monomers of nucleic acids DNA and RNA. Many types of complex proteins are seen in association with DNA and ribosomes, for example, basic proteins (histones) and acidic or neutral proteins (non-histones). The nucleoplasm also contains many enzymes which are necessary for the synthesis of DNA and RNA, for example, DNA polymerase, RNA polymerase, nucleoside triphosphatase, NAD synthetase, adenosine diaminase, guanase, aldolase, enolase, etc. The minerals like Mn^{++} and Mg^{++} act as cofactors.

CHROMATIN FIBRES

Chromatin is a complex of DNA and protein found inside the nucleus. These chromatin fibres are observed only during interphase.

History

The term chromatin was coined by W. Flemming in 1879.

Structure

The chromatin comprises of DNA as the main nucleic acid while small quantity of RNA may also be present that accounts for nearly 5% of the total chromatin present. The nucleic acids are generally in the form of double-stranded DNA (a double helix). The major proteins involved in chromatin are histone proteins, but other chromosomal proteins are prominent too. DNA is packaged into chromatin thereby constraining the size of the molecule and allowing the cell to control gene expression of the chromatin-packaged genes.

Two types of chromatin material have been recognized: Euchromatin and Heterochromatin.

Euchromatin: It is a lightly packed form of chromatin that is rich in gene concentration, and is often (but not always) under active transcription. Euchromatin participates in the active transcription of DNA to mRNA products. The unfolded structure allows gene regulatory proteins and RNA polymerase complexes to bind to the DNA sequence, which can then initiate the transcription process. There is, therefore, a direct link between the amount of euchromatin and the productive activity of a cell. It is thought that the cell uses transformation from euchromatin into heterochromatin as a method of controlling gene expression and replication.

Heterochromatin: It is (usually but not always) a tightly packed form of DNA. Its major characteristic is that it is not transcribed. It is a genetically inactive region of chromosomes that either lack genes or contain genes that are repressed. Heterochromatin also replicates later in S phase of the cell cycle, and is found only in eukaryotes. The centromeres and telomeres are heterochromatic, as is the Barr body of the second inactivated X chromosome in a female.

Heterochromatin is generally stably inherited when a cell divides the two daughter cells. Both the daughter cells will typically contain heterochromatin within the same regions of DNA, resulting in epigenetic inheritance.

NUCLEOLUS

Nucleolus (L. *Nucleolus* = a small nucleus) is an RNA-rich intranuclear organelle in the nucleus of eukaryotic cells, produced by a nucleolar organizer. It represents the storage place for ribosomes and ribosome precursors. The nucleolus consists primarily of ribosomal precursor RNA, ribosomal RNA, their associated proteins, and enzymatic equipment like polymerase, RNA methylase, RNA cleavage enzymes which are required for synthesis, conversion and assembly of ribosomes.

History

Nucleoli were first observed by Fontana in 1781 and were studied by M.J. Schleiden in 1838.

Structure

Nucleoli are made of protein and ribosomal DNA (**rDNA**). The ribosomal DNA serves as a template for transcription of ribosomal RNA for inclusion into new ribosomes. The nucleolus includes fibrillar centers (FC), dense fibrillar components (DFC), granular components (GC) and rDNA.

Fibrillar center (FC): It is made up by a network of fine (4-5 nm thick) fibrils. It is roughly globular, with the diameter ranging from about 50 nm to 1 μm. The number and size of FCs per nucleolus is variable, and changes with cellular activity and the need for ribosome production. Cells with lower cellular activity usually have fewer FC than others.

Dense fibrillar component (DFC): This component is also made up by very fine (3-5nm) and densely packed fibrils. DFCs usually surround FCs when they are present and form a meshwork. The amount of DFC roughly reflects the nucleolar engagement in ribosome biogenesis.

Granular component (GC): The granular component appears to consist of small granules with a diameter of about 15 nm. They typically form a mass surrounding the fibrillar complexes and embed the FCs and DFC. Thus a transition zone between DFC and GC can be observed.

Ribosomal DNA (rDNA): rDNA is a set of tandemly-repeated genes coding for preribosomal RNA, these regions of the chromosome are called 'nucleolus organizer regions' or NORs.

NUCLEO-CYTOPLASMIC RELATIONSHIP

In 1943, a Danish biologist named Joachim Hammerling experimentally proved the significance of nucleus. He carried out an important experiment in which he searched for the part of a cell that directs its physical appearance or phenotype. Hammerling used large unicellular green algae called *Acetabularia* which is composed of one single large cell about 6 cm long having three main body parts: a foot or base containing the nucleus and anchoring the cell to a rock or other support, a stalk resembling a plant stem, and a cap that carries out the process of photosynthesis.

For his experiment, Hammerling used two different species of *Acetabularia*, *A. mediterranea* that has a disk-shaped cap, and the other species, *A. crenulata*, which has a branched cap, more like a flower. He cut the stalk and cap off of an *A. mediterranea* cell and grafted a stalk from *A. crenulata* in its place. It was observed that a modified branched cap, with marked similarity to *A. crenulata* caps grew on that cell. Hammerling then removed the new cap from the

A. mediterranea base to see what type of cap would now grow on the grafted stalk. This time a disk-shaped cap, just like the *A. mediterranea* caps, grew.

This experiment suggested to Hammerling that the factor directing the growth of the algal cap was located in the base of the cell. He felt that the growth of the modified branched cap resulted from the presence of a message of some sort in the grafted stalk. That message was used up in directing the production of the modified *A. mediterranea* cap, so that when a second cap was regenerated, it grew according to instructions derived from the nucleus contained in the base of the cell (Fig. 16.4a, b and c).

Hammerling concluded from these experiments that the nucleus of the cell was directing its development and somehow also specifying its hereditary characteristics. Later experiments by other scientists confirmed this conclusion and broadened its implications to suggest that nucleus not only directed the synthesis of new parts of an individual cell, but also directed the growth and development of entire multicellular organisms.

Hammerling's experiment on *Acetabularia*

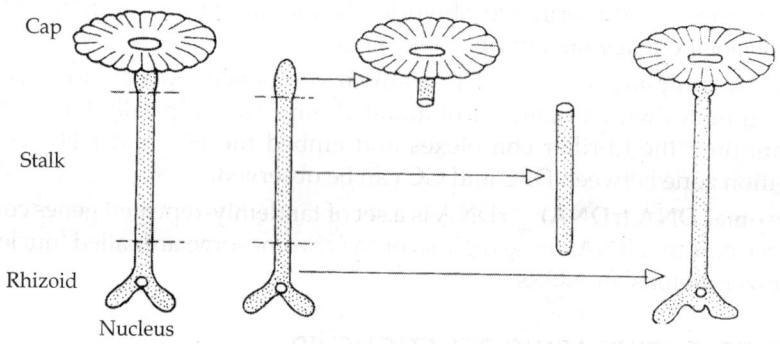

FIG. 16.4a The morphological structure of the cell with hat, stem and rhizoid with nucleus

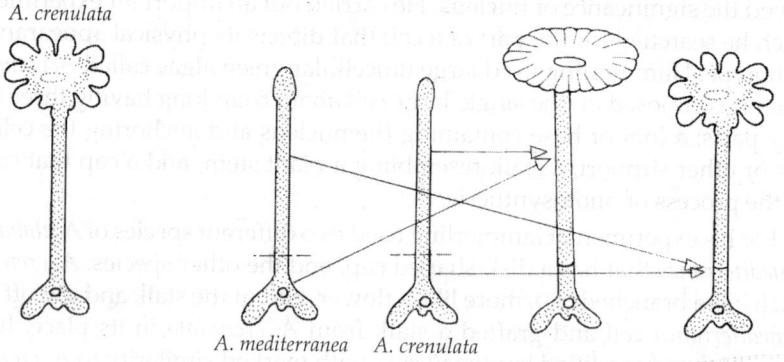

FIG. 16.4b The influence of the nucleus on the hat morphology of *Acetabularia*

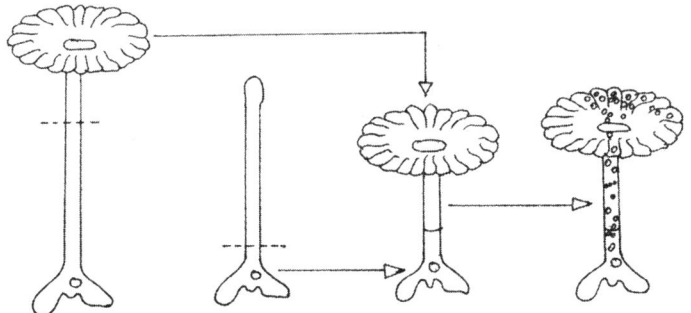

FIG. 16.4c Influence of the plasma on the gamete formation

OBJECTIVE TYPE QUESTIONS

1. Which structure includes all of the others?
   - A. Nucleolus
   - B. Nucleus
   - C. Chromosomes
   - D. Genes

2. _____ is located in the middle of the cell, is the center for cell reproduction, contains the hereditary material of the cell.
   - A. Chromosome
   - B. DNA
   - C. Nucleus
   - D. Nucleolus

3. _____ is threadlike material in the cell nucleus that forms the chromosomes.
   - A. DNA
   - B. Chromatin
   - C. Histones
   - D. Chrommomeres

4. _____ is a thin wall enclosing the nucleus, double membrane, controls what goes in and out of nucleus.
   - A. Nucleoplasm
   - B. Endoplasmic reticulum
   - C. Nuclear membrane
   - D. Nuclear lamina

5. Nuclear membrane is mainly composed of
   - A. Lipids and carbohydrates
   - B. Lipids and proteins
   - C. Proteins and fatty acids
   - D. Proteins and starch

Answers

1. B   2. C   3. B   4. C   5. B

SHORT ANSWER TYPE QUESTIONS

1. What structures can not be seen in interphase but can be seen in prophase? Why?
2. Write a short note on nuclear pore complex.
3. Comment upon nucleo-cytoplasmic relationship.

LONG ANSWER TYPE QUESTIONS

1. What are the major roles of the nucleus, and what parts of the nucleus carry out these roles?
2. Write an account on the structure and functional organization of the cell nucleus.
3. The nucleus is influential but is also influenced by factors beyond its control. Comment.

Chromosomes 17

INTRODUCTION

Chromosome (Greek, *chroma*=colour; *soma*=body) is long, threadlike association of genes in the nucleus of all eukaryotic cells consisting of DNA and protein. It forms the basis of heredity and carries genetic information in DNA in the form of a sequence of nitrogenous bases. Each chromosome has two arms, the shorter one called 'p' arm (from the French, *petit*- small) and the longer 'q' arm (*q* following *p* in the Latin alphabet). Prokaryotes do not possess histones or nuclei.

HISTORY

Chromosomes were first observed in plant cells by Swiss botanist Karl Wilhelm von Nägeli (1817-1891) in 1842, and independently, in *Ascaris*, by the Belgian scientist, Edward Van Beneden (1846-1910). Their behavior in animal (Salamander) cells was later described in detail by German anatomist, Walther Flemming (1843-1905), the discoverer of mitosis in 1882. The name was invented later by another German anatomist, Heinrich von Waldeyer (Fig. 17.1).

CHROMOSOMES IN BACTERIA

Bacterial chromosomes are often circular but sometimes linear. When linear, bacterial chromosomes tend to be attached to the plasma membrane of the bacteria. Some bacteria have one chromosome, while others have more than one. Bacterial DNA is not bound to proteins as in eukaryotes.

FIG. 17.1 Heinrich Wilhelm Von Waldeyer (1836-1921)

The area where the chromosome in bacteria is located is having some amount of cytoplasm surrounding the chromosome forming nucleoid. Bacterial chromosomes initiate replication at one origin of replication. Bacterial DNA also exists as plasmids which are circular pieces of DNA that can be transmitted between bacteria. Plasmids often carry the genes for antibiotic resistance and can thus spread between different bacteria (Fig. 17.2).

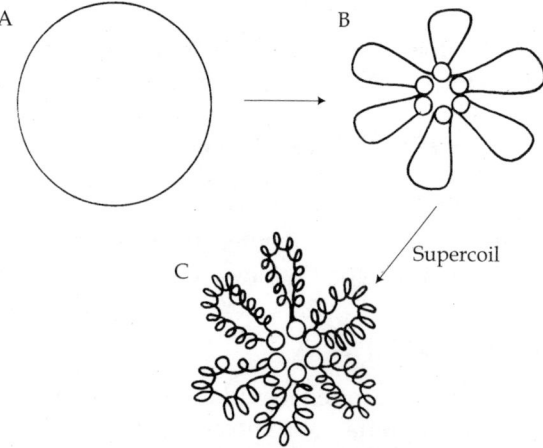

FIG. 17.2 Structure of chromosome in *E. coli*

CHROMOSOME NUMBER

All normal members of a particular species have the same number of chromosomes. Sexually reproducing species have somatic or body cells, which are diploid [2n] having two sets of chromosomes, one from the mother and one from the father. Gametes, that is, sperms and ova are haploid [n] and have one set of chromosomes. The gametes are produced by meiosis of a diploid germ line cell. During meiosis, the matching chromosomes of father and mother can exchange small parts of themselves (a process known as chromosomal crossing over), and thus create new chromosomes that are not inherited solely from one parent. When a male and a female gamete merge during fertilization a new diploid organism is formed as a result.

Some animal and plant species are polyploid [Xn] and have more than two sets of chromosomes. Agriculturally important plants such as tobacco or wheat are often polyploid compared to their ancestral species. Wheat has a haploid number of seven chromosomes, still seen in some wild progenitors. The chromosome number of some animal and plant species is tabulated below:

Chromosome numbers in some common organisms

Common name	Species	Diploid number
Human	Homo sapiens	46
Monkey	Macaca mulatto	42
Dog	Canis familiaris	78
Cat	Felis domesticus	38
Mouse	Mus musculus	40
Frog	Rana pipiens	26
Fruitfly	Drosophila melanogaster	08
Flatworm	Planaria torva	16
Corn	Zea mays	20
Green alga	Acetabularia mediterannea	20

Smallest chromosome number: The female of a subspecies of the ant, *Myrmecia pilosula*, has one pair of chromosomes per cell. The male ant has only one chromosome in each cell.

Largest chromosome number: In the fern family of plants, the species *Ophioglossum reticulatum* has about 630 pairs of chromosomes, or 1260 chromosomes per cell.

DUPLICATED AND UNDUPLICATED CHROMOSOMES

Chromosomes either have one or two molecules of DNA along with associated proteins. A chromosome with one molecule of DNA is called an unduplicated chromosome. On the other hand, a duplicated chromosome contains two identical daughter DNA molecules that have come from an original DNA molecule during synthetic (S) phase of interphase at the time of division. These chromosomes are also referred as unreplicated and replicated chromosomes, respectively. In the case of a duplicated chromosome, each molecule of DNA and associated proteins is called a sister chromatid.

CHROMATID AND CHROMOSOME

Remember that when two DNA molecules are joined together at mitotic metaphase, each molecule is called a chromatid and the two of the molecules are called a duplicated chromosome. Both chromatids are attached to one another by a centromere and become separated at the beginning of anaphase when the sister chromatids move to opposie poles (Fig. 17.3).

CENTROMERE

When a chromosome is examined during mitosis or meiosis there is a constriction in the region somewhere along the length of the chromosome. This constriction is called the centromere. The centromere is a region where the spindle fibers attach to the chromosome during cell division and it is in a characteristic position that is constant for different types of chromosomes. Thus the centromere is important for studying and identifying chromosomes. The centromere also contains a small ring of protein called a kinetochore which is important in the movement of chromosomes during mitosis and meiosis.

TELOMERE

The ends of the chromosome in eukaryotes are called telomeres. This region is important because during DNA replication, the telomere does not always get

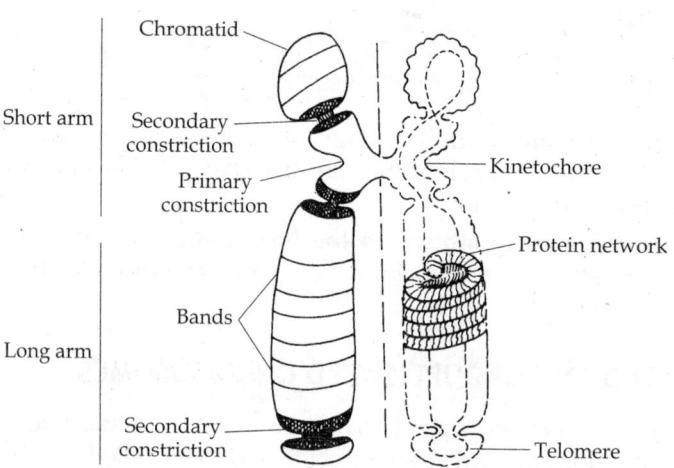

FIG. 17.3 Metaphasic chromosome

duplicated properly and the chromosome shortens slightly. The telomere contains many repeating sections of DNA rather than regions of DNA that code for specific genes. The telomeres provide stability to the chromosomes because the broken ends of chromosomes are sticky, whereas the normal ends are not sticky, suggesting the ends of chromosomes have unique features. Usually, but not always, the telomeric DNA is heterochromatic and contains direct tandemly repeated sequences.

CHROMATIN

During certain period of the cell's life cycle chromosomes are not visible. This is because the chromosomes are stretched out very thin to allow surfaces for the various chemical reactions that involve chromosomes to take place. When the nucleus is stained and examined, it appears uniformly colored and the chromosomes collectively are termed chromatin. It is critical to remember that even though individual chromosomes cannot be seen in the chromatin, the chromosomes still exist. Two types of chromatin can be distinguished:

Euchromatin: It consists of DNA that is active and is expressed as protein.

Heterochromatin: It consists of mostly inactive DNA. It seems to serve structural purposes during the chromosomal stages. Heterochromatin can be further distinguished into two types:

(i) *Constitutive heterochromatin* is never expressed. It is located around the centromere and usually contains repeated DNA sequence.

(ii) *Facultative heterochromatin* is sometimes expressed.

AUTOSOMES AND SEX CHROMOSOMES

Some organisms such as birds and mammals have a chromosomal system of sex determination. For example, in humans, individuals which have two X chromosomes (XX) typically develop into females and individuals which have an X chromosome and a Y chromosome (XY) develop into males.

The term autosome refers to those chromosomes that are not involved in sex determination. Human diploid cells have 22 pairs of autosomes and 1 pair of sex chromosomes.

Sex chromosomes are chromosomes involved in a major way in sex determination. All species do not follow the same XY chromosome system as in human beings. For example, in birds the females are XY and the males are XX.

Autosomes and sex chromosomes together form a complete set of chromosomes, called genome, that is the complete DNA component of an individual.

CHROMOSOME MORPHOLOGY

Under the microscope chromosomes appear as thin, thread-like structures. They all have a short arm and long arm separated by a primary constriction called the centromere. The short arm is designated as p and the long arm as q. The centromere is the location of spindle attachment and is an integral part of the chromosome. It is essential for the normal movement and segregation of chromosomes during cell division. Human metaphase chromosomes (Fig. 17.4a) are present in three basic shapes and can be categorized according to the length of the short and long arms and also the location of the centromere (Fig. 17.4b).

Metacentric chromosomes: They have short and long arms of roughly equal length with the centromere in the middle. The metacentric chromosomes appear V-shaped during anaphasic movement.

Sub-metacentric chromosomes: They have short and long arms of unequal length with the centromere more towards one end. Such chromosomes appear L-shaped during anaphasic movement.

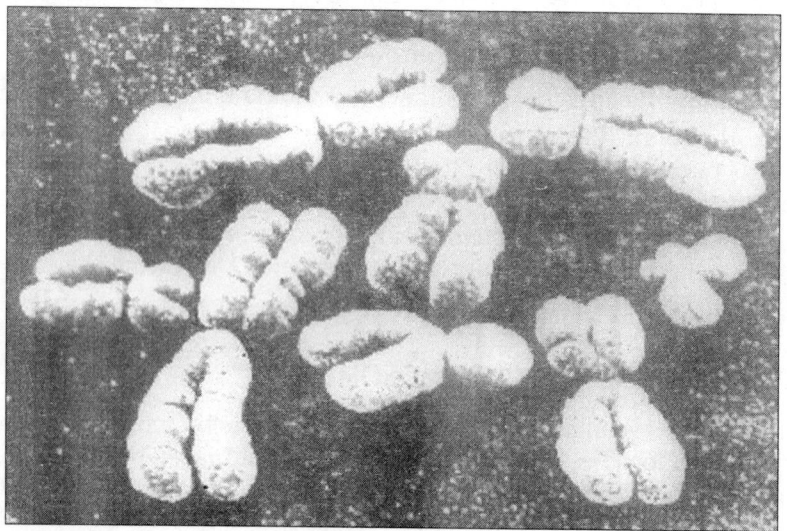

FIG. 17.4a Human chromosomes

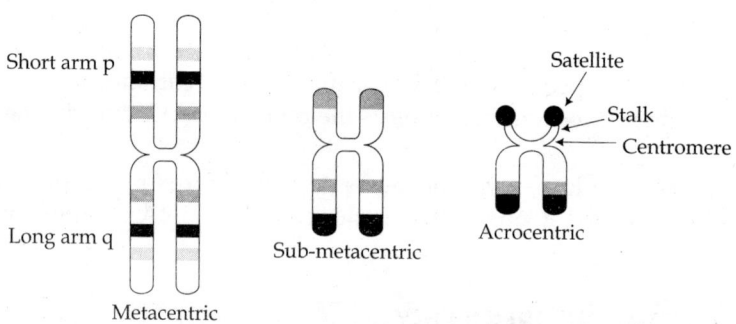

FIG. 17.4b Types of chromosomes

Acrocentric chromosomes: These have a centromere very near to one end and have very small short arms. These chromosomes appear J-shaped during anaphasic movement. They frequently have secondary constrictions on the short arms that connect very small pieces of DNA, called stalks and satellites, to the centromere. The stalks contain genes which code for ribosomal RNA.

Sub-telocentric chromosomes: These are with a more terminally placed centromere, forming very unequal chromosome arms and appear I-shaped during anaphasic movement.

EUKARYOTIC CHROMOSOME KARYOTYPE

The eukaryotic chromosomes are detected to occur during cell division and are in their most condensed form during metaphase when the sister chromatids are attached. This is the primary stage when cytogenetic analysis is performed. The eukaryotic species possess at least one pair of chromosomes but most of the organisms have more than one pair. Each species is characterized by a karyotype. The karyotype is a photographic representation of the number of chromosomes in the normal diploid cell, as well as their size distribution. For example, the human chromosome complement has 23 pairs of chromosome, 22 somatic pairs and one pair of sex chromosomes (Fig. 17.5, 17.6). One important aspect of genetic research is correlating changes in the karyotype with changes in the phenotype of the individual.

To further distinguish among chromosomes, they are treated with a dye that stains the DNA in a reproducible manner. After staining, some of the regions are lightly stained and others are heavily stained. The lightly stained regions are called euchromatin, and the dark stained region is called heterochromatin. The current dye of choice is the Giemsa stain, and the resulting pattern is called the G-banding pattern.

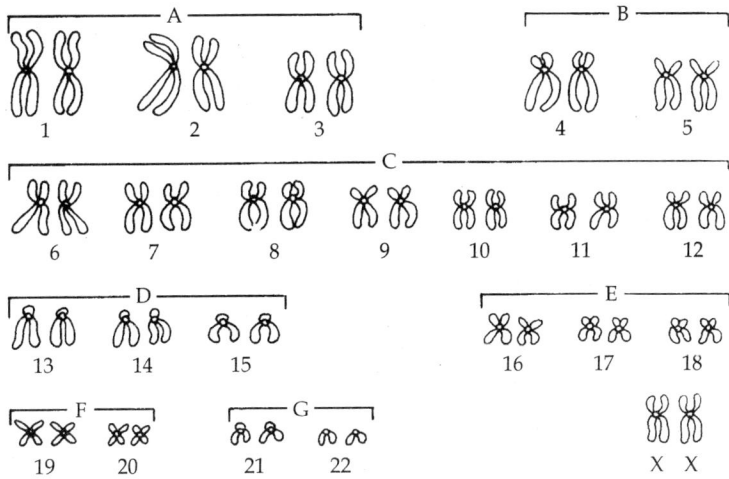

FIG. 17.5 Karyotype of human chromosomes (normal female)

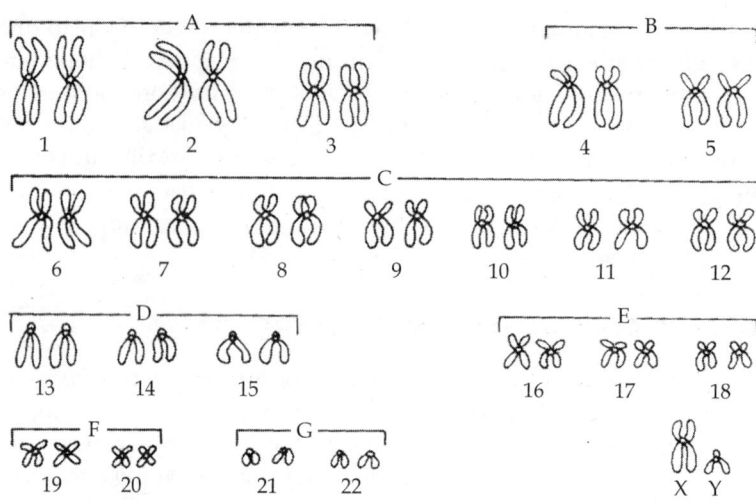

FIG. 17.6 Karyotype of human chromosomes (normal male)

DOSAGE COMPENSATION AND LYONIZATION

Dosage compensation is a genetic regulatory mechanism which operates to equalize the phenotypic expression of characteristics determined by genes on the X chromosome so that they are equally expressed in the human XY male and the XX female. Because females have twice as many copies of X-linked genes as males, one copy of each must be turned off. This occurs by inactivating one X chromosome in every cell of a female which is done by converting it to heterochromatin. This highly condensed inactive X chromosome is called a Barr body.

We have seen that the gender is determined by the sex chromosomes. Humans usually have 46 chromosomes per diploid cell consisting of 22 sets of autosomes and one set of sex chromosomes—either XX or XY. In the usual course of events, individuals with the karyotype 46 + XX are female and individuals with 46 + XY are male. The male being the heterogametic sex (XY) produces two types of spermatozoa in equal proportions, i.e., one carries the X-chromosome and the other carries the Y-chromosome. The female, on the other hand, being the homogametic sex (XX) produces only one kind of gametes. However, this is not true for all organisms, for example, in birds and reptiles the female is heterogametic (ZW) and the male is homogametic (ZZ). But largely, the type of sex determination mechanism found to exist in all mammals and some insects is the common XY system wherein the individuals who possess XX chromosomes are female while male animals possess an X and a Y chromosome. But the question arises as to how the sex of an individual is determined. Is sex determined by the number of X chromosomes— with one X you are male or with 2 X's you are female? Or is sex determined by the presence or absence of the Y chromosome—the presence of a Y makes for a male or the absence of a Y produces a female? The answer was provided by individuals resulting from non-disjunction of the sex chromosomes. Some individuals have 45 chromosomes with only one X chromosome, such individuals are phenotypically female while other individuals have 47 chromosomes with two X chromosomes and a Y chromosome and are phenotypically male. If having two X chromosomes was the determining factor, then these individuals should be female. The table below indicates the sex of these individuals.

Chromosome Constitution	Name of Syndrome	Sex of Individual
46, XX	Normal	Female
46, XY	Normal	Male
45, XO	Turner Syndrome	Female
47, XXY	Klinefelter Syndrome	Male

Clearly, the presence or absence of a Y chromosome is more important to sex determination in humans than the number of X chromosomes present. Further research has shown that the gonads (which early in embryonic development can differentiate into either testes or ovaries) are directed to become testes if there is a Y chromosome present. If no Y chromosome is present, then the gonads differentiate as ovaries.

The region of the Y chromosome which is important for sex determination has been identified, and called SRY (for sex-determining region of Y). This region was identified in phenotypic males who were found to have two X chromosomes and no Y, although a small amount of Y chromosome DNA (the SRY) was found to have been transferred to one of the X chromosomes. Phenotypic females who were XY were found to be missing this region in their Y chromosomes. Proof of the role of this region in sex determination came from work on mice, in which the SRY was inserted into the genome of XX embryos. These mice developed as perfectly normal males.

X-Chromosome Inactivation

This strategy of sex determination creates a situation of inequity. Females, because of their two X chromosomes, have two copies of all X-linked genes, while males only have one. Therefore, females should produce twice as much of the proteins encoded by these genes as males. This would mean that females and males would have differences in color vision, and other things. (In fact, the overall effect would probably be lethal to female embryos.) To compensate for the extra X chromosome (the X chromosome 'dose'), one X chromosome (and all of its genes) is inactivated in every cell of early female mammalian embryos. The X chromosome is inactivated by converting it to hetero-chromatin. This highly condensed inactive X chromosome is called a Barr body and this phenomenon of bringing about equality in products synthesized under the control of genes carried on X chromosomes is termed dosage compensation (Fig. 17.7). Which chromosome will be inactivated is

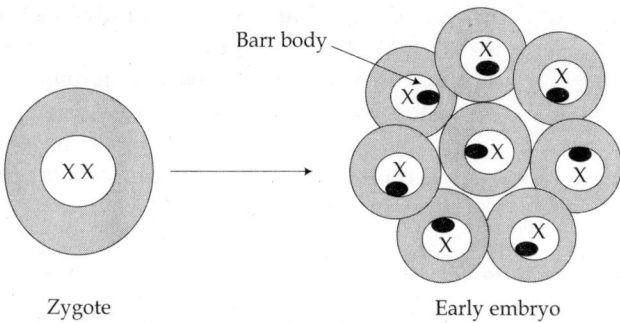

Zygote   Early embryo

FIG. 17.7 Diagram illustrating inactivation of X-chromosome and formation of barr body

randomly determined in every cell, and once inactivated, the same X chromosome will continue to be inactivated in all the cells which are derived from it. This phenomenon of inactivation of X chromosomes was first described by Mary Lyon (Fig. 17.8) and was known as the Lyon Hypothesis, now considered as the Lyon Principle. So, if a female is heterozygous for an X-linked trait, the dominant allele will be expressed in some cells, while the recessive allele will be expressed in others. Female mammals are therefore mosaics with regard to X-linked genes. This explains calico cats, females which have patches of different colored fur (one coat color gene for cats is found on the X chromosome). It turns out that during the first few days of embryonic life the X chromosome inactivation occurs. In any one female cell

FIG. 17.8 Mary F Lyon

the inactivated X chromosome may be either the paternal (X^P) or maternal (X^m). Which one is inactivated is a matter of chance but once that X chromosome has been inactivated in the cell, all that nucleus' descendants (clones through mitosis) will have the same inactive X chromosome. X inactivation is randomly determined but once it happens it is permanent. Each cell's descendants contribute to the overall body of the organism. Because some cells have inactivated paternal Xs and others have inactivated maternal Xs, hence, the embryo and the organism become a mixture of two types of cells — expressing different alleles (Fig. 17.9).

It has now been shown that only parts of the inactivated X chromosome are in fact inactive, and that certain areas crucial to development remain active. This X is usually late replicating, highly coiled and condensed and can be visualized microscopically as a clear dense mass against the inner aspect of the nuclear membrane.

There are two classes of heterochromatin. Unlike the first, constitutive heterochromatin that is never transcriptionally active, located in the centromeric region of each chromosome and also found in the telomeres, the other class is called facultative heterochromatin. This type is sometimes active and sometimes not, depending upon the cell type in an organism and on the

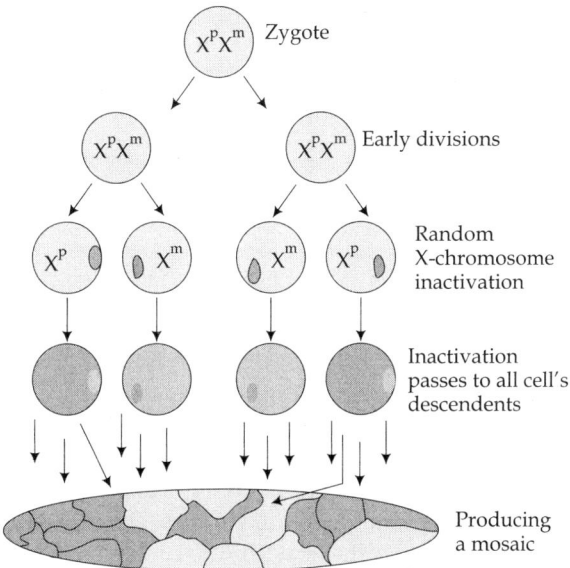

FIG. 17.9 Diagram on Lyon hypothesis and how it causes female mammals, including women, to be mosaic. Early in development the X chromosomes are randomly inactivated in each cell. In any particular cell it could be the paternal (X^P) or the maternal (X^m) that remains active. The inactivated X chromosome becomes a tiny blob of dense, useless chromatin called the Barr body-shown here as a dark oval. These cells pass their inactivation patterns to all their descendants. These descendants contribute to the body plan of the organism, often in a complicated way that produces a mottled pattern. The complete organism is a mosaic of cells expressing alleles from either the X^P or the X^m chromosome, depending on which one has escaped inactivation.

organism's stage of development. There is little facultative heterochromatin in embryonic cells; whereas some highly specialized cell types have much more. It means that as cells differentiate and no longer need access to all of their genes, the unnecessary genes are put into the equivalent of long-term storage by taking the form of heterochromatin. Heterochromatin is replicated late in the S period of the cell cycle. For example, the active X chromosome replicates before the inactive X chromosome. The less condensed DNA is more accessible to the replication machinery. Therefore, the less condensed DNA replicates early. The housekeeping genes, which are essential for the moment-to-moment functioning of the cell, are replicated first. If, in a given species, a particular gene is active in one type of cell but not in another type of cell, it will replicate early in the first type, but later in the second type. In humans, the X chromosome that is inactivated is determined by chance. In marsupials, however, the paternal X chromosome is always inactivated. Inactivation is achieved by methylation of the X chromosome DNA, a common way in which the cell silences particular genes.

One might wonder, "If women with Turner syndrome (45, with XO) do not make a barr body, they do not inactivate their single X chromosome. So why do they have problems? Why aren't they as healthy as any male? On the other hand, why is Klinefelter syndrome (47, XXY) a problem since these men inactivate the extra X chromosome anyway (producing a barr body like a normal female)?"

The fact is that a few of the thousands of genes on the X chromosome escape inactivation. Also, there are some delicate interactions between the X chromosome genes and other genes on the autosomes (and perhaps the Y chromosome too). All together, these poorly understood effects contribute to the unusual medical problems caused by an 'over dose' or 'under dose' effect.

Dosage Compensation in Non-mammalian Mechanisms

In both the fruit fly *Drosophila melanogaster* and the nematode *Caenorhabditis elegans,* the primary sex determination mechanisms and the molecular cascades controlling sexual differentiation have been studied extensively. Primary sex determination in these animals does not involve the Y chromosome but instead is determined by the ratio of the number of X chromosomes to autosomal (non-sex) chromosomes. By examining individuals with unusual numbers of various chromosomes it has been determined that in *Drosophila* those with one or fewer X chromosomes per diploid autosome set develop as males while those with two or more X chromosomes develop as females. Individuals with intermediate ratios such as those with two X chromosomes and a triploid set of autosomes develop as intersexes with both male and female characteristics. Although this ratio serves as the primary determinant of sex in both of these organisms, the specific gene products that influence this ratio assessment are different, demonstrating that different molecular mechanisms can be used for a similar purpose.

The phenomenon of dosage compensation in *Drosophila* has been studied extensively by many workers and it has been shown that the X chromosome in male *Drosophila* shows hyperactivity rather than the X chromosome undergoing inactivation in female *Drosophila*.

C-VALUE PARADOX

The C-value is the amount of DNA in the haploid genome of an organism. It varies over a very wide range with a general increase in C-value with complexity of organism from prokaryotes to eukaryotes. This term was coined by R. Vendrely and C. Vendrely in 1948 pointing to a 'remarkable constancy in the nuclear DNA content of all the cells in all the individuals within a given animal species'. However, it was soon found that C-values, i.e., genome sizes vary enormously among species of the same genus as well as among genera of the same family and that this bears no relationship to the presumed number of genes as reflected by the complexity of the organism. For example, the cells of some salamanders may contain 40 times more DNA than those of humans. But, man is considered more complex in terms of genetic development. This lack of correlation between genome size and genetic complexity is considered paradoxical and is thus, referred to as C-value paradox. Fig. 17.10 summarizes the range of C-values found in different evolutionary phyla. There is an increase in the minimum genome size found in each group as the complexity

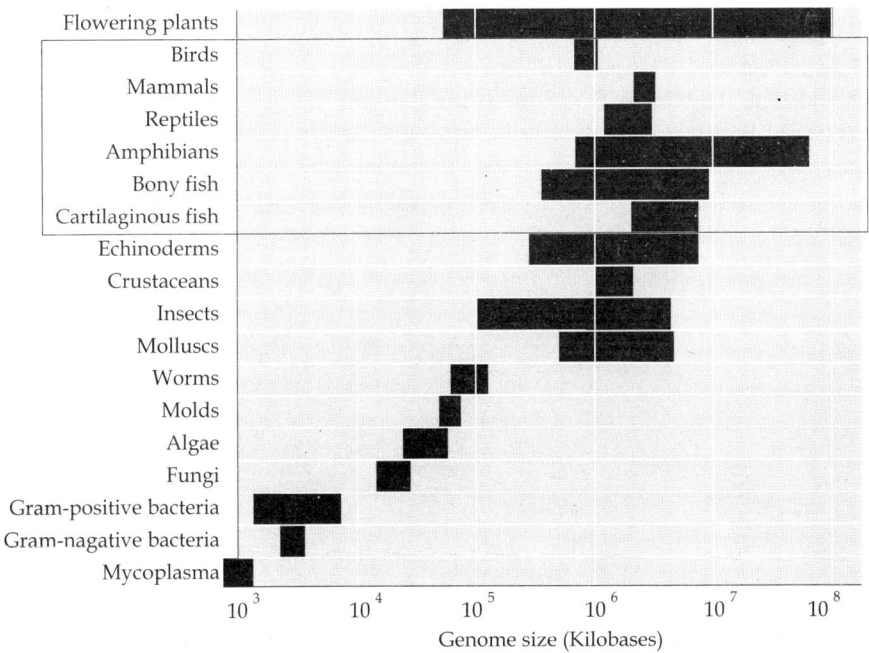

FIG. 17.10 Variations in genomic DNA in a variety of animals and plants

increases. But as absolute amounts of DNA increase in the higher eukaryotes, some wide variations in the genome sizes are seen within some phyla.

The term C-value paradox was coined by C.A. Thomas in 1971. There is yet another aspect of C-value paradox involving huge variation in C-values between species, whose apparent complexity does not vary correspondingly. This is especially seen in case of insects, amphibians and plants (Fig. 17.11).

The discovery of non-coding DNA resolved the C-value paradox, that is, why the genome size does not reflect gene number in eukaryotes. Most eukaryotic (but not prokaryotic) DNA is non-coding and therefore does not consist of genes, and as such total DNA content is not determined by gene number in eukaryotes. The human genome, for example, is comprised of only about 1.5% protein-coding genes, with the other 98.5% being various types of non-coding DNA. This large proportion of non-coding DNA in most eukaryotes has been shown to be present in the form of multiple copies of identical DNA sequences; this multiplicity ranges from a few copies to millions of copies within the same cell. These DNA sequences are together described as repetitive DNA. The remaining DNA is described as Unique DNA and is found in the form of single copy DNA sequences. These sequences are dispersed throughout the genome and are responsible for coding proteins. The repetitive DNA, on the other hand, may be clustered at one place or interspersed with unique sequences. Though they do not carry any genetic information, but they are believed to play some other structural or regulatory role.

Recently, it has been suggested that the term C-value paradox may be abandoned and the usage of a more appropriate term C-value enigma be considered. This term was coined by T. Ryan Gregory in 2001. This term is preferred because it explicitly includes all the questions that will need to be

Group	Approximate Size of haploid Genome (bp)
Viruses & Bacteria	$10^5 - 5 \times 10^6$
Algae & Fungi	$5 \cdot 10^6 - 5 \times 10^6$
Worms	$5 \times 10^7 - 10^8$
Insects	$10^8 - 5 \times 10^9$
Echinoderms	$5 \times 10^8 - 5 \times 10^9$
Fish	$10^9 - 5 \times 10^{10}$
Amphibians	$5 \times 10^9 - 5 \times 10^{11}$
Birds	$5 \times 10^9 - 10^{10}$
Reptiles	$5 \times 10^9 - 5 \times 10^{10}$
Mammals	$5 \times 10^9 - 5 \times 10^{10}$
Flowering plants	$5 \times 10^9 - 10^{12}$

FIG. 17.11 Range of DNA content of haploid genomes of representatives groups of organisms

answered if a complete understanding of genome size evolution is to be achieved. It was endorsed in preference to C-value paradox at the 'Second Plant Genome Size Discussion Meeting and Workshop' held at the Royal Botanic Gardens, UK.

Repetitive DNA

The chromosomes of eukaryotes contain certain base sequences that are repeated many times in the haploid chromosome complement. DNA containing such repeated sequences, called repetitive DNA, represents a major component of the eukaryotic genome. In order to detect the presence of repetitive DNA in the genome, the DNA solution is heated resulting in the denaturation of the double stranded DNA into single stranded DNA. This is accompanied with increase in the optical density, a phenomenon described as hyperchromicity. In the next step, the solution of single stranded DNA is allowed to cool slowly so that the double stranded DNA forms again resulting in decrease in the optical density, i.e., hypochromicity. The amount of repetitive DNA is determined on the basis of extent of the formation of double stranded DNA in a definite period of time keeping concentration constant. The formation of double stranded DNA is actually measured over different values of *Cot*, that is, concentration × time. The patterns of reassociation in a number of organisms are shown by *Cot* curves.

DNA Renaturation Kinetics

The relationships among the rates of renaturation, complexity and repetition frequency in eukaryotes can be studied by DNA renaturation kinetics. If a long DNA molecule with no repeated sequences is broken into fragments of a particular length and denatured, and is then allowed to renature under appropriate conditions, the rate of renaturation will depend on (1) the concentration of DNA in solution and (2) the complexity of the DNA, that is, the number of different fragments of a particular length. Let us consider a particular fragment composed of two complementary strands A and B. Reassociation of A and B will require a specific collision between these two strands and a collision of either of the two strands with any other single strand will not lead to hybridization. For any given concentration, the larger the DNA molecule, the more non-identical base-pair fragments of a particular length are seen and hence it results in the slower reassociation reaction because of a smaller proportion of the random collisions between the complementary single strands such as A and B. Hence, the rate of renaturation should be a function of the square of the concentration of single strands, which follows the second order or bimolecular reaction kinetics.

The time required for half-renaturation is inversely proportional to the rate constant. Let C = concentration of single stranded DNA at time t (expressed as moles of nucleotides per litre). The rate of loss of single-stranded DNA during renaturation is given by the following expression for a second-order rate process:

$$\frac{-dC}{dt} = kC^2 \quad \text{or} \quad \frac{dC}{C^2} = -kdt$$

That is, the decrease in concentration of single stranded DNA ($-dC$) with time (dt) is equal to the proportionality constant (k) times the square of the concentration of single stranded DNA.

Integration from the initial conditions ($t = 0$ seconds and $C = C_0$, where C_0 equals the concentration of single stranded DNA at $t = 0$) shows:

$$\frac{C}{C_0} = \frac{1}{1 + kC_0^2}$$

Thus, at half-renaturation when

$$\frac{C}{C_0} = 0.5 \quad \text{and} \quad t = t_{1/2}$$

One obtains:

$$C_0 t_{1/2} = \frac{1}{k}$$

where k is the rate constant in litres (mole nt)$^{-1}$ sec^{-1}

The rate constant for renaturation is inversely proportional to sequence complexity. The rate constant k shows the following proportionality:

$$ka\frac{\sqrt{L}}{N}$$

where L = length; N = complexity

Empirically, the rate constant k has been measured as

$$K = 3 \times 10^5 \frac{\sqrt{L}}{N}$$

in 1.0 M Na$^+$ at T = T_m – 25°C

The time required for half-renaturation (and thus $C_0 t_{1/2}$) is directly proportional to sequence complexity

Thus, $$C_0 t \times a\frac{N}{\sqrt{L}}$$

For a renaturation measurement, one usually shears DNA to a constant fragment length L (e.g., 400 bp). Then L is no longer a variable, and

$$C_0 t \times aN$$

The data for renaturation of genomic DNA are plotted as **Cot curves**.

DNA renaturation kinetics experiments not only show what proportion of a genome consists of a particular class of repetitive DNA, but also how many copies of that particular sequence are present.

Satellite DNA

There is remarkable variability in genome size among eukaryotes that has little correlation with organismal complexity, ploidy or number of coding genes. For example, a newt has six times the genome size of a human. Much of this variation is due to non-coding, tandemly repeated DNA.

Indeed, a substantial fraction of the genomes of many eukaryotes is composed of repetitive DNA in which short sequences are tandemly repeated in small to huge arrays.

Tandemly repetitive sequences, commonly known as 'satellite DNAs' are classified into three major groups:

- **Satellites:** They are highly repetitive with repeat lengths of one to several thousand base pairs. These sequences are typically organized as large (up to 100 million bp) clusters in the heterochromatic regions of chromosomes, near centrosomes and telomeres; these are also found abundantly on the Y chromosome.

- **Minisatellites:** They are moderately repetitive, tandemly repeated arrays of moderately sized (9 to 100 bp, but usually about 15 bp) repeats, generally involving mean array lengths of 0.5 to 30 kb. They are found in euchromatic regions of the genome of vertebrates, fungi and plants and are highly variable in array size.

- **Microsatellites:** They are moderately repetitive, and composed of arrays of short (2-6 bp) repeats found in vertebrate, insect and plant genomes. The human genome contains at least 30,000 microsatellite loci located in euchromatin. Copy numbers are characteristically variable within a population, typically with mean array sizes on the order of 10 to 100.

In general, satellite DNAs show exceptional variability among individuals, particularly with regard to the number of repeats at a given locus. Minisatellite loci are the most highly polymorphic sequence elements yet discovered in the human genome, and delineating the repeat lengths of these loci is the basis of most DNA often consists of tandem highly repeated DNA.

The first evidence for repetitive DNA came from density-gradient analysis of eukaryotic DNA. In eukaryotes, since G-C base-pairs are denser than A-T base-pairs, the density of DNA increases as the %GC content increases. When such DNA is isolated, fragmented and centrifuged to equilibrium into a cesium chloride (CsCl) density gradient, the result is the presence of one large band of DNA (Mainband DNA) and one to several small bands. These small bands are called satellite bands and the DNA contained in these bands is called satellite DNA (Fig. 17.12).

Therefore, 'satellite' actually refers to any DNA population whose density differs from that of the bulk of DNA.

Human DNA has five satellites. I to IV have short reiterated sequences of less than 100 bp containing divergent GGAAT repeats.

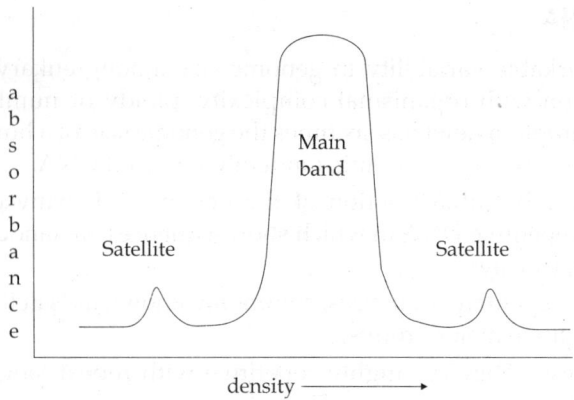

FIG. 17.12 Density configuration of embryonic DNA of *Drosophila melanogaster*, showing a main band and satellite bands

Two remarkable features of satellite DNA are (1) relative uniformity within the same species and (2) the great variability between otherwise closely related species.

The chromosomal locations of several satellite DNAs have been determined by a technique called *in situ* hybridization. It involves annealing single strands of isolated radioactive satellite DNA directly to denatured DNA in chromosome squash preparations. After washing out the non-hybridized radioactive material, the locations of the satellite DNA sequences in chromosomes are determined by autoradiography.

Satellite DNAs are usually in heterochromatic regions of the chromosomes often around centromeres and near telomeres and are not transcribed into RNA. Other kinetically distinguished classes of DNA are the moderately repetitive and unique DNA.

EUKARYOTIC CHROMOSOME STRUCTURE

The length of DNA in the nucleus is far greater than the size of the compartment in which it is contained. To fit into this compartment the DNA has to be condensed in some manner. The degree to which DNA is condensed is expressed as its packing ratio.

Packing ratio is the length of DNA divided by the length into which it is packaged. For example, the shortest human chromosome contains 4.6×10^7 bp of DNA (about 10 times the genome size of *E. coli*). This is equivalent to 14,000 μm of extended DNA. In its most condensed state during mitosis, the chromosome is about 2 μm long. This gives a packing ratio of 7000 (14,000/2).

To achieve the overall packing ratio, DNA is not packaged directly into final structure of chromatin. Instead, it contains several hierarchies of organization. The first level of packing is achieved by the winding of DNA around a protein core to produce a 'bead-like' structure called a nucleosome.

This gives a packing ratio of about 6. This structure is invariant in both the euchromatin and heterochromatin of all chromosomes. The second level of packing is the coiling of beads in a helical structure called the 30 nm fiber that is found in both interphase chromatin and mitotic chromosomes. This structure increases the packing ratio to about 40. The final packaging occurs when the fiber is organized in loops, scaffolds and domains that give a final packing ratio of about 1000 in interphase chromosomes and about 10,000 in mitotic chromosomes (Fig. 17.13 a).

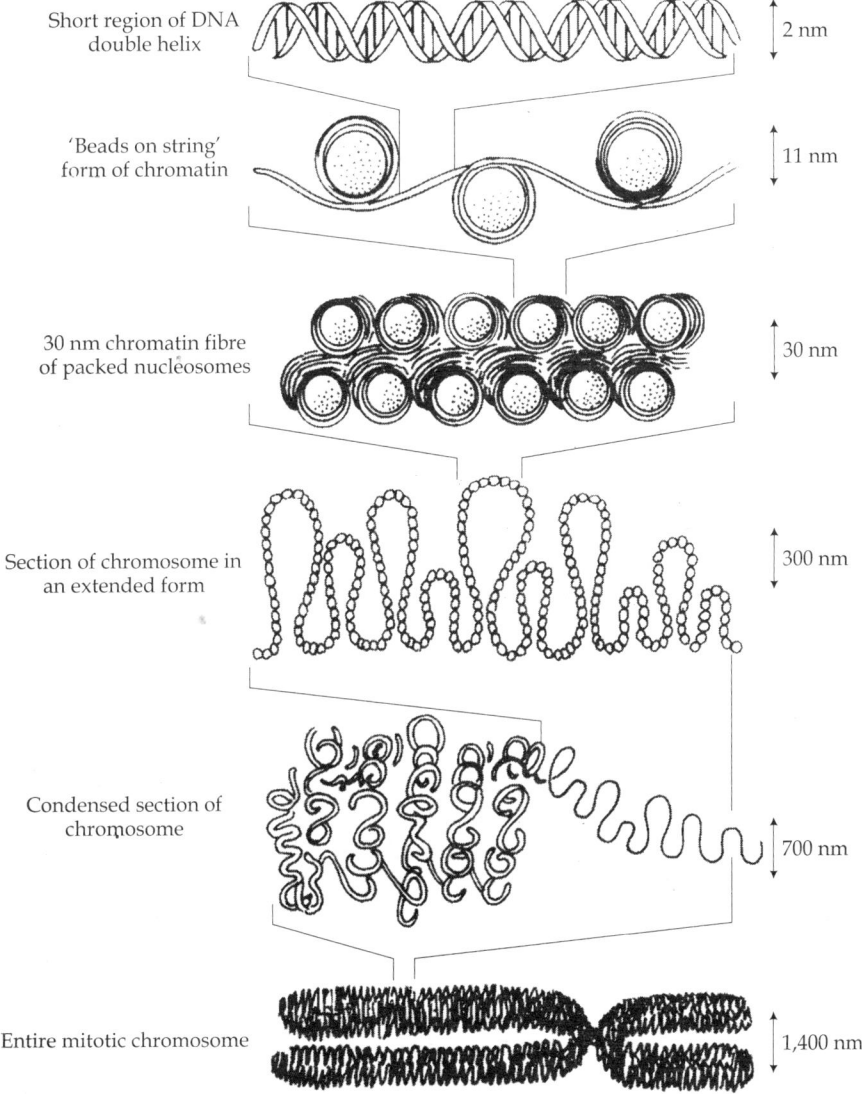

Short region of DNA double helix — 2 nm

'Beads on string' form of chromatin — 11 nm

30 nm chromatin fibre of packed nucleosomes — 30 nm

Section of chromosome in an extended form — 300 nm

Condensed section of chromosome — 700 nm

Entire mitotic chromosome — 1,400 nm

FIG. 17.13a Levels of packaging of DNA

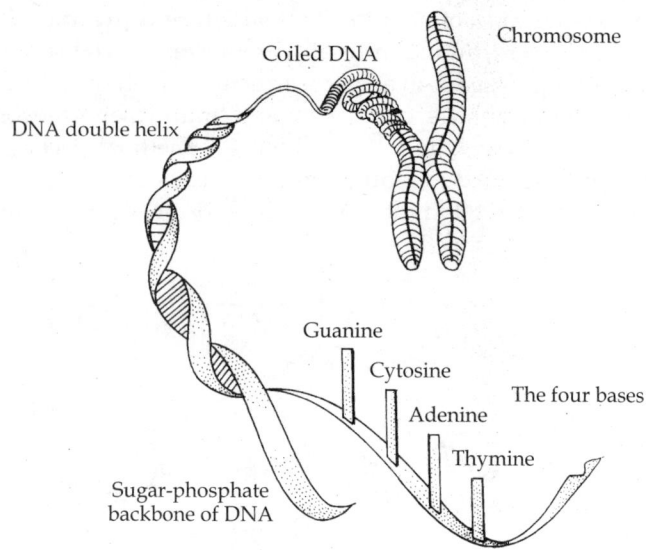

FIG. 17.13b Organization of Eukaryotic chromosome

Eukaryotic chromosomes consist of a DNA-protein complex that is organized in a compact manner which permits the large amount of DNA to be stored in the nucleus of the cell (Fig. 17.13 b). The subunit designation of the chromosome is chromatin. The fundamental unit of chromatin is the nucleosome.

NUCLEOSOME MODEL

Nucleosomes are the simplest packaging structure of all eukaryotic chromatin having fundamental repeating subunits. They package DNA into chromosomes inside the cell nucleus and control gene expression. They are made up of DNA and four pairs of proteins called histones, and resemble 'beads on a string' (Fig. 17.14). The nucleosome concept was proposed by Roger Kornberg in 1974.

The structure of nucleosome consists of 200 bp DNA wrapped around an octamer of small basic proteins called histones H2A, H2B, H3 and H4, all of which exist in two copies known as core histones. 146 bp of DNA is wrapped around the core and the remaining bases link to the next nucleosome called linker DNA; this structure causes negative supercoiling.

According to the crystal structure, the histone octamer likely interacts with the dsDNA around it roughly every 10 bp. Each of the four histone dimers contain three regions of interaction with the dsDNA. The central interaction site for each dimer is formed by an α-helix from each histone in the pair pointing at a single phosphate group on the dsDNA to which they hydrogen bond.

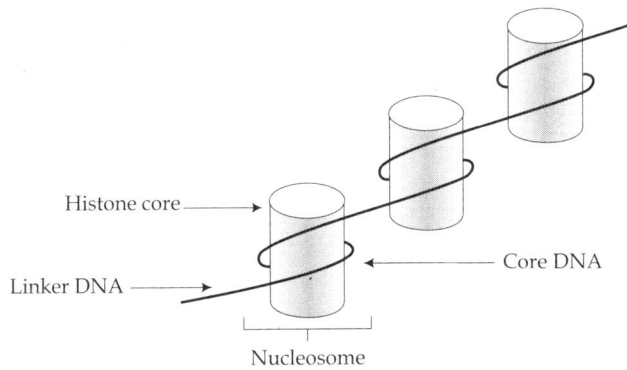

FIG. 17.14   Nucleosome model

Analysis of the structure of dsDNA wrapped around the histone octamer suggests that it is predominantly B-form, although more tightly constrained than free DNA due to its interaction with the octamer.

SOLENOID MODEL

In the condensed form, nucleosomes are packed into a 30 nm fiber with about 6 nucleosomes per turn. A string of nucleosomes is coiled into a solenoid configuration by the fifth histone, called H1. One molecule of H1 binds to the site at which DNA enters and leaves each nucleosome, and a chain of H1 molecules coils the string of nucleosomes into the solenoid structure of the chromatin fibre.

Nucleosomes not only neutralize the charges of DNA, but they have other consequences. First, they are an efficient means of packaging. DNA becomes compacted by a factor of six when wound into nucleosomes and by a factor of about 40 when the nucleosomes are coiled into a solenoid chromatin fibre. The winding into nucleosomes also allows some inactive DNA to be folded away in inaccessible conformations, a process that probably contributes to the selectivity of gene expression.

The final level of packaging is characterized by the 700 nm structure seen in the metaphase chromosome. The condensed piece of chromatin has a characteristic scaffolding structure that can be detected in metaphase chromosomes. This appears to be the result of extensive looping of the DNA in the chromosome.

GIANT CHROMOSOMES

In certain organisms some cells at particular stages contain large nuclei with giant or large-sized chromosomes. Such giant chromosomes are of two types, viz., Polytene chromosomes and Lampbrush chromosomes.

The polytene chromosomes are giant multistranded chromosomes commonly occurring in many dipteran flies. The enormous size of the polytene chromosomes is achieved by the process of endoreduplication in which the repeated rounds of DNA replication occur without any cell division. Another characteristic feature of these chromosomes is the occurrence of somatic synapsis as a result of which their number always appears to be half as compared to that in other normal somatic cells.

They have a distinct pattern of transverse bands and interbands along their length. The bands become enlarged at certain points forming chromosome puffs or Balbiani rings which are associated with differential gene activation. The polytene chromosomes are used as a good model to examine the structural aberrations in chromosomes due to their large size and characteristic banding pattern (Fig. 17.15).

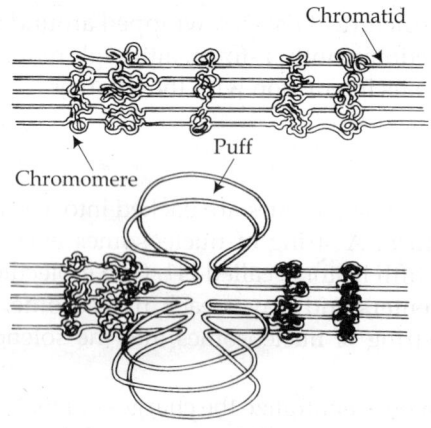

FIG. 17.15 Structure of Polytene Chromosome

The lampbrush chromosomes are meiotic or 'germ-line' chromosomes, present during prophase I of oogenesis in many vertebrates particularly amphibians. Due to their large size they provide favourable material for cytological studies. Each lampbrush chromosome consists of a central axial region composed of two bivalent chromosomes and numerous pairs of lateral loops that project from the chromomeres present along the length of the axial filaments. The main axis is rich in DNA and protein whereas the loops mainly consist of RNA. The lampbrush chromosomes are associated with synthetic activity in the growing oocyte (Fig. 17.23).

POLYTENE CHROMOSOMES

The polytene chromosomes were observed for the first time by E.G. Balbiani in 1881 in the salivary gland nuclei of midges of the genus *Chironomus*. These structures were also studied separately by Korschelt and Carnoy in 1884, but the early workers did not attach any genetic significance to their observations.

Later, various cytologists like Heitz and Bauer (1933), Painter (1933) and King and Beams (1934) worked out the cytological importance of these chromosomes.

Polytene chromosomes although widespread in the order Diptera, they have also been observed in several species of protozoans and plants. In order Diptera, besides occurring in salivary glands, they are also present in several other tissues such as Malphigian tubules, gut, hepatic cecae and muscle cells.

The term 'Polytene chromosome' was coined by Kollar in 1935 which is suggestive of the presence of many chromonemal fibrils in them.

The polytene chromosomes are larger in size in comparison to other somatic chromosomes. In *Drosophila melanogaster*, these chromosomes are over 100 times the length of somatic metaphase chromosomes. The work of Heitz and Bauer showed that these chromosomes exist in haploid number, each being formed by intimate side-by-side pairing of two homologous chromosomes as a result of somatic pairing phenomenon.

Structure

From structural point of view polytene chromosomes can be defined as a special case of chromosomes, present during interphase, characterized by (i) a *high content of DNA* and (ii) *particular chromatin structures* related to their high DNA contents. The large size of polytene chromosomes is attributed to the type of growth that occurs in the larval glandular tissues of dipterous insects. Salivaries and other glands grow by enlargement rather than by division. As a larva develops, the synapsed elements replicate nine to ten times giving rise to a polytene strand of about 1000 DNA double helix molecules. The process of recurrent duplication cycle without consequent mitosis is called endoreduplication. The probable function of polyteny is suggested to be gene amplification. Multiple copies of genes permits a high level of gene expression; that is, abundant transcription and translation to produce the gene products in large, metabolically active cells (like salivary glands). For unknown reasons, the centromeric regions of the chromosomes do not endoreplicate very well. As a result, the centromeres of all the chromosomes bundle together in a mass called the chromocenter.

The polytene chromosomes bear alternate transverse dark and light regions along their length. The dark bands observed in the polytene chromosomes are chromomeres, densely packed with chromatin, while the lighter interband regions are chromonemata made up by chromatin that is packed much more loosely (Fig. 17.15). The bands are the regions of high DNA concentration, Feulgen positive, stain intensely with basic stains and absorb ultraviolet light at 2,600Å. The interbands, on the other hand, are the regions of low DNA concentration, Feulgen negative, do not stain with basic dyes and absorb very little ultraviolet light. The polytene chromosomes of plants, however, do rarely display a clear banding pattern. A possible explanation is that chromomeres in this case are not lined up exactly in parallel and as a consequence the denser chromatin parts do not lie side-by-side to form a clear-cut banding pattern.

The work of C.B. Bridges (1936) showed that the bands of the polytene chromosomes are related to genes and the interbands are relatively inert linker regions. It is now believed that the interbands also contain many genes, though chromomeres contain much folded or spiralized DNA while in interband regions the DNA is in extended state. Du Praw and Rae (1961) suggested that a gene may overlap from band to interband or may even lie entirely in the interband. Polytene chromosomes thus correspond in linear structure with other chromosomes of the same species. The difference being that the polytene chromosomes are formed of about 1000 to 4000 unit chromatids which do not separate out to new cells through cell division. About 5,000 bands have been observed on the four chromosomes of *Drosophila melanogaster*. The banding pattern is distinctive for each homologous pair of chromosomes in any given species (Fig. 17.16).

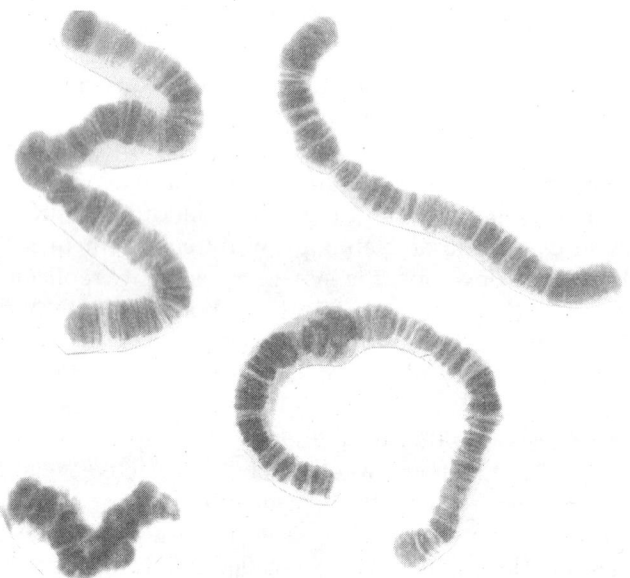

FIG. 17.16 Polytene chromosome complement of *Chironomus*

Puffs and Gene Activity

At certain stages in development, some genetic loci in the polytene chromosomes become enlarged into swellings or puffs which have been called Chromosome puffs or Balbiani rings. The process of puff formation has been studied in detail by Beerman and Bahr (1954) and it has been suggested that puffing involves unfolding or unraveling of DNA in the chromomeres of a particular band in the form of a loop.

The process of puffing is due to the biosynthetic activity of the particular loci concerned and hence plays an important role in the understanding of gene action. Thus, puffs are the active genes and represent the sites of RNA synthesis.

Majority of puffs are visible only at certain times of the development of an organism and appear at different positions on the polytene chromosomes. This is explained by the fact that at different times during the development different genes are active. Furthermore, the activity of genes can be different from one tissue to another. For example, the three huge Balbiani rings in the fourth chromosome of *Chironomus tentans* are only seen in the salivary nuclei and not at all in other tissues.

It has been shown by a number of workers that numerous puffs can be experimentally induced at an earlier developmental stage than that at which they would normally occur by administration of the insect steroid moulting hormone ecdysone. The degree of puffing and its duration depend on the concentration of the hormone. A number of other environmental conditions such as heat treatment (Barettino *et al.*, 1988) and galactose treatment (Diez *et al.*, 1990) are also capable of inducing specific puffs.

The phenomenon of puffing is associated with the synthesis of RNA has been demonstrated by Pelling in 1966 by tritiated uridine autoradiography. It has been suggested that the RNA formed in the puff is messenger RNA (mRNA). This mRNA is transferred to the cytoplasm as ribonucleoprotein and acts as a template for the synthesis of special proteins. The mRNA formed in one puff varies from the one made in another puff.

Importance in Cytogenetics

Polytene chromosomes have been extremely important in cytogenetics due to their large size and characteristic banding pattern. In the normal somatic chromosomes the identification of the chromosomal aberrations simply by observation is a major problem due to lack of visible markers for identifying different areas along the length of the chromosome. In the giant polytene chromosomes such studies are greatly facilitated due to the presence of alternate dark and light bands. They provide a useful tool for comparing the individual chromosomes within a chromosome set or genome, as well as whole sets from different organisms. Comparison of banding sequences has been of great significance in the analysis of evolutionary cytogenetic processes. Also, the chromosomal rearrangements such as deletions, duplications, inversions and translocations can be studied in detail since any structural rearrangement is expressed in corresponding alteration in somatic synapsis.

Deletions: A single break near the end of the chromosome results in a terminal deficiency. In case two breaks occur, a section may be deleted and an intercalary deficiency is created. In this case a loop like structure is observed in the polytene chromosomes due to alteration in the pairing of two homologues.

By identifying the part of the polytene chromosome in which the formation of such a loop like structure is seen and then studying the phenotype of the organism carrying a recessive gene in the homologous chromosome opposite the deletion or deficiency, the gene can be spatially positioned on the chromosome. The physical location of many genes is now precisely known in *Drosophila melanogaster* and some other species of Diptera because of this technique (Fig. 17.17).

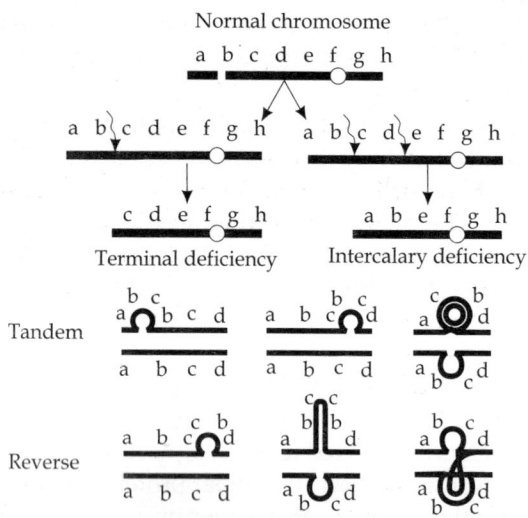

FIG. 17.17 Deletion

Duplications: Duplications occur due to the addition of chromosome parts. If duplication occurs only on one of the two homologues, the cytological condition comparable to that of deficiency will be obtained at meiosis (Fig. 17.18).

In *Drosophila* the first case of duplication to be critically analyzed was the one involving the B (bar) locus in the X-chromosome. Bar eye is a character where eyes are narrower compared to the normal eye shape and this condition appears due to the duplication in the region 16A in the X chromosome. The wild type eye of *Drosophila* is large and has an average of 779 facets. It has two Bar regions on each X chromosome. If this region undergoes duplication, the size of the eye is reduced and number of facets decreases. Though, there is no change in the amount of genetic material but the order of the genes gets altered. This gave rise to the concept of Position effect. Barred individuals give rise to double bar and ultra bar conditions due to unequal crossing over. This was suggested by the fact that this type of condition occurred only in females (Fig. 17.19).

Inversions: An inversion occurs when a part of the chromosome becomes detached and reunites in reverse order after rotating through 180°. Inversions are called pericentric when the segment includes the centromere and

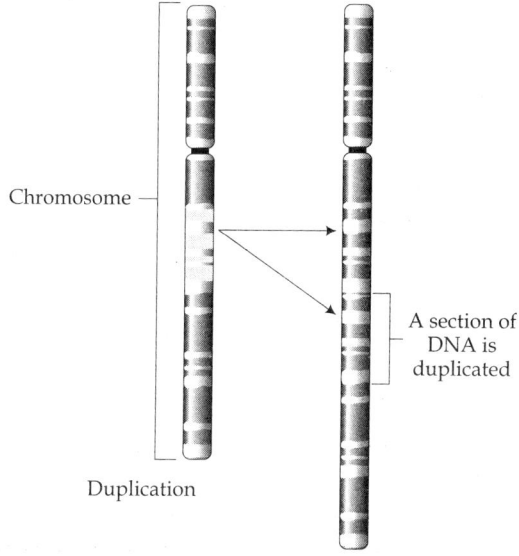

Chromosome

A section of
DNA is
duplicated

Duplication

FIG. 17.18 Duplication

Chromosome Structure	Phenotype
X-16A X-16A	Normal Eye Female
X-16A-16A X-16A	Bar Eye female
X-16A-16A X-16A-16A	Bar Eye Female (more restricted, 68 facets)
X-16A-16A-16A X-16A	Double Bar Eye Female (45 facets)
X-16A-16A-16A X-16A-16A-16A	Double Bar Eye Female (further reduced)

FIG. 17.19 Effects of different arrangements of duplicated sections of 16A in the
X-chromosome of *Drosophila melanogaster* on size of the eye

paracentric if the centromere is located outside the inverted segment. When
crossing over occurs within the inverted segment of a paracentric inversion, it
leads to the formation of dicentric and acentric chromatids during meiosis.
The dicentric chromatids have two centromeres and are connected together by
a bridge (Fig. 17.20). If, on the other hand, crossing over occurs within the loop
of a pericentric inversion then the resulting chromatids with duplication and
deficiency are formed (Fig. 17.21).

Inversions are known to have played a significant role in evolution of
different species and races of dipterans. It has been found that this condition is

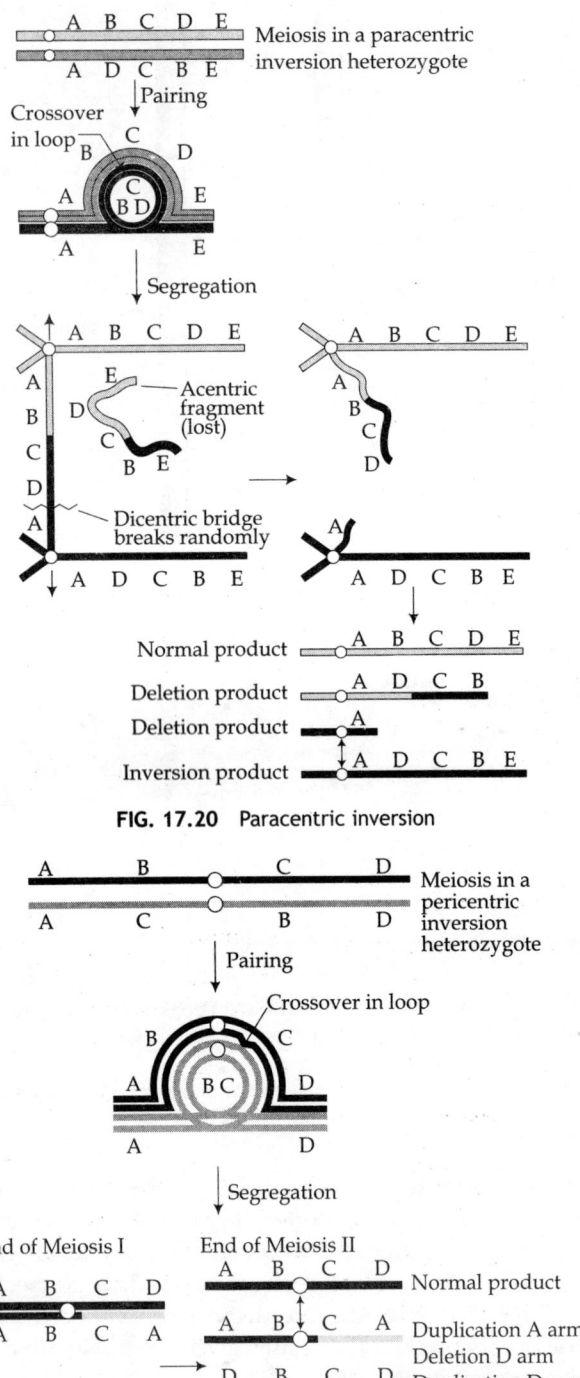

FIG. 17.20 Paracentric inversion

FIG. 17.21 Pericentric inversion

maintained in nature by balancing selection due to selective advantage of inversion heterozygotes. As a result, many inversions have become part of the stable chromosome system in *Chironomus* (Beerman, 1953; Martin, 1974). In *Drosophila* it has been seen that inversions occurred spontaneously in nature and became established in populations due to the adaptive value. These inversions are restricted to different localities and the different races actually derived their names after these localities.

Translocations: In translocation, a segment of a chromosome becomes attached to a non-homologous chromosome. In reciprocal translocation, two non-homologous chromosomes exchange segments. Translocations occur due to irregularities during crossing over. They may not involve a loss or an addition of chromosome material, but frequently they do become associated with deficiencies, duplications and unbalanced combinations of genetic units (Fig. 17.22).

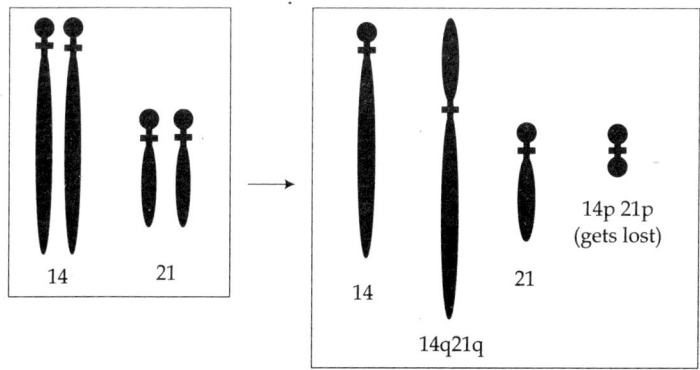

FIG. 17.22 Robertsonian translocation

LAMPBRUSH CHROMOSOMES

Lampbrush chromosomes are transitory structures that exist during an extended diplotene of the first meiotic division in female gametocytes of most animals, except mammals. They are found in the oocyte nuclei of vertebrates like sharks, amphibians, reptiles and birds and in invertebrates like *Sagitta, Sepia, Echinaster* and in several species of insects. The name lampbrush is given to these giant chromosomes due to their resemblance to the brushes used to clean old-fashioned oil lamp chimneys (Fig. 17.23).

Lampbrush chromosomes were first seen in sections of salamander (*Ambystoma mexicanum*) oocytes by Flemming in 1882. Ten years later, in 1892, they were described in the oocytes of a dogfish by J. Ruckert. They have also been reported in plants (Grun, 1958).

Structure

Since, lampbrush chromosomes exist in diplotene stage of meiosis, they are bivalents, i.e., they consist of two homologues (two pairs of sister chromatids)

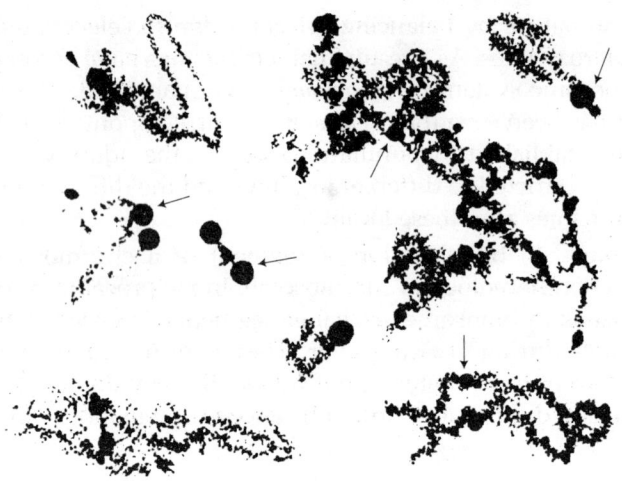

FIG. 17.23 Lampbrush chromosomes, arrows indicate unstained protein bodies

held together by chiasmata at one or more points. They are about 800 µm long. At maximum development, they are even larger than the largest salivary gland chromosomes (Fig. 17.24).

Chromosomal axis: Lampbrush chromosomes are generally described as having a central chromosomal axis formed of four chromatids. The axis varies from 30-50 Å in diameter and is extensible and elastic, composed of DNA and protein.

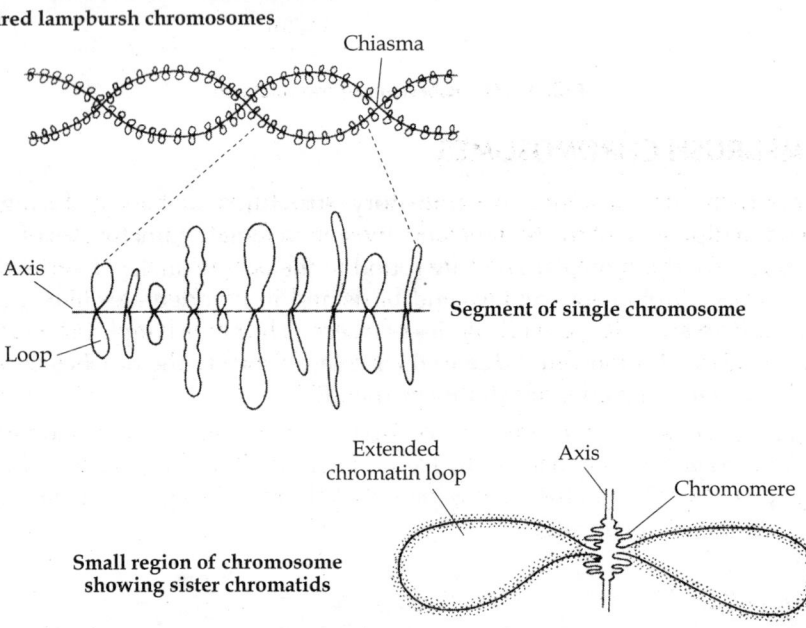

FIG. 17.24 Structure of lampbrush chromosome

Chromomeres: The chromosomal axis is dotted with a row of Feulgen-positive granules called chromomeres which are about 1 μm in diameter and 1-5 μm in length, connected by an extremely thin continuous chromonema. In addition to DNA, the chromomeres contain basic protein (Gall, 1954). These chromomeres are found in pairs, one for each homologue.

Loops: Most of the chromomeres bear pairs of lateral loops. It is the presence of hundreds of loops arising from each bivalent which give the chromosome the characteristic hairy 'lampbrush' appearance. Members of each pair of loops are similar in length and morphology but loops arising from different non-homologous chromomeres differ in length and appearance. The size of the loops varies from 5 μm to 100 μm in length. Loops are of two types: (a) Typical and (b) Special.

Each typical loop consists of a central axis which gives off RNA fibrils all along its length. The RNA fibrils on this axis increase progressively in length. This makes the loop markedly thicker on one side. The special loops bear granules at the ends of these RNA fibrils.

One to nine loops may arise from a single chromomere. Each loop is thought to be composed of a loop axis which is covered with matrix. The loop axis is composed of DNA about 30-50 Å in thickness whereas, matrix has protein combined with RNA. The loops appear to be Feulgen-negative in stained preparations. The difference between the individual loops depends on the amount and kind of matrix.

As meiosis proceeds towards metaphase I, the loops undergo regression and are eventually reabsorbed into the chromosomes which are finally converted into normal compact meiotic bivalents.

Master and Slave Hypothesis

Callan and Lloyd in 1960 postulated the 'master and slave hypothesis'. According to this, each loop represents a long operon, consisting of a series of duplicate copies of the same structural gene (cistrons). Each gene locus produces an RNA molecule which interacts with protein to form ribonucleoprotein particles. There is a master copy of a particular gene in the chromomere which transfers information to several identical gene copies on the same loop (slave genes). Between the gene loci matrix free regions of 'spacer' DNA are present which do not produce RNA. It has been observed that only the slave genes take part in RNA synthesis. A high rate of synthesis is possible because of the presence of repeating similar genes.

Functional Significance

Lampbrush chromosomes are thought to assist in fulfilling the high demand for transcripts during oogenesis. Studies on oogenesis in vertebrates reveal that extensive transcription occurs during prophase I (diplotene) of meiosis. The loops of the lampbrush chromosomes have been shown to be the regions of active transcription by means of pulse-labeling with [^{3}H] uridine and autoradiography. The transcribed genes are those whose products are required during early stages of embryogenesis.

OBJECTIVE TYPE QUESTIONS

1. A human being has _____ autosomes and _____ sex chromosomes
   A. 23, 1   B. 22 pairs, 1 pair
   C. 23, 23   D. 2, 2
2. A section of chromosomes that codes for a trait can be called a
   A. Nucleotide   B. Base-pair
   C. Gene   D. Nucleus
3. Somatic cells of a human have _____ chromosomes and are called
   _____ .
   A. 10, haploid   B. 92, diploid
   C. 23, haploid   D. 46, diploid
4. Each chromosome consists of two identical
   A. Genes   B. Nuclei
   C. Chromatids   D. Bases
5. If a person receives an X and a Y chromosome, that person is
   A. Female   B. Male
   C. Red eyed   D. Mentally challenged
6. Two alleles for pea plant height are designated T (tall) and t (dwarf). These alleles are found on
   A. Genes   B. Sex chromosomes
   C. Ribosomes   D. Homologous chromosomes
7. An animal has 40 chromosomes in its gametes, how many chromosomes would you expect to find in this animal's brain cells?
   A. 1   B. 20
   C. 40   D. 80
8. A picture of a person's chromosomes is called a
   A. Karyotype   B. Syndrome
   C. Chromatin   D. Fingerprint
9. The histone protein that joins with H2B to form one tetramer subunit of a nucleosome is
   A. H4   B. H2A
   C. H1   D. H3

10. The histone protein that joins with H3 to form another tetramer subunit of a nucleosome is

    A. H2B   B. H2A

    C. H1   D. H4

11. The histone protein that attaches to DNA strands between nucleosomes is

    A. H3   B. H1

    C. H4   D. H2A

12. Chromosomes must condense to approximately 1/500th of their length for cell division. The first reduction is

    A. forming a coiled solenoid fiber

    B. coiling around nucleosomes

    C. looping of solenoid fibers to form a 300 nm fiber

    D. Looping of 300 nm fibers

13. Data are consistent with a model that shows nucleosomes packing together to form a coil. This is called a

    A. 300 nm fibre   B. Chromatid

    C. Solenoid   D. Chromosome

14. If a small chromosome consists of 500,000,000 nucleotides, how many nucleosomes will attach to that chromosome?

    A. 2,500,000   B. 500,000,000

    C. 250,000,000   D. none

15. Are histone proteins the only proteins associated with chromosomes?

    A. Depends on the life cycle   B. Yes

    C. No

Answers

1. B	2. C	3. D	4. C	5. B
6. A	7. D	8. A	9. B	10. D
11. B	12. B	13. C	14. A	15. B

SHORT ANSWER TYPE QUESTIONS

1. What is heterochromatin? Write its significance?

2. What is karyotype? Why is it prepared?

3. How many sex chromatin bodies are expected to occur in cell nuclei with each of the following chromosome arrangements: 1. XY 2. XXY 3. XXXX 4. XXYY 5. XX

4. What roles are played by centromeres and telomeres?

5. What is chromatin? Discuss its major structural components?

LONG ANSWER TYPE QUESTIONS

1. Clarify the difference between genes, chromosomes and DNA.
2. As genome size increases in eukaryotes, the percentage of the genome composed of genes becomes smaller. Comment.
3. Describe the successive levels of compaction that must occur to compress an extended double helical DNA sufficiently to fit in a condensed mitotic chromosome.
4. What is a polytene chromosome? What importance does it have in the study of genetics?
5. Distinguish between the following pairs.
    a. Repititive and unique sequence DNA
    b. Heterochromatin and euchromatin
    c. Acrocentric and telocentric chromosomes.
    d. Kinetochore and centromere
    e. Karyotype and genome

Cell Reproduction: Cell Growth and Cell Cycle

18

INTRODUCTION

Cell growth refers to increase in size of a developing embryo. It is often used in the context of reproduction of living cells and is considered as growth in cell populations by means of cell reproduction. In other contexts, cell growth refers to increase in the cell size.

The cell cycle is the series of events in a eukaryotic cell between one cell division and the next. It consists of four distinct phases: G_1 phase, S phase, G_2 phase and M phase. G_1, S and G_2 phase are collectively known as interphase. M phase is further composed of two tightly coupled processes, viz., mitosis, in which the cell's chromosomes are divided between the two daughter cells, and cytokinesis, in which the cytoplasm of the cell divides. Cells that have temporarily or reversibly stopped dividing are said to have entered a state of quiescence called G_0 phase, while cells that have permanently stopped dividing due to age or accumulated DNA damage are said to be senescent.

HISTORY

The history of cell division dates back to 1846 when K. Nageli suggested that plant cells arise from the division of pre-existing cells. Later, Rudolf Virchow in 1855 confimed Nageli's principle and stated in Latin, "*Omnis cellula e cellula*" meaning cells arise only from pre-existing cells. E. Strasburger in 1875 described mitosis in plant cells. The term 'mitosis' was coined by W. Flemming in 1882. J. B. Farmer and J. E. Moore in 1905 coined the term 'meiosis' for the reductional cell division.

CELL CYCLE

The normal cell cycle consists of 3 major stages. The first is interphase, during which the cell lives and grows larger; the second is mitosis, when the cell divides and the final is cytokinesis, which helps the two daughter cells to complete their separation. However, the cell cycle is regulated in several ways. Some cells divide rapidly (for example, beans take 19 hours for the complete cycle; red blood cells must divide at a rate of 2.5 million per second). Others, such as nerve cells, lose their capability to divide once they reach maturity, liver cells do not normally utilize their capacity for division, and they divide only if part of the liver is removed. This division continues until the liver reaches its former size. Cancer cells however, undergo a series of rapid divisions such that the daughter cells divide before they have reached functional maturity. Environmental factors such as changes in temperature and pH, and declining nutrient levels lead to declining cell division rates. When cells stop dividing, they stop usually at a point late in the G1 phase known as the R point (for restriction).

Interphase

Throughout the interphase, the cell carries out its normal metabolic activities and is actively engaging in transcription and translation of its genome (Fig. 18.1).

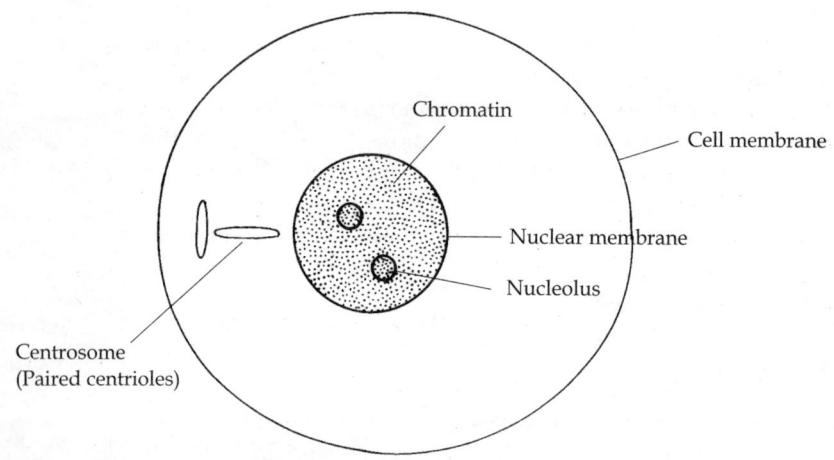

FIG. 18.1 Interphase

G₁ phase is the first growth phase. During this phase, the cell carries on its usual metabolic activities while preparing to duplicate its DNA. These preparations often include growing by increasing the amount of cytoplasm and the number of important organelles such as mitochondria.

S phase or synthesis phase is the stage during which the DNA is replicated, where S stands for the Synthesis of DNA.

G₂ phase is the second growth phase, in this phase the cell continues its growth and metabolism in preparation for undergoing mitosis.

M phase or mitotic phase is the stage that facilitates the equal partitioning of replicated chromosomes into two identical groups. Before partitioning can occur, the chromosomes must become aligned so that the separation process can occur in an orderly fashion. The alignment of replicated chromosomes and their separation into two groups is a process that can be observed in virtually all eukaryotic cells (Fig. 18.2)..

MITOSIS

The process of mitosis is divided into four stages, viz., prophase, metaphase, anaphase and telophase.

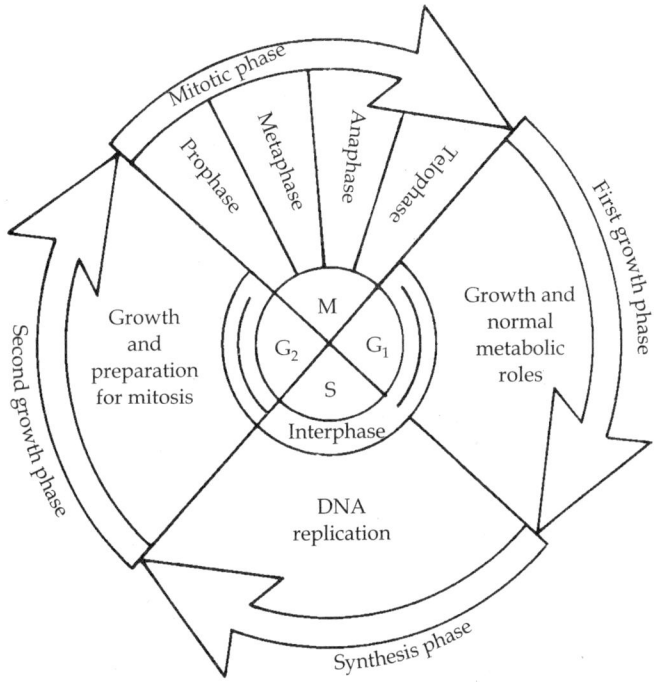

FIG. 18.2 The cell cycle

Prophase (Fig. 18.3)

Prophase is the first stage of mitosis in which chromatin condenses (remember that chromatin/DNA replicates during interphase) and undergoes extensive condensation (*i.e.,* coiling). The chromosomes are greatly thickened and shortened but are still contained within the nuclear envelope. During late prophase, the nuclear envelope dissolves, nucleolus disappears; the centrioles divide and migrate to opposite poles, the kinetochore fibers attach to the kinetochores of centromeres and the spindle apparatus forms.

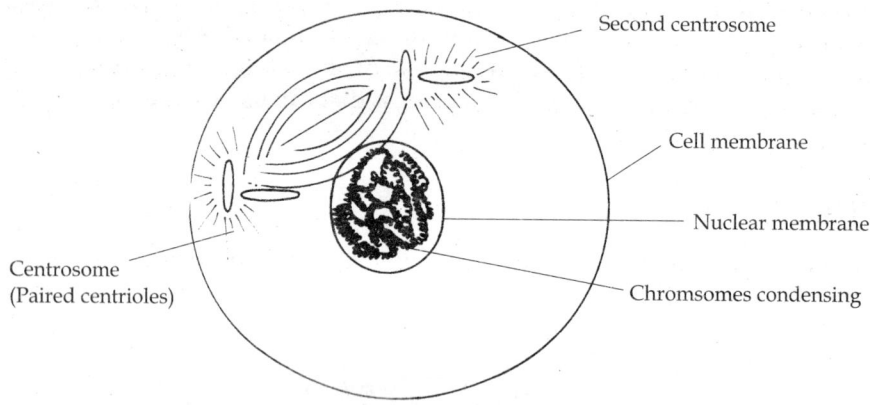

Second centrosome

Cell membrane

Nuclear membrane

Centrosome
(Paired centrioles)

Chromsomes condensing

FIG. 18.3 Prophase

Metaphase (Fig. 18.4)

Metaphase follows prophase. Once the nuclear envelope has broken down, the spindle microtubules and the chromosomes are no longer separated by a (double) membrane boundary. The microtubules begin to interact with the

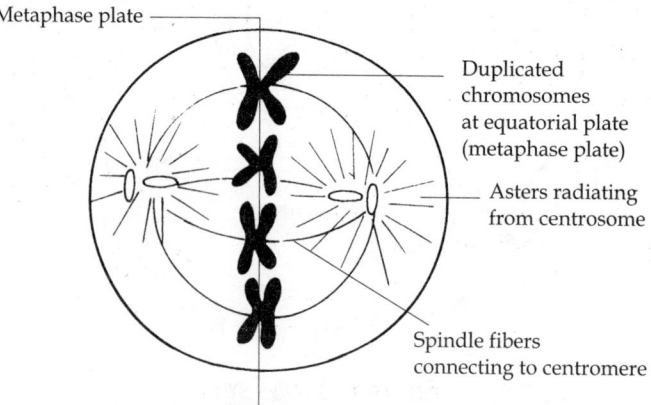

Metaphase plate

Duplicated
chromosomes
at equatorial plate
(metaphase plate)

Asters radiating
from centrosome

Spindle fibers
connecting to centromere

FIG. 18.4 Metaphase

chromosomes, and the chromosomes undergo what is known as congressional movement, where they ultimately end up with their centromeres all situated in middle of the spindle, at a site known as the metaphase plate. Each kinetochore of the replicated chromosome is pointed toward one side of the spindle. The replicated chromosomes converge toward the center of the spindle, and once they get there, significant movements cease. At several points during metaphase, the chromatid arms may move away from each other (Fig. 18.5).

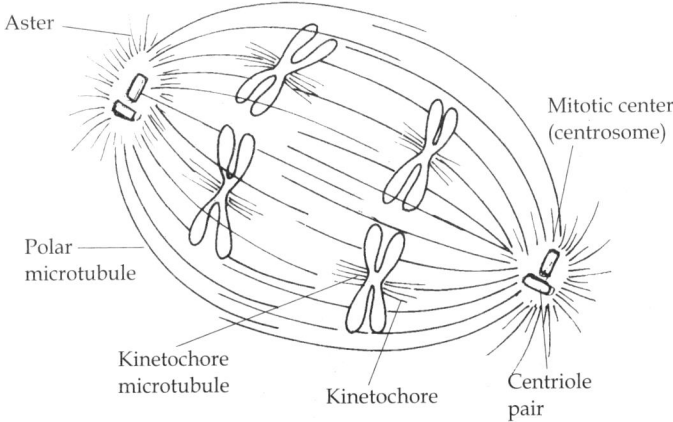

FIG. 18.5 Structure and main features of the spindle apparatus

Anaphase (Fig. 18.6)

Anaphase starts with the splitting of sister chromatids at their centromeres. These daughter chromosomes then begin to separate from each other, each

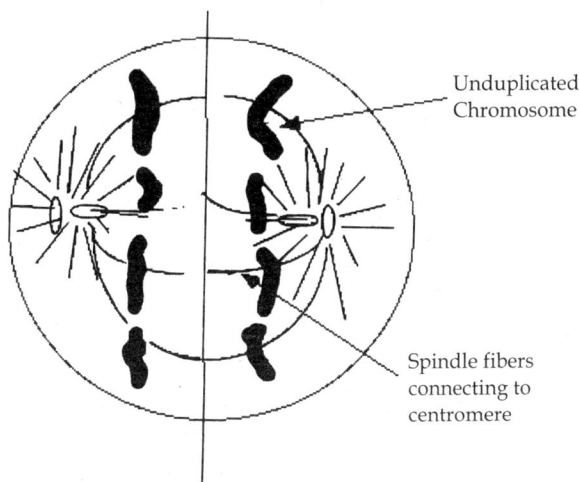

FIG. 18.6 Anaphase

moving away from the metaphase plate and toward one of the two spindle pole regions. The mechanisms that control chromosome separation clearly involve the interactions between microtubules and components in or near the kinetochore. Anaphase begins with the separation of the centromeres, and the pulling of chromosomes (we call them chromosomes after the centromeres are separated) to opposite poles of the spindle.

Telophase (Fig. 18.7)

Telophase occurs when the chromosomes reach the poles of their respective spindles, the nuclear envelope appears again, chromosomes uncoil into chromatin form, and the nucleolus (which had disappeared during prophase) reappears. Where there was one cell there are now two smaller cells each with exactly the same genetic information. These cells may then develop into different adult forms via the processes of development.

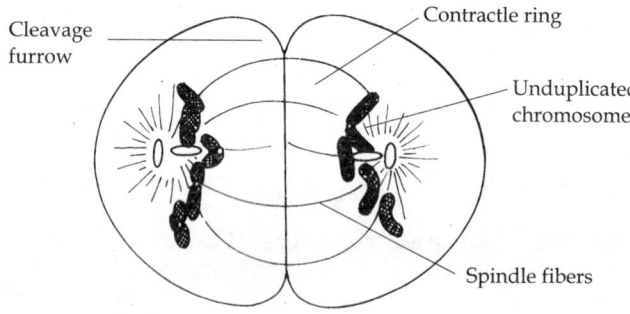

FIG. 18.7 Telophase

Cytokinesis

Cytokinesis is the process of splitting the daughter cells apart. Whereas mitosis is the division of the nucleus, cytokinesis is the splitting of the cytoplasm and allocation of the organelles like Golgi bodies, mitochondria, plastids and cytoplasm into each new cell.

SUMMARY OF MITOSIS (Fig. 18.8)

1. Mitosis results in the formation of two daughter cells with identical nuclei and chromosomes.
2. The stages in the division of the nucleus are called karyokinesis (karyon = nucleus and kinesis = splitting or dividing).
3. The division of the cytoplasm is called cytokinesis.
4. At the end of mitosis, the two new daughter cells are half the size of the original parent cell, which have to grow to reach the size of the parent cell.

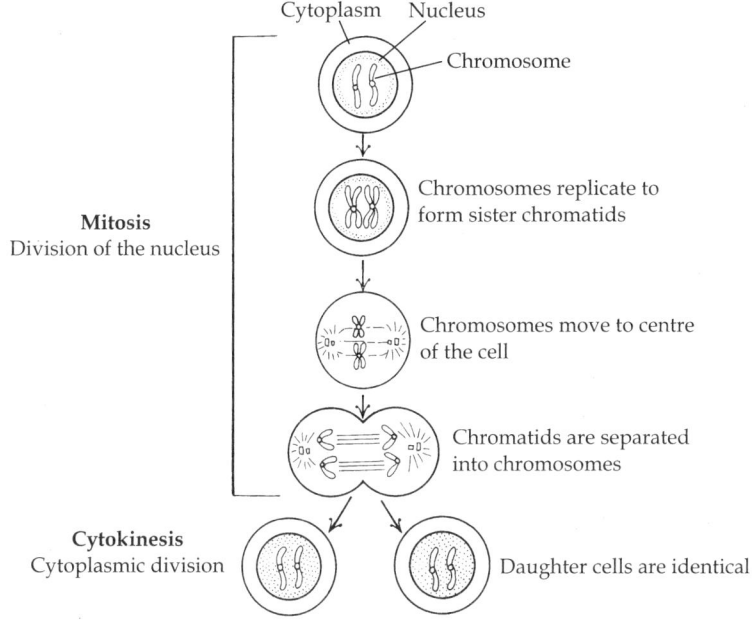

Cytoplasm   Nucleus

Chromosome

Mitosis
Division of the nucleus

Chromosomes replicate to form sister chromatids

Chromosomes move to centre of the cell

Chromatids are separated into chromosomes

Cytokinesis
Cytoplasmic division

Daughter cells are identical

FIG. 18.8   Stages of mitosis

SIGNIFICANCE OF MITOSIS

1. The significance of mitosis is that it involves duplication of the genetic material and its equal distribution to each of two daughter cells. There is little or no variation.

2. The process provides the cells needed for growth. An example would be a zygote developing into a functioning multicellular organism. For this to occur, there must be an increase in the number of cells from one cell to millions of cells in a human body.

3. Mitosis supplies the cells for repair of worn or damaged tissues, e.g., in the human alimentary canal, the skin, lung lining and blood cells.

4. It maintains the chromosome number. Daughter cells have an identical set of chromosomes compared to the parent. This set of chromosomes function together as part of a tissue or an organ or an organism.

5. Asexual reproduction produces offspring that are genetically similar to the parent.

MEIOSIS

Meiosis or the reductional division is the process of formation of germ cells. The germ cells produced are haploid; the maternal and paternal germ cells fuse during fertilization and thus generate a diploid fusion product, the zygote. It represents nature's solution of the problem of chromosome doubling that would occur, if two diploid cells, i.e., two cells with a double set of chromosomes would fuse. Therefore, meiosis is a special type of nuclear division which segregates one copy of each homologous chromosome into each new gamete. It reduces the number of sets of chromosomes by half, so that when gametic recombination (fertilization) occurs the chromosome count of the parents is reestablished.

The process of meiosis consists of two successive nuclear divisions, Meiosis I and Meiosis II. At the end of meiosis 4 haploid cells are produced. Meiosis I reduces the ploidy level from 2n to n (reduction) while Meiosis II divides the remaining set of chromosomes in a mitosis-like process (division). Hence, the process is named as reductional division.

Meiosis I: It starts after an interphase which is similar to the interphase of mitosis (Fig. 18.1) and comprises of the following stages:

Prophase I (Fig. 18.9)

It is the longest stage of meiosis and includes five substages (Fig. 18.10):

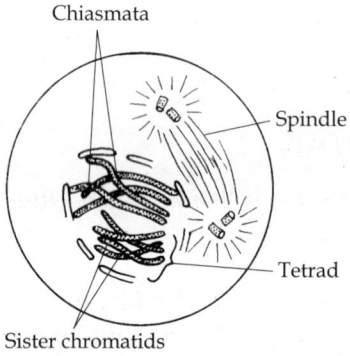

FIG. 18.9   Prophase I

Leptotene or leptonema (Gr., *leptas*=thin; *nema*= thread):   In this stage the chromosomes appear as long, thin, thread like structures with linear series of darkly stained bead like structures called chromomeres. At this stage, the chromosomes are oriented in such a way inside the nucleus that they converge towards one side forming the shape of a bouquet (bouquet stage). The centriole duplicates and each daughter centriole starts moving toward the opposite poles of the cell.

Zygotene or zygonema (Gr., *zygon*=to pair; *nema*=thread):   During this phase the pairing of homologous chromosomes begins, the process of pairing is called synapsis and results in the formation of bivalents. Each bivalent consists of one maternal and one paternal chromosome and the resulting structure is called synaptonemal complex (Fig. 18.11). Directly after initiation

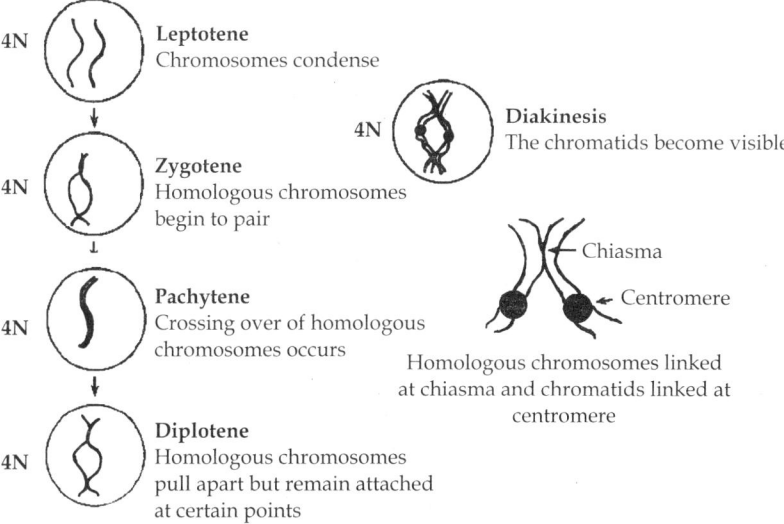

FIG. 18.10 The five phases of meiotic prophase

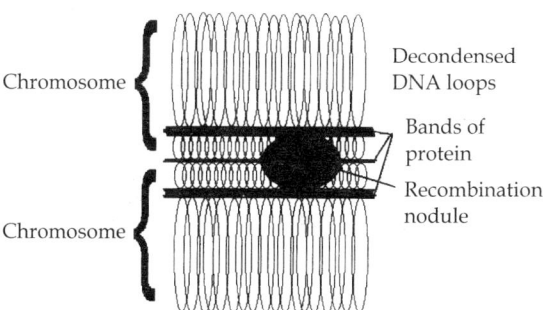

FIG. 18.11 Synaptonemal Complex

of the process the pairing spreads like a zipper across the whole length of the chromosome.

Pachytene or pachynema (Gr., *pachus*=thick; *nema*=thread): During the pachytene the pairing stabilizes. Each chromosome now consists of two sister chromatids joined at centromere called dyad, while the bivalent consists of two dyads and is called a tetrad. During this stage an important phenomenon called crossing over takes place in which the genetic material of two homologous chromosomes gets exchanged by the formation of chiasmata between two non-sister chromatids (Fig. 18.12). The number of synaptic complexes corresponds to the number of chromosomes in a haploid set of the respective species.

Diplotene or diplonema (Gr., *diplos*=double; *nema*=thread): During this stage the bivalents separate again and it emerges that each chromosome is built of two chromatids, so that the whole complex harbours four strands

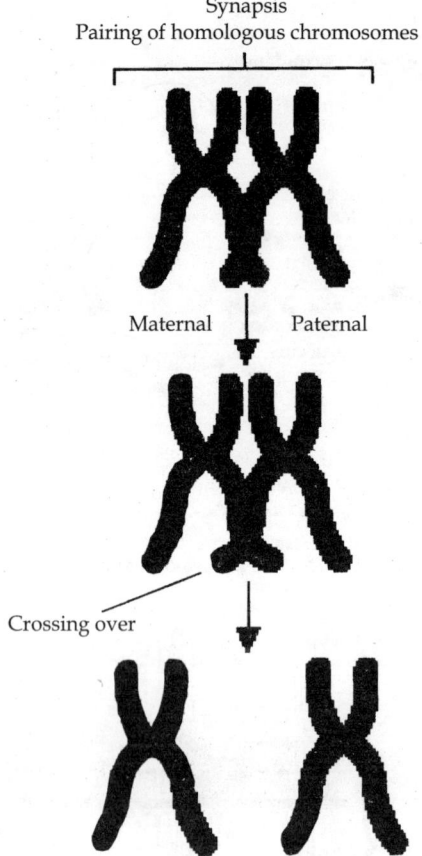

Synapsis
Pairing of homologous chromosomes

Maternal   Paternal

Crossing over

FIG. 18.12 Crossing over

during the separation. Normally the separation is not accomplished, but the homologous chromosomes stick together at certain points, the chiasmata (singular- chiasma) where the crossing over took place. This state is marked by the formation of cross-like structures, single or multiple loops.

Diakinesis (Gr., *dia*=across; *kinesis*=movement): This is continuation of the diplotene stage. It is usually difficult to demarcate both the stages. The chromosomes condense and become more compact. The nucleolus disappears, nuclear envelope breaks down, the chiasmata move away from the centromere towards the terminal ends (terminalization).

Metaphase I (Fig. 18.13)

In metaphase I the centrioles are at opposite poles of the cell. Spindle fibers attach to the centromere region of each homologous chromosome pair. The pairs of homologous chromosomes (the bivalents), get condensed and become arranged on the metaphase plate in a plane at equal distance from the poles.

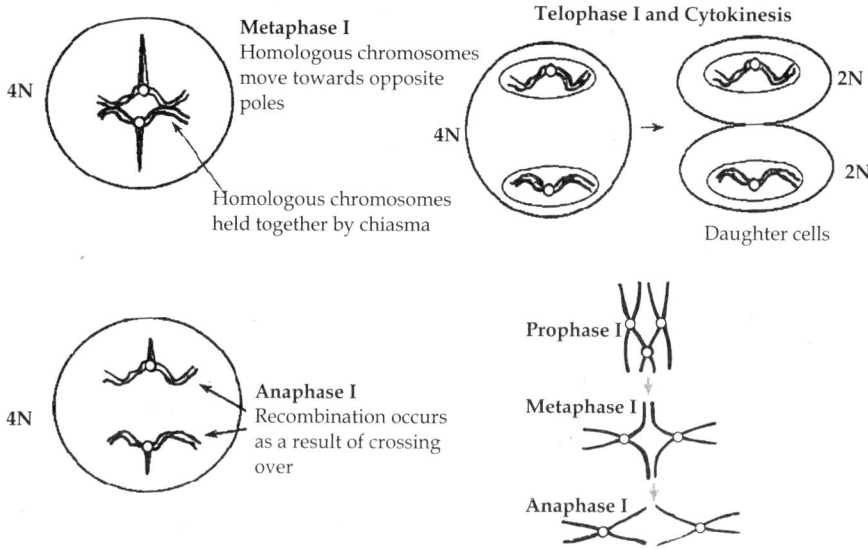

FIG. 18.13   Metaphase I, Anaphase I and Telophase I

The spindle fibers from each pole of the cell attach to one chromosome of each pair (called sister chromatids).

Anaphase I (Fig. 18.13)

In anaphase I the tetrads separate, and are drawn to opposite poles by the spindle fibers. The centromeres in anaphase I remain intact and the sister chromatids do not separate but move together toward the poles. A key difference between mitosis and meiosis is that the sister chromatids remain joined after metaphase in meiosis I, so that meiosis involves reduction in chromosome number whereas in mitosis they get separated.

Telophase I (Fig. 18.13)

Telophase I is similar to the telophase of mitosis, except that only one set of (replicated) chromosomes is in each cell. Depending on species, new nuclear envelopes may or may not form. Some animal cells may have division of the centrioles during this phase.

Interkinesis

It represents the interphase between meiosis I and meiosis II. It involves only protein and RNA synthesis but no DNA synthesis.

Meiosis II: This is actually similar to mitosis and divides each meiotic daughter cell into two haploid cells. It includes the following stages:

Prophase II

During prophase II nuclear envelopes dissolve and spindle fibers reform. The chromosomes with two chromatids each become thick and short.

Metaphase II (Fig. 18.14)

The chromosomes arrange on the equatorial plate with spindle fibres attaching to the opposite sides of the centromeres in the kinetochore region. The centromere divides into two and produces two monad chromosomes.

Anaphase II (Fig. 18.14)

During anaphase II, the centromeres split and the former chromatids (now chromosomes) are segregated into opposite sides of the cell.

Telophase II (Fig. 18.14)

Telophase II is identical to the telophase of mitosis. The nuclear envelope forms around the chromosomes, nucleolus reappears. The spindle apparatus disappears.

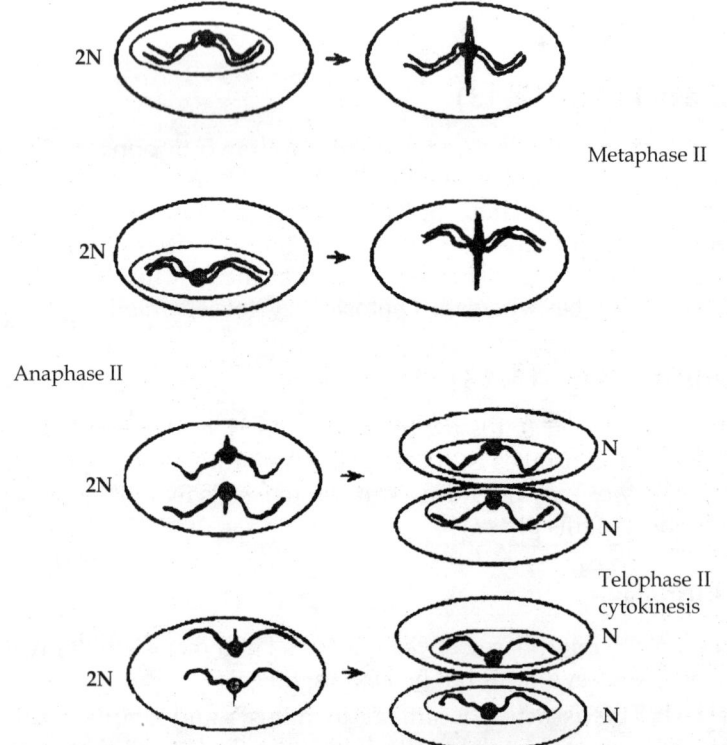

Metaphase II

Anaphase II

Telophase II
cytokinesis

FIG. 18.14 Metaphase II, Anaphase II and Telophase II

Cytokinesis

Cytokinesis separates the cells by the formation of cell furrow in animal cells and cell plate in plant cells.

SUMMARY OF MEIOSIS

1. In meiosis cell divides two times, the first division is called meiosis I and the second is called meiosis II.
2. The different stages of meiosis have the same names as those of mitosis. However, the numerals I and II indicate the division number (1st or 2nd).
3. Meiosis I consists of stages prophase I, metaphase I, anaphase I, and telophase I and meiosis II consists of prophase II, metaphase II, anaphase II, and telophase II.
4. The prophase I is an elaborate stage consisting of different substages, viz., leptotene, zygotene, pachytene, diplotene and diakinesis.
5. In the first meiotic division, the number of cells is doubled but the number of chromosomes is not. This results in 1/2 as many chromosomes per cell.
6. The second meiotic division is like mitosis. In this the number of chromosomes remains the same in each cell as the separation of chromatids takes place.

SIGNIFICANCE OF MEIOSIS

1. Meiosis is necessary to ensure that the diploid chromosome number is reduced to a haploid number. The number of chromosomes must be halved during meiosis.
2. Meiosis consists of two successive cell divisions, but only one cycle of chromosome replication. There is much variation due to each parent contributing half of their genetic material to the new organism.
3. Meiosis separates chromosomes, halves the diploid (2n) number of chromosomes and introduces some variation into the haploid (n) cells that are produced. The doubling of the chromosome number at the time of fertilization is avoided.
4. Crossing over occurs during prophase I. This is when all four chromatids are exactly aligned with each other causing non-sister chromatids to cross over, break and reassemble so that parental gene combinations are replaced by recombinants.
5. Crossing over, independent assortment of chromosomes and random fertilization are important sources of genetic variation. New genetic material is thus provided for natural selection and later evolution.

Comparison of Mitosis and Meiosis

Mitosis	Meiosis
1. Nucleus divides once.	1. Nucleus divides twice.
2. In prophase, chromosomes remain as separate units	2. In prophase I homologous chromosomes pair to form bivalents of homologous chromosomes.
3. In metaphase, each chromosome splits into two sister chromatids.	3. In metaphase I each bivalent splits into two separate chromosomes.
4. In anaphase sister chromatids now called daughter chromosomes are pulled to opposite poles.	4. In anaphase I chromosomes are pulled to opposite poles.
5. Two daughter cells formed.	5. Four daughter cells formed.
6. Each daughter cell has the same chromosome number as the parent cell, i.e. diploid.	6. Each daughter cell has half the number of chromosomes present in the parent cell, i.e. haploid.

OBJECTIVE TYPE QUESTIONS

1. _____ only occur(s) in the gonads to produce gametes.
   - A. Mitosis
   - B. Meiosis
   - C. Mitosis and meiosis
   - D. Sporogony

2. Which of the following mitosis is not used for?
   - A. Repair (of a wound) in multicellular organisms
   - B. Asexual reproduction in unicellular organisms
   - C. Development (e.g., baby in mother's womb)
   - D. Production of gametes

3. During which stage of mitosis do the centromeres split?
   - A. Prophase
   - B. Interphase
   - C. Anaphase
   - D. Telophase

4. During which stage of mitosis does cytokinesis usually occur in animals?
   - A. Prophase
   - B. Metaphase
   - C. Anaphase
   - D. Telophase

5. What is the correct order of the stages of mitosis? 1-Metaphase 2-Telophase 3-Anaphase 4-Prophase
   - A. 4,1,2,3
   - B. 2,3,1,4
   - C. 1,2,3,4
   - D. 4,1,3,2

6. During which stage of meiosis do the sister chromatids begin to move toward the poles?
   - A. Prophase I
   - B. Telophase I
   - C. Anaphase II
   - D. Anaphase I

7. During which stage of meiosis do tetrads line up at the equator?
   - A. Metaphase I
   - B. Telophase I
   - C. Metaphase II
   - D. Anaphase II

8. In both mitosis and meiosis, sister chromatids seperate during anaphase, but there are _____ haploid daughter nuclei produced by meiosis compared to _____ diploid nuclei by mitosis.
   - A. 6,3
   - B. 4,2
   - C. 2,4
   - D. 3,6

9. During which stage of mitosis does the nuclear envelope begin to disappear?
   - A. Metaphase I
   - B. Telophase I
   - C. Anaphase II
   - D. Prophase I

10. When _____ occurs between non-sister chromatids genetic exchange between chromosomes provides new combination of genes that are different from either parent.
    - A. Cytokinesis
    - B. Crossing-over
    - C. Mitosis
    - D. Cell division

11. In meiosis the chromosome number is reduced to the haploid number in a diploid animal.
    - A. Yes, and this is a normal part of mitosis too.
    - B. Yes
    - C. No, it remains at the diploid number.

12. In which organs does meiosis occur?
    - A. The thyroid and parathyroid.
    - B. The ovary and testis.
    - C. The testis and seminiferous tubules.
    - D. Ovary and Graffian follicle.

13. Which of the following statements about human reproduction is true?
    - A. Mitosis in males is also known as spermatogenesis.
    - B. Sperm and ova are zygotes.
    - C. Oögenesis takes place in the ovaries of females.

14. Which of the following statements is true about mitosis in humans?
    - A. All cells of the body go through mitosis more or less constantly from conception until death.
    - B. Each cell undergoing mitosis divides into two complete new cells that are usually identical to the cell form which they originated.
    - C. It takes roughly two weeks for a cell to go through all six phases of mitosis.

15. Which of the following statements is true about meiosis in humans?
    A. Sperm and ova are not identical to the parent cells that produced them.
    B. Females produce far more gametes than do males.
    C. The process begins in males and females at puberty.

Answers

1. B	2. D	3. C	4. D	5. D
6. C	7. A	8. B	9. D	10. B
11. B	12. B	13. C	14. B	15. A

SHORT ANSWER TYPE QUESTIONS

1. How do sister chromatids differ from chromosomes?
2. What is the centromere?
3. A diploid human cell contains 46 unreplicated chromosomes in early interphase. How many sister chromatids will be present in the human cell during prophase of mitosis?
4. Explain the significance of meiosis.
5. Define homologous chromosome.

LONG ANSWER TYPE QUESTIONS

1. How is the genetic blueprint transmitted faithfully from one cell to the next?
2. Describe the need and the mechanism of conservation of hereditary material.
3. Explain the chemical composition of a chromatid.
4. Contrast extended and condensed chromosomes.
5. Do you think that the homologous replicated chromosomes will pair with one another during mitosis? Explain.

Gametogenesis

19

INTRODUCTION

Gametogenesis is the production of haploid gametes by diploid multicellular organisms through the process of meiosis. The production of female gametes or ova (eggs) is called oogenesis. The production of male gametes or spermatozoa (sperms) is called spermatogenesis. Gametes are derived from the primordial germ cells, which enter the gonads during development. When the primordial germ cells become established in the gonad, they become stem cells that divide to produce the supply of gametes that the organism requires for reproduction. When they enter the gonads, the germ cells may associate with specific somatic cells that support, nurture and protect them. In the female, these somatic cells are called follicle cells. Various names are applied to the comparable somatic cells in the male gonad, in mammals, they are called sertoli cells. When the organism reaches maturity, germ cells acquire the ability to differentiate into functional gametes and undergo meiosis to reduce the chromosome number from 2n to 1n.

SPERMATOGENESIS (Fig. 19.1)

Spermatogenesis is the process by which stem cells develop into mature spermatozoa. There are three phases: (1) Spermatocytogenesis (Mitosis), (2) Meiosis and (3) Spermiogenesis.

Spermatocytogenesis: Each stem cell divides by mitosis to produce two daughter cells with 2n chromosomes each. One daughter cell, known as a type A spermatogonium, does not

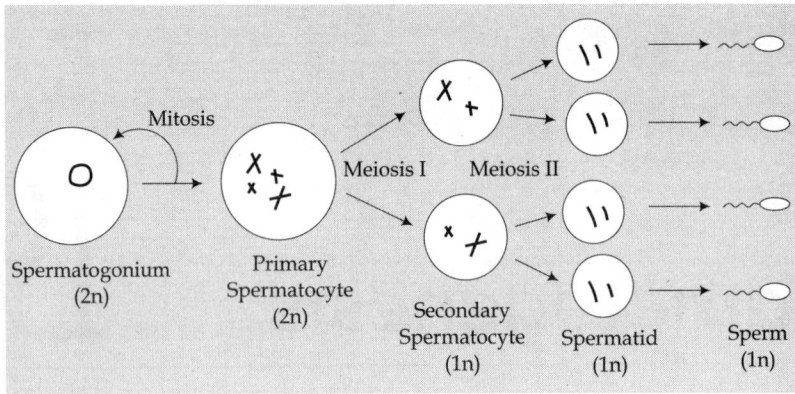

FIG. 19.1 Spermatogenesis

participate in spermatogenesis and is needed to ensure that stem cells are in a continous supply as they are needed in large quantities since the average male produces trillions of sperm cells throughout his lifetime. The other daughter cell, known as type B spermatogonium, is the one which actually enters into spermatogenesis.

The two daughter cells produced initially have chromosomes with just one chromatid. These cells enter into a period of growth, known as interphase in order to help the cells to replenish their chromosomal material. The end result is two cells with double-chromatid chromosomes. These cells are known as primary spermatocytes.

Maturation or meiosis: The primary spermatocytes enter into meiosis, which is a reductional division. The first division, meiosis I produces two daughter cells each with 'n' chromosomes. These daughter cells are called secondary spermatocytes. It should be noted that the chromosomes of these cells have two chromatids.

The second division, meiosis II, is called the equational division, and is nearly identical to mitosis. However, it is very important to note that there is no interphase between the meiosis I and meiosis II. The each of the two previously obtained daughter cells produce two new daughter cells, which brings the total number of daughter cells to four. The newly formed daughter cells have 'n' chromosomes which consist of one chromatid. They are called spermatids.

Differentiation or spermiogenesis: During differentiation, the spermatid is slowly molded into an elongated shape, and large portions of the cytoplasm (the residual cytoplasm) are shed off. Most notably, the spermatid develops an acrosome, produced by the Golgi apparatus, and a flagellum that looks roughly like a long tail and is responsible for motility. The flagellum is produced by the centrioles present in the spermatid. The nucleus is squeezed into an elongated shape and forms with the acrosome the head of the sperm, the acrosome occupying the topmost part of the head. The mitochondria which act as power plants for cells are arranged in the middle piece of the sperm along with the centriole. Once differentiation is complete, the sperm cell separates from the sertoli cell to which it was bound during spermatogenesis and migrates into the lumen of the seminiferous tubules.

STRUCTURE OF MATURE SPERMATOZOON

The spermatozoon is a single specialized cell that is modified for swimming and fertilizing the ovum. It is small in size with a long tail or flagellum. Typically, each sperm consists of an anterior head, an intermediate middle piece, and a posterior tail. The head contains a nucleus with haploid chromosomes, and the acrosome which effects penetration of the sperm into the egg. The middle piece has proximal and distal centrioles, the axial filament and numerous mitochondria which provide energy for locomotion of the sperm. The tail is long and consists of the main piece and end piece, with central axial filament, fibrils, and partly covered by protoplasmic membrane. The human sperm is about 0.05 mm in length , and a single ejaculation (about 4 ml) contains some 300 million cells. To fertilize the egg, more than 60 to 80 million sperm cells are required per ejaculation. Such high sperm count is necessary to induce pregnancy. Head is often conical in shape. The nucleus is contained within the head, which in most of the mammals has a flattened, oval shape. During spermiogenesis, the haploid sperm cell develops a tail or flagellum, and all of its mitochondria become aligned in a helix around the first part of the tail, forming the middle piece. The entire cell is, of course, enveloped by a plasma membrane (Fig. 19.2).

The acrosome is, in essence, a gigantic lysosome that forms around the anterior portion of the nucleus. It is bounded by a membrane that is considered to have two faces the *inner acrosomal membrane* faces the nucleus, while the *outer acrosomal membrane* is in close contact with the plasma membrane. It is formed from the Golgi complex and helps the spermatozoon to penetrate through the egg membranes.

OOGENESIS (Fig. 19.3)

Oogenesis is the process when haploid ova are formed within the ovary. In the typical mammalian situation a diploid primary oocyte divides by meiosis producing a secondary oocyte and a polar body after the first division. In the second division the secondary oocyte divides giving rise to an ootid and another polar body. This second division does not occur unless fertilization of the secondary oocyte by a sperm occurs. The process is completed in three stages as follows:

Multiplication phase: The cells of germinal epithelium divide repeatedly to produce oogonia; the oogonia undergo mitotic divisions and form diploid primary oocytes which enter the growth phase.

Growth phase: The growth phase of oogenesis is very long and elaborate due to the fact that egg cytoplasm accumulates food substances needed during

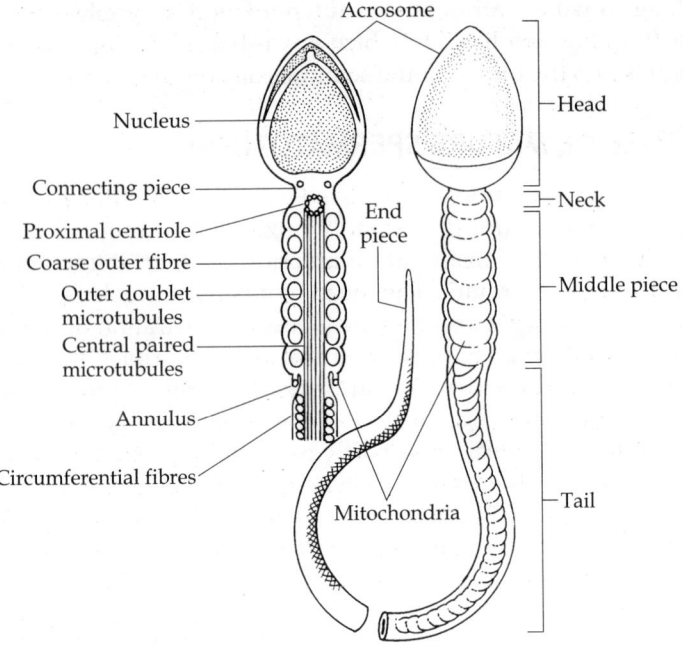

FIG. 19.2 Structure of spermatozoon

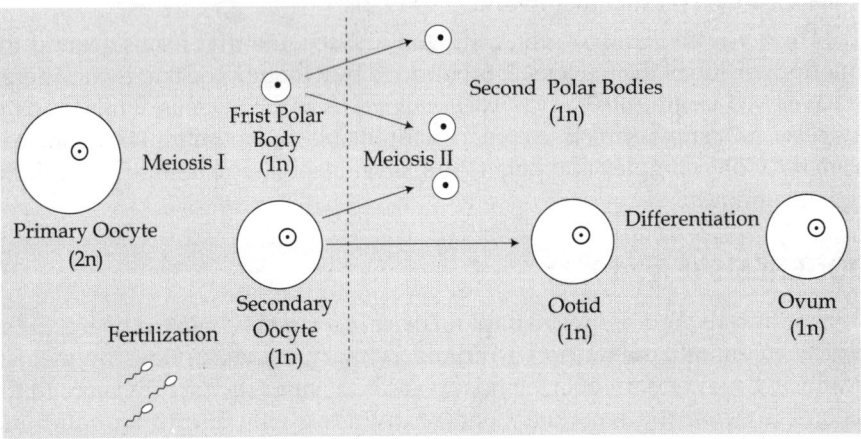

FIG. 19.3 Oogenesis

development. First of all, the size of the primary oocyte increases enormously and then large amount of fats and proteins become accumulated in the form of yolk.

The yolk comprises of (1) the cytoplasm of the ordinary animal cell with its spongioplasm and hyaloplasm; this is frequently termed the formative yolk; (2) the nutritive yolk or deutoplasm, which consists of numerous rounded granules of fatty and albuminoid substances embedded in the cytoplasm. In the mammalian ovum the nutritive yolk is extremely small in amount, and is of service in nourishing the embryo in the early stages of its development only, whereas in the egg of the bird there is sufficient to supply the chick with nutriment throughout the whole period of incubation. The nutritive yolk not only varies in amount, but in its mode of distribution within the egg; thus, in some animals it is almost uniformly distributed throughout the cytoplasm; in some it is centrally placed and is surrounded by the cytoplasm; in others it is accumulated at the lower pole of the ovum, while the cytoplasm occupies the upper pole. Yolk is distributed in the amphibian oocyte in an asymmetrical pattern, with most platelets and the largest ones located in the vegetal hemisphere. The other hemisphere is the animal hemisphere. This polarity foreshadows the polarity of the embryo itself. The establishment of this polarity has been studied by monitoring the deposition of fluorescent-labeled vitellogenin. Yolk is formed in one of two ways: It is either synthesized within the oocyte (autosynthesis) or exogenously and transported into the oocyte (heterosynthesis). Vertebrate vitellogenesis is predominantly heterosynthetic. A centrosome and centriole are present and lie in the immediate neighborhood of the nucleus.

The germinal vesicle or nucleus is a large spherical body which at first occupies a nearly central position, but becomes eccentric as the growth of the ovum proceeds. The cytoplasm becomes rich in RNA, DNA, ATP and enzymes. In large yolky eggs of reptiles, birds, and bony fishes the chromosomes become large in size and develop side loops resembling the bristles of a lampbrush and are called lampbrush chromosomes. These chromosomes are related to the increased transcription of mRNA molecules and active protein synthesis in the cytoplasm.

In the mammalian follicle, each oocyte is surrounded by a multilayered cohort of follicle cells, often referred to as granulosa cells. Between the oocyte and granulosa cells is the acellular zona pellucida, which is secreted by the oocyte. The zona is penetrated by many short microvilli from the surface of the oocyte and cytoplasmic processes from the follicle cells. Desmosomes and gap junctions form at the points where the cytoplasmic processes contact the oocyte surface. The gap junctions function in transfer of nutrients and regulatory molecules into the oocyte.

After the growth period, the primary oocyte becomes ready for the maturation phase.

Maturation phase: This takes place previous to or immediately after its escape from the follicle, and consists essentially of an unequal subdivision of the ovum by the process of meiosis first into two and then into four cells. Three of the four cells are small, incapable of further development, and are termed polar bodies or polocytes, while the fourth is large, and constitutes the mature ovum. The process of maturation has not been observed in the human ovum, but has been carefully studied in the ova of some of the lower animals, to which the following description applies. During the first meiotic division, the primary oocyte divides to produce one small polar body and one secondary oocyte. The latter enters the second meiotic division to produce the second polar body and the haploid ovum, which is the only functional sex cell to result from meiotic reduction of an oogonium.

STRUCTURE OF A MATURE OVUM

Human ova are extremely minute, measuring about 0.2 mm in diameter, and are enclosed within the egg follicles of the ovaries; as a rule each follicle contains a single ovum, but sometimes two or more are present. By the enlargement and subsequent rupture of a follicle at the surface of the ovary, an ovum is liberated and conveyed by the uterine tube to the cavity of the uterus. Unless it is fertilized it undergoes no further development and is discharged from the uterus, but if fertilization take place it is retained within the uterus and is developed into a new being.

In appearance and structure the ovum differs little from an ordinary cell, but distinctive names have been applied to its several parts; thus, the cell substance is known as the yolk or ooplasm, the nucleus as the germinal vesicle, and the nucleolus as the germinal spot. The ovum is enclosed within a thick, transparent envelope, the zona striata or zona pellucida, adhering to the outer surface of which are several layers of cells, derived from those of the follicle and collectively constituting the corona radiata.

OBJECTIVE TYPE QUESTIONS

1. Eggs and sperms are genetically very similar, but structurally very different. Why is this so?
    A. Both contain a haploid chromosome number, but eggs must provide nutrients for early development, while sperm must be able to move efficiently.
    B. Both contain a diploid chromosome number, but eggs must provide nutrients for early development, while sperm must be able to move efficiently.
    C. Both contain maternal chromosomes, but only sperm can control which chromosomes are passed on.
    D. None of the above
2. How would mammalian reproduction be affected if the meiotic strategy of spermatogenesis and oogenesis were reversed?
    A. Not enough eggs would be made each month to ensure reproductive success.
    B. Sperm production would decrease to one-fourth.
    C. Eggs would be diploid while sperm would be haploid.
    D. Sperm would be diploid while eggs would be haploid.

3. To mature, germ cells must be surrounded by specialized supportive cells, called _____ in the ovary and _____ in the testis.
    A. Granulosa cells, sertoli cells
    B. Sertoli cells, follicle cells
    C. Granulosa cells, nurse cells
    D. Sertoli cells, granulosa cells
4. A primary spermatocyte is
    A. Diploid
    B. Haploid
    C. Triploid
    D. Tetraploid
5. Which piece of the sperm is called power house?
    A. Head piece
    B. Neck piece
    C. Middle piece
    D. Tail piece

Answers

1. A   2. B   3. A   4. A   5. C

SHORT ANSWER TYPE QUESTIONS

1. How is haploidy achieved?
2. Why do the gametes have to be haploid?
3. Differentiate between spermatocytogenesis and spermiogenesis.

LONG ANSWER TYPE QUESTIONS

1. Describe the structure of the mature spermatozoon.
2. Determine the differences in egg and sperm production.
3. Explain the process of spermatogenesis/oogenesis.

Fertilization

20

INTRODUCTION

Fertilization is the fusion of gametes to form a new organism of the same species. In animals, the process involves a sperm fusing with an ovum, which eventually leads to the development of an embryo. Depending on the animal species, the process can occur within the body of the female in internal fertilization, or outside in the case of external fertilization.

MECHANISM OF FERTILIZATION

The process of fertilization involves two aspects (Fig. 20.1):

Activation of the egg: The penetration of sperm into the egg initiates a series of changes in the egg cortex and activates the egg cytoplasm by enhancing its metabolic activities.

Amphimixis: The fusion of male and female pronuclei is known as amphimixis. Though, it is not very essential for the development because even the single set of chromosomes of female pronucleus may control the development of a haploid embryo, however, fertilization restores the original diploid chromosome number of the parents.

Fertilization can be described as the following steps:

Sperm Capacitation

Freshly ejaculated sperm are unable or poorly able to fertilize the eggs. Rather, they must first undergo a series of changes collectively known as capacitation. Capacitation is associated with removal of adherent seminal plasma proteins,

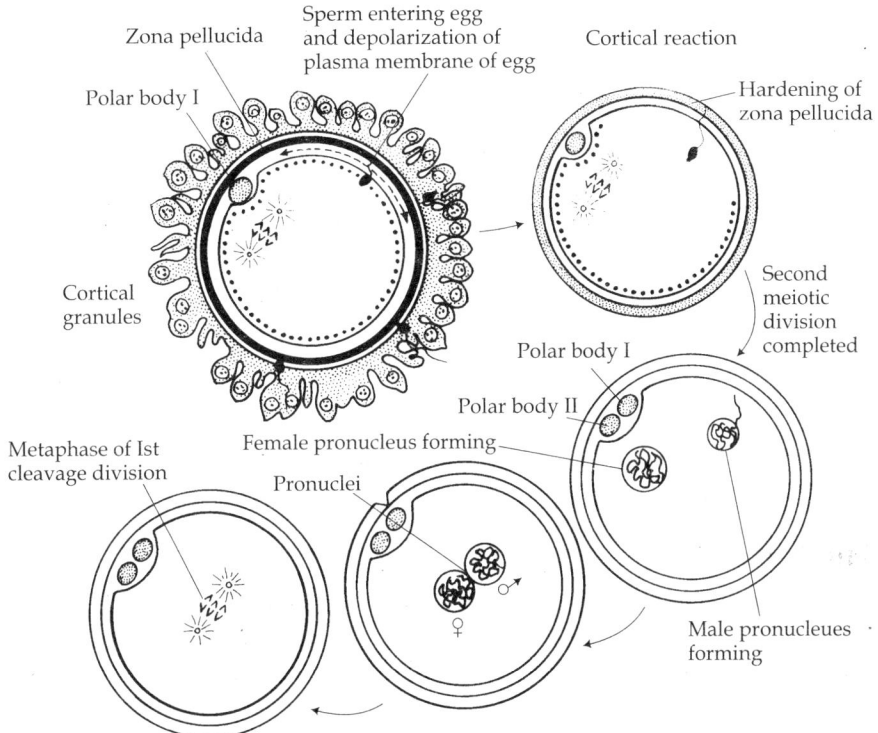

FIG. 20.1 Summary of main events of fertilization

reorganization of plasma membrane, lipids and proteins. It also seems to involve an influx of extracellular calcium, increase in cyclic AMP, and decrease in intracellular pH. The molecular details of capacitation appear to vary somewhat among species (Fig. 20.2).

Capacitation occurs while sperm reside in the female reproductive tract for a period of time, as they normally do during gamete transport. The length of time required varies with species, but usually requires several hours. The sperm of many mammals, including humans, can also be capacitated by incubation in certain fertilization media.

Sperm that have undergone capacitation are said to become hyperactiviated, and display hyperactivated motility. Most importantly however, capacitation appears to destabilize the sperm's membrane to prepare it for the acrosome reaction, as described below.

Sperm-Zona Pellucida Binding

Binding of sperm to the zona pellucida is a receptor-ligand interaction with a high degree of species specificity. The carbohydrate groups on the zona pellucida glycoproteins function as sperm receptors. The sperm molecule that

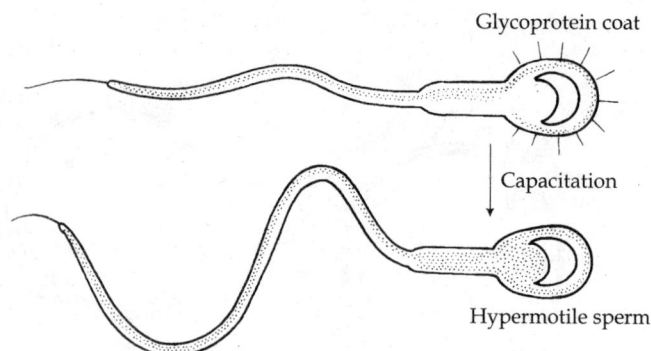

Glycoprotein coat

Capacitation

Hypermotile sperm

FIG. 20.2   Sperm capacitation

binds this receptor is not known with certainty, and indeed, there may be several proteins that can serve this function.

The Acrosome Reaction

After binding of sperm to the zona pellucida, the sperm penetrates the zona pellucida to get to the oocyte. This process is accomplished with the help of the acrosome—a huge modified lysosome that is packed with zona-digesting enzymes and located around the anterior part of the sperm's head.

The acrosome reaction provides the sperm with an enzymatic drill to get through the zona pellucida. The same zona pellucida protein that serves as a sperm receptor also stimulates a series of events that lead to many areas of fusion between the plasma membrane and outer acrosomal membrane. Membrane fusion (actually an exocytosis) and vesiculation expose the acrosomal contents, leading to leakage of acrosomal enzymes from the sperm's head.

As the acrosome reaction progresses and the sperm passes through the zona pellucida, more and more of the plasma membrane and acrosomal contents are lost. By the time the sperm traverses the zona pellucida, the entire anterior surface of its head, down to the inner acrosomal membrane, is exposed. Sperm that lose their acrosomes before encountering the oocyte are unable to bind to the zona pellucida and thereby unable to fertilize.

Penetration of the Zona Pellucida

The constant propulsive force from the sperm's flagellating tail, in combination with acrosomal enzymes, allows the sperm to create a tract through the zona pellucida. These two factors that is motility and zona-digesting enzymes allow the sperm to traverse the zona pellucida.

Sperm-oocyte Binding

Once a sperm penetrates the zona pellucida, it binds to and fuses with the plasma membrane of the oocyte. Binding occurs at the posterior (post-acrosomal) region of the sperm head.

The molecular nature of sperm-oocyte binding is not completely resolved. A chemical substance in some species is a dimeric sperm glycoprotein called fertilizin, which binds to a protein in the oocyte plasma membrane and may also induce fusion. Interestingly, humans and apes have inactivating mutations in the gene encoding one of the subunits of fertilizin, suggesting that they use a different molecule to bind oocytes.

Egg Activation and the Cortical Reaction

Prior to fertilization, the egg is in a quiescent state, arrested in metaphase of the second meiotic division. Upon binding of a sperm, the egg rapidly undergoes a number of metabolic and physical changes that collectively are called egg activation. Prominent effects include a rise in the intracellular concentration of calcium, completion of the second meiotic division and the so-called cortical reaction.

The cortical reaction refers to a massive exocytosis of cortical granules seen shortly after sperm-oocyte fusion. Cortical granules contain a mixture of enzymes, including several proteases, which diffuse into the zona pellucida following exocytosis from the egg. These proteases alter the structure of the zona pellucida, inducing the zona reaction. Components of cortical granules may also interact with the oocyte plasma membrane.

Zona Reaction

The zona reaction refers to an alteration in the structure of the zona pellucida catalyzed by proteases from cortical granules. The critical importance of the zona reaction is that it represents the major block to polyspermy in most mammals. This effect is the result of two measurable changes induced in the zona pellucida:

1. **The zona pellucida hardens:** The runner-up sperm that have not finished traversing the zona pellucida by the time the hardening occurs are stopped in their tracks.

2. **Sperm receptors in the zona pellucida are destroyed:** Therefore, any sperm that have not yet bound to the zona pellucida will no longer be able to bind and hence not fertilize the egg.

Post-fertilization Events

Following fusion of the fertilizing sperm with the oocyte, the sperm head is incorporated into the egg cytoplasm. The nuclear envelope of the sperm disperses, and the chromatin rapidly loosens from its tightly packed state in a process called decondensation. In vertebrates, other sperm components, including mitochondria, are degraded rather than incorporated into the embryo. Chromatin from both the sperm and egg are soon encapsulated in a nuclear membrane, forming pronuclei. Each pronucleus contains a haploid genome. They migrate together, their membranes break down, and the two genomes condense into chromosomes, thereby reconstituting a diploid organism.

OBJECTIVE TYPE QUESTIONS

1. The cell formed through fertilization is called a(n)
   - A. Gamete
   - B. Sperm cell
   - C. Zygote
   - D. Egg cell

2. What is fertilization?
   - A. The fusion of male and female gametes.
   - B. The division of the zygote into a larger and larger number of smaller cells.
   - C. The continued division of cells that move inward to form three cellular layers.
   - D. The development of pattern, shape, and form.

3. The fusion of male and female pronuclei is known as _____ .
   - A. Automixis
   - B. Hemimixis
   - C. Amphimixis
   - D. Heteromixis

4. The acrosome reaction helps the sperm to penetrate through the _____ .
   - A. Zona pellucida
   - B. Egg cortex
   - C. Egg cytoplasm
   - D. All of the above

5. Prior to fertilization, the egg is in a quiescent state, arrested in _____ stage of the second meiotic division.
   - A. Prophase
   - B. Metaphase
   - C. Anaphase
   - D. Telophase

Answers

1. C   2. A   3. C   4. A   5. B

SHORT ANSWER TYPE QUESTIONS

1. Explain the significance of fertilization.
2. Write short note on sperm capacitation.
3. Explain the phenomenon of acrosome reaction.

LONG ANSWER TYPE QUESTIONS

1. Discuss the key embryological phenomena of fertilization.
2. Write a short note on amphimixis.
3. Describe the function of acrosome.

Genetic Material: Properties and Replication

21

INTRODUCTION

Genes are the units of heredity in living organisms. They are encoded in the organism's genetic material that is composed of DNA or RNA and is responsible for directing the physical development and behaviour of the organism. The total complement of genes in a given organism is known as its genome. During reproduction, the genetic material is passed on from the parent(s) to the offsprings; in asexual reproduction the offspring is a genetic clone of its parent and in sexual reproduction the offspring inherits half of its genetic material from each parent. Especially in discussions of classical genetics, the genes of an organism are known as its genotype and its physical appearance or behavior as its phenotype.

Until the early 1950s most biologists were inclined to believe that the proteins were the chief carriers of hereditary traits. Nucleic acids contain only four different unitary building blocks, but proteins are made up of 20 different amino acids. Therefore, proteins appeared to have a greater diversity structure, and the diversity of the genes seemed likely to rest on the diversity of the proteins. However, the evidences in support of DNA as the genetic material came from three experimental studies, viz., the transformation of bacteria, transduction and conjugation.

Principle of Transformation—Evidence for DNA as the Genetic Material (Fig. 21.1)

In 1928, Frederick Griffith, an English army doctor, wanted to make a vaccine against a bacterium named *Streptococcus pneumoniae*, which causes a type of pneumonia. About 50 years ago, since the time of Pasteur, vaccines had been made using killed microorganisms which could be injected into patients to elicit the immune response of live cells without the risk of disease. Though Griffith did not succeed in making the vaccine he accidentally discovered the transmission of genetic informations by a process now called the 'transformation principle'.

He found that the bacterium, when grown on agar plates, had two forms, a smooth (S) and a rough (R) form. The R bacteria were harmless, but the S bacteria were lethal when injected into mice. The heat-killed S cells were also harmless; this effect was also seen by Pasteur. However, surprisingly when live R cells were mixed with killed S cells and injected into mice, the mice died, and the bacteria obtained from the mice were found to be 'transformed' into the S type.

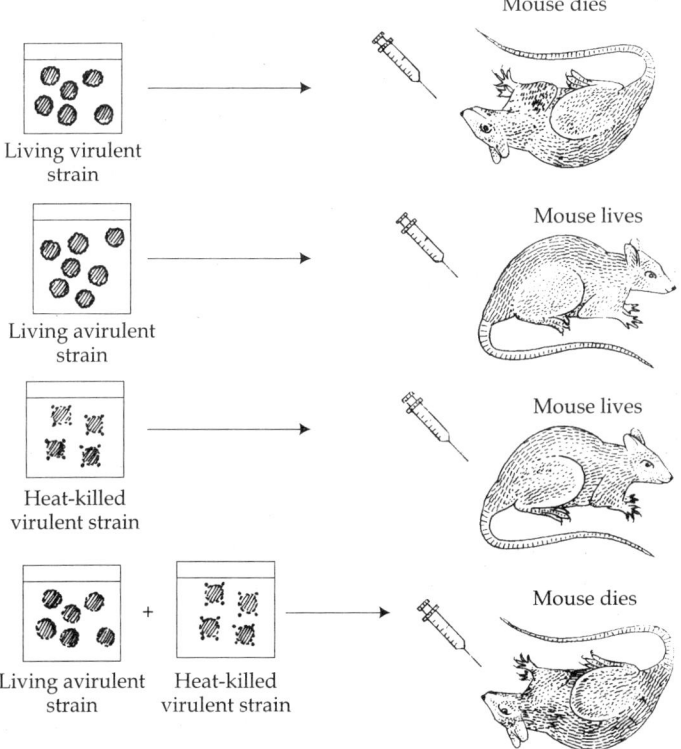

Mouse dies

Living virulent
strain

Mouse lives

Living avirulent
strain

Mouse lives

Heat-killed
virulent strain

Mouse dies

Living avirulent   Heat-killed
strain   virulent strain

FIG. 21.1   The transforming pinciple

This experiment strongly implied that the genetic material had been transferred from the dead to the live cell. It was hard to be certain of this, or to know what genetic material was transferred and was responsible for the transformation process.

Sixteen years later, in 1944, the team of Oswald Avery, Colin Macleod, and Maclyn McCarty, at the Rockefeller Institute revisited this experiment and attempted a more definitive experiment. They extracted purified DNA, proteins and other materials from S bacteria and mixed R bacteria with these different materials, and it was found that only those bacteria that were mixed with DNA were transformed into S bacteria.

This experiment strongly implied that DNA is the 'transforming factor' and not proteins or other materials and the genes are made of DNA.

The Hershey-Chase Blender Experiment (Fig. 21.2)

In 1952, American biologists, Alfred Hershey and Martha Chase performed an experiment that showed DNA to be the genetic material of a bacteriophage. They knew that a bacterial virus was an extremely simple organism,

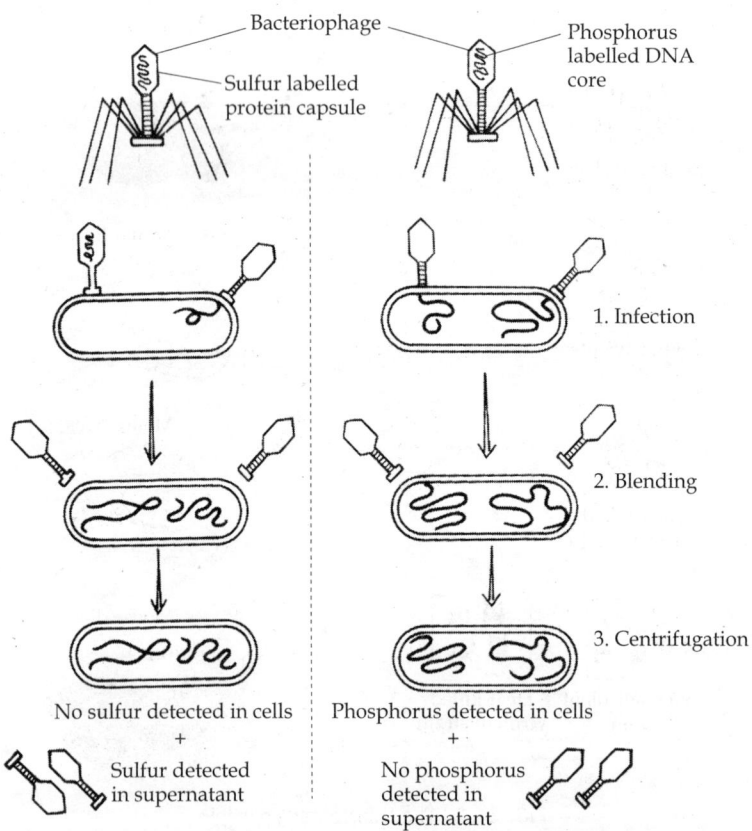

FIG. 21.2 The Hershey-Chase experiment

composed only of protein and DNA. The protein makes up the exterior of the virus, and the DNA is contained within it. When a bacterium is infected by a bacteriophage, the bacterium's internal machinery falls under the control of the virus, which uses the bacterium to produce more viruses. The question arose that which substance directed this takeover—DNA or protein?

The experiment that Hershey and Chase devised to differentiate between these possibilities was simple and took advantage of the differences in the composition of protein and DNA. Protein contains sulfur, DNA doesn't. Protein contains a small amount of phosphorus; DNA contains a lot of phosphorus.

Hershey and Chase added bacteriophage to cultures containing either radioactive sulfur or radioactive phosphorus. The bacteriophages grown in the cultures with radioactive sulfur picked it up and incorporated it into their protein. The bacterial viruses grown in the culture with radioactive phosphorus picked that up, incorporating a little of it into the protein, but most of it into their DNA. Hershey and Chase now had two types of bacteriophages: one with a radioactive external protein coat, the other with highly radioactive DNA. They were ready to begin their experiment.

Each of the two types of radioactive bacteriophage was added to a separate culture of bacteria. The bacteriophages were allowed to infect the bacteria, and then the cultures were centrifuged, causing any part of the bacteriophages that had not entered the bacteria to fall off. Next, the cultures were again centrifuged and this separated materials suspended in liquid according to their weights. The heavier bacterial cells settled to the bottom and formed a pellet while the lighter bacteriophages and loose phage parts remained floating.

Now looking for radioactivity it was found that in the cultures infected by bacteriophages with radioactive sulfur (with labeled protein), most of the radioactivity was in the liquid with the phages, whereas the cultures infected by bacteriophages with radioactive phosphorus (with most of the label in their DNA), most of the radioactivity was in the pellet of infected bacteria. Thus, it was concluded that the radioactively labeled DNA had entered the bacterial cells, but not the proteins.

The final proof that DNA, not protein, was the genetic material was provided by the progeny of the phosphorus-labeled bacteriophages. They had radioactive DNA, passed down from their parents, but no radioactive protein. These experiments convinced the scientific community that DNA alone was the genetic material, and inspired Watson and Crick to begin their efforts to discover its structure.

Bacterial Conjugation (Fig. 21.3)

Lederberg and Tatum in 1946 studied the phenomenon of bacterial conjugation. It is the transfer of genetic material between bacteria through cell-to-cell contact. This process is often regarded as equivalent to sexual

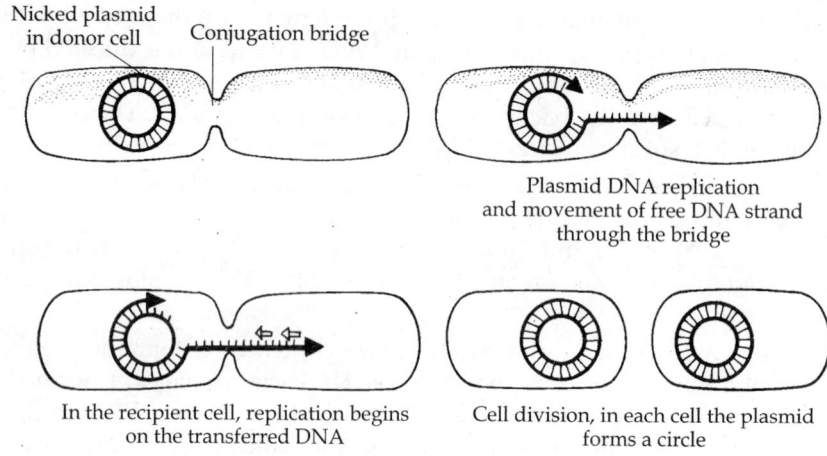

FIG. 21.3 Bacterial conjugation

reproduction or mating; however it is merely the transfer of genetic information from a donor cell to a recipient. In order to perform conjugation, one of the bacteria called the donor, plays as host to a conjugative or mobilizable genetic element which is most often a conjugative plasmid. Most conjugative plasmids have systems ensuring that the recipient cell does not already contain a similar element.

These elements are best viewed as genetic parasites on the bacterium, and conjugation as a mechanism evolved by the element to spread itself into new hosts.

The prototype for conjugative plasmids is the F-plasmid, also called the F-factor. The F-plasmid is an episome (a plasmid that can integrate itself into the bacterial chromosome by genetic recombination). It carries its own origin of replication, called *oriV*. There can only be one copy of the F-plasmid in a bacterium (which is then called F-positive), either free or integrated.

Among other genetic information, the F-plasmid carries a *tra* and a *trb* locus, which together consist of about 40 genes. The *tra* locus includes the *pilin* gene and regulatory genes, which together form pili on the cell surface, polymeric proteins that can attach themselves to the surface of F-negative bacteria and initiate the mating. When conjugation is initiated, via a mating signal, a complex of proteins called the relaxosome creates a nick in one strand of plasmid DNA at the origin of transfer, or *oriT*. In the F-plasmid system, the relaxosome consists of proteins TraI, TraY, TraM, and the integrated host factor, IHF. The transferred, or *T-strand*, is unwound from the duplex and transferred into the recipient bacterium in a 5'-terminus to 3'-terminus direction. The remaining strand is replicated, either independent of conjugative action (vegetative replication, beginning at the *oriV*) or in concert with conjugation (conjugative replication similar to the rolling circle replication of lambda phage).

If the F-plasmid becomes integrated into the host genome, donor chromosomal DNA may be transferred along with plasmid DNA. The amount of chromosomal DNA that is transferred depends on how long the bacteria remain in contact; for common laboratory strains of *Escherichia coli* the transfer of the entire bacterial chromosome takes about 100 minutes. The transferred DNA can be integrated into the recipient genome via recombination.

A culture of cells containing non-integrated F plasmids usually contains a few that have accidentally become integrated, and these are responsible for the low-frequency of chromosomal gene transfer by such cultures. Strains of bacteria with an integrated F-plasmid can be isolated and grown in pure culture. Because such strains transfer chromosomal genes very efficiently, they are called Hfr (high frequency of recombination). The *E.coli* genome was originally mapped by interrupted mating experiments, in which various Hfr cells in the process of conjugation were sheared from recipients after less than 100 minutes (initially using a Waring blender) and investigating which genes were transferred.

DNA REPLICATION

DNA was proven as the hereditary material and Watson and Crick had deciphered its structure. The Watson-Crick Model of DNA structure suggested a possible mechanism for replication of DNA molecules. It had become rather important to determine how DNA copied its information and how that was expressed in the phenotype. Three models of replication were considered likely (Fig. 21.4):

1. **Conservative replication:** This method of replication would leave the original DNA molecule intact and generate an entirely new DNA strand during replication.

2. **Semiconservative replication:** This would produce two DNA molecules, each of which was composed of one-half of the parental DNA along with an entirely new complementary strand. In other words, the newly synthesized DNA would consist of one new and one old strand of DNA. The existing strands would serve as complementary templates for the new strand. The nature of base pairing meant that if the two strands of a DNA molecule were separated, they could each serve as a template for the creation of a complementary strand by bringing in individual nucleotides to base pair with their complementary base on the template, and joining the new nucleotides together. Thus, each DNA molecule after replication would consist of one of the original strands plus one newly synthesized strand.

3. **Dispersive replication:** This involved the breaking of the parental strands during replication, and somehow, a reassembly of molecules that were a mix of old and new fragments on each strand of DNA so that each DNA molecule after replication might consist of segments of new and old DNA interspersed.

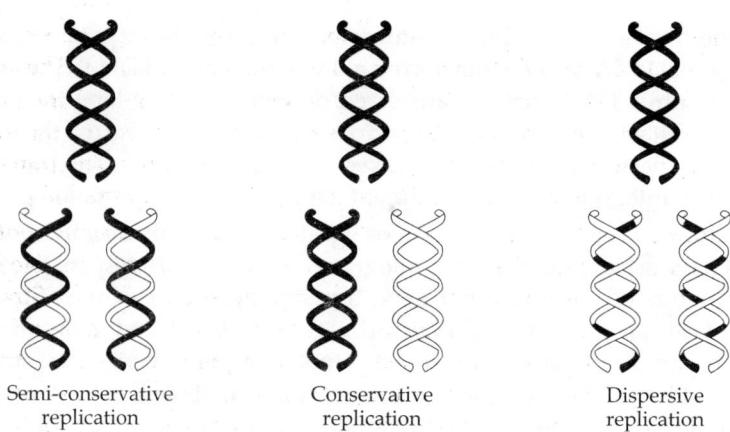

Semi-conservative
replication

Conservative
replication

Dispersive
replication

FIG. 21.4 Three modles for DNA replication

Meselson-Stahl's Experiment

Matthew Meselson and Franklin W. Stahl designed an experiment to determine the method of DNA replication. They proved that DNA replication was semiconservative by using nitrogen which is a major constituent of DNA. It is commonly found in the ^{14}N isotope, but it can also be found in the heavier ^{15}N isotope. Semiconservative replication means that when the double stranded DNA helix was replicated, each of the two double stranded DNA helices consisted of one strand coming from the original helix and one newly synthesized.

E. coli were grown for several generations in a medium with ^{15}N. The DNA of the resulting cells had a higher density (was heavier). After that, E. coli cells with only ^{15}N in their DNA were put back into a ^{14}N medium and were allowed to divide only once. DNA was then extracted from a cell and was compared to DNA from ^{14}N DNA and ^{15}N DNA. It was found to have exactly an intermediate density. Since conservative replication would result in equal amounts of DNA of the higher and lower densities (but no DNA of an intermediate density), conservative replication was excluded. However, this result was consistent with both semiconservative and dispersive replication. Semiconservative replication would result in double-stranded DNA with one strand of ^{15}N DNA, and one of ^{14}N DNA, while dispersive replication would result in double-stranded DNA with both strands having mixtures of ^{15}N and ^{14}N DNA, either of which would have appeared as DNA of an intermediate density.

DNA was then extracted from cells which had been grown for several generations in a ^{15}N medium, followed by two divisions in a ^{14}N medium. DNA from these cells was found to consist of equal amounts of two different densities, one corresponding to the intermediate density of DNA of cells grown for only one division in ^{14}N medium, the other corresponding to cells grown exclusively in ^{14}N medium. This was inconsistent with dispersive replication, which would have resulted in a single density, lower than the intermediate density of the one-generation cells, but still higher than cells grown only in ^{14}N DNA medium, as the original ^{15}N DNA would have been split evenly among all DNA strands. The result was consistent with semiconservative replication, in that half of the second-generation cells would have one strand of the original ^{15}N DNA along with one of ^{14}N DNA, accounting for the DNA of intermediate density, while the DNA in the other half of the cells would consist entirely of ^{14}N DNA one synthesized in the first division, and the other in the second division (Fig. 21.5).

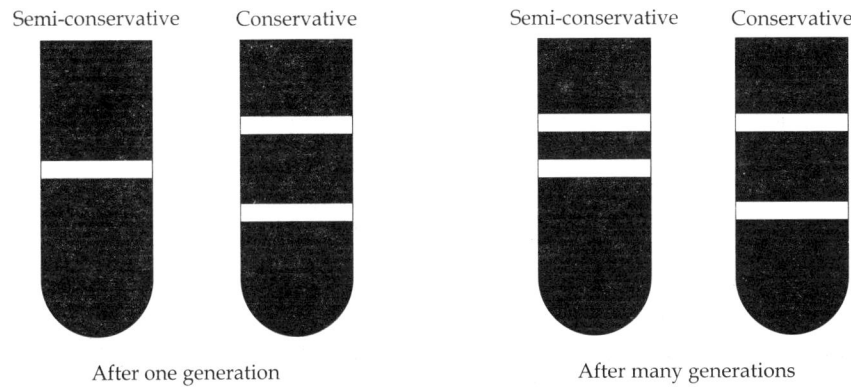

FIG. 21.5

Mechanism of DNA Replication (Fig. 21.6)

The process of DNA replication involves the following steps in *E. coli*:

Initiation of replication: DNA replication begins with a partial unwinding of the double helix at an area known as the replication fork. The process of replication begins in the DNA molecules at thousands of sites called origins of replication. At these sites, the hydrogen bonds between the bases are broken and the paired bases separate.

Unwinding the DNA: The helix begins to pull apart or unwind. This unwinding is accomplished by an enzyme known as DNA helicase. This unwound section appears under electron microscopes as a bubble and is thus known as a replication bubble. As the strands unwind, another enzyme, gyrase, removes the positive supercoils into the DNA helix by making temporary single-strand nicks. Where the DNA is nicked, the helix can unwind around the connected strand. The nick will be sealed later by another enzyme called ligase.

Replication complex contains single-strand binding proteins to help keep the two template strands apart and a family of enzymes of vital importance to replication called DNA polymerases.

RNA priming: As the two DNA strands separate or unzip and the bases are exposed, the enzyme DNA polymerase III moves into position at the point

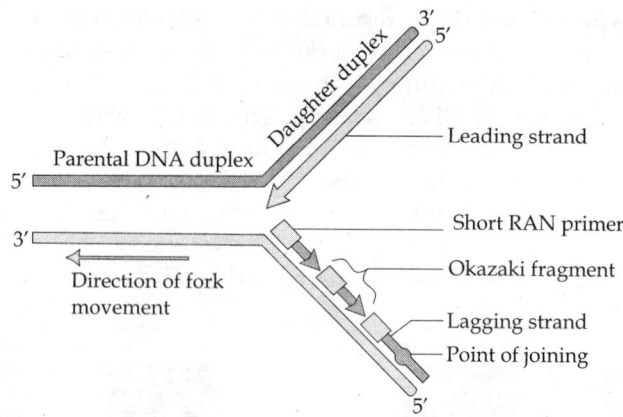

FIG. 21.6 Schematic representation of DNA replication at a growing fork

where synthesis will begin. The start point for DNA polymerase is a short segment of RNA known as RNA primer. The term 'primer' is indicative of its role which is to prime or start DNA synthesis at certain points. The primer is laid down complementary to the DNA template by an enzyme known as RNA polymerase or Primase.

Elongation of DNA chain: The DNA polymerase III (once it has reached its starting point as indicated by the primer) starts adding nucleotides one by one in an exactly complementary manner, A to T and G to C. DNA polymerase is described as being 'template dependent' in that it will 'read' the sequence of bases on the template strand and then 'synthesize' the complementary strand. The template strand is always read in the 3' to 5' direction (that is, starting from the 3' end of the template and reading the nucleotides in order toward the 5' end of the template). The new DNA strand (since it is complementary) must be synthesized in the 5' to 3' direction (remember that both strands of a DNA molecule are described as being antiparallel). DNA polymerase III catalyzes the formation of the hydrogen bonds between each arriving nucleotide and the nucleotides on the template strand.

In addition to catalyzing the formation of hydrogen bonds between complementary bases on the template and newly synthesized strands, DNA polymerase III also catalyzes the reaction between the 5' phosphate on an incoming nucleotide and the free 3'OH on the growing polynucleotide (called a phosphodiester bond). As a result, the new DNA strands can grow only in the 5' to 3' direction, and strand growth must begin at the 3' end of the template. Again, note that a phosphodiester bond is formed between the 3' OH group of the sugar and the 5' phosphate group of the incoming nucleotide.

Since the original DNA strands are complementary and run antiparallel, only one new strand can begin at the 3' end of the template DNA and grow continuously as the point of replication (the replication fork) moves along the template DNA. The other strand must grow in the opposite direction because it is complementary, not identical to the template strand. The result of this side's discontinuous replication is the production of a series of short sections of new DNA called Okazaki fragments (after their discoverer, a Japanese researcher). To make sure that this new strand of short segments is made into a continuous strand, the sections are joined by the action of an enzyme called DNA ligase which ligates the pieces together by forming the missing phosphodiester bonds.

Termination: Termination occurs when DNA replication forks meet one another or run to the end of a linear DNA molecule. Also, termination may occur when a replication fork is deliberately stopped by a special protein, called a replication terminator protein, that binds to specific sites on a DNA molecule.

Before the DNA replication is finally complete, enzymes are used to proofread the sequences to make sure the nucleotides are paired up correctly in a process called DNA repair. If mistake or damage occurs, enzymes such as a nuclease will remove the incorrect DNA. DNA polymerase I will then fill in the gap.

Nucleoside Triphosphates

Replication involves large numbers of free nucleoside triphosphates. A nucleoside triphosphate is much like a nucleoside except that instead of having one phosphate molecules, it has three.

There are eight types of nucleoside triphosphates involved in replication. They are:

dATP - deoxyribose adenosine triphosphate

dTTP - deoxyribose thymine triphosphate

dCTP - deoxyribose cytosine triphosphate

dGTP - deoxyribose guanine triphosphate

ATP - ribose adenosine triphosphate

TTP - ribose thymine triphosphate

CTP - ribose cytosine triphosphate

GTP - ribose guanine triphosphate

These molecules will pair with the bases on the template strands during the replication process. The high energy of the triphosphates is used to form bonds between the nucleotides. The two outermost phosphates are liberated, leaving the innermost group still attached. Once the two outermost phosphates have been released, the nucleoside triphosphate has become a nucleotide.

This remaining phosphate forms a link between the two deoxyribose subunits. The free 3'-OH group of the deoxyribose in the first nucleotide then reacts with the first 5' phosphate of the second nucleotide to form a sugar-phosphate-sugar linkage. The 3'-OH group of the second nucleotide is free to, and will, react with the 5' phosphate of the third nucleotide and so on.

Enzymes of DNA Replication

DNA exists condensed and compact structure. In order to prepare DNA for replication, a series of enzymes are required in the unwinding and separation of the double-stranded DNA molecule. These proteins are required because DNA must be single-stranded before replication can proceed.

DNA Helicases: These enzymes bind to the double stranded DNA and initiate the separation of the two strands.

DNA single-stranded binding proteins: These proteins bind to the DNA as a tetramer and stabilize the single-stranded structure that is generated by the action of the helicases. Replication is 100 times faster when these proteins are attached to the single-stranded DNA.

DNA Gyrase: This enzyme catalyzes the formation of negative supercoils that is thought to aid with the unwinding process.

In addition to these proteins, several other enzymes are involved in bacterial DNA replication.

DNA Polymerase: DNA Polymerase I (Pol I) was the first enzyme to be discovered with polymerase activity, and it is the best characterized enzyme. It was isolated by Arthur Kornberg, in 1960, from bacteria. Although this was the first enzyme to be discovered that had the required polymerase activities, it is not the primary enzyme involved with bacterial DNA replication. The primary enzyme is DNA Polymerase III (Pol III). Three activities are associated with DNA polymerase I;

$5' \rightarrow 3'$ elongation (polymerase activity)

$3' \rightarrow 5'$ exonuclease (proof-reading activity)

$5' \rightarrow 3'$ exonuclease (repair activity)

The second two activities of DNA Pol I are important for replication, but DNA Polymerase III (Pol III) is the enzyme that performs the $5' \rightarrow 3'$ polymerase function. DNA polymerase II is a minor enzyme involved in DNA repair.

Primase: The requirement for a free 3' hydroxyl group is fulfilled by the RNA primers that are synthesized at the initiation sites by these enzymes.

DNA Ligase: Nicks occur in the developing molecule because the RNA primer is removed and synthesis proceeds in a discontinuous manner on the lagging strand. The final replication product does not have any nicks because DNA ligase forms a covalent phosphodiester linkage between 3'-hydroxyl and 5'-phosphate groups.

Origins of Replication

Although the semiconservative nature of DNA replication had been confirmed, many questions about replication still remain unsolved. One of these questions was: Is the DNA replication initiated at a specific site on the chromosome, or is it initiated at random sites, or even multiple sites?

The answer to this question depends somewhat on the organism being considered. The bacteria have a single specific origin of replication; in other words, bacterial replication begins at the same spot on the chromosome every time. In *E. coli*, this site is called *OriC*. The *OriC* is a 9 base-pair (bp) sequence that is repeated four times within the region.

Eukaryotes also have specific sites at which replication is originated. However, because eukaryotic cells contain much more DNA than bacteria (humans have approximately 1500 times as much DNA as *E. coli*), there must be multiple origins of replication on each chromosome in order to replicate the entire DNA in a timely fashion. The amount of DNA replicated from a single origin is called a replicon.

Other research has revealed that DNA replication proceeds bidirectionally from an origin of replication. This means that replication proceeds in opposite directions away from the origin:

Note in the diagram how each original DNA molecule branches, or forks, at the point where replication is occurring. These branch points are called replication forks. Because replication is bidirectional, two replication forks form at each origin of replication. The open area of the chromosome between the replication forks is called a replication bubble.

Eukaryotic DNA Replication

Synthesis of DNA in eukaryotes is less well understood, but the process appears to be basically the same as in prokaryotes, with a few notable exceptions. For one thing, eukaryotic DNA is complexed with histones to form chromatin. Every round of replication, therefore, requires that the histones be removed and then replaced after replication is complete. This requirement however slows down the whole replication process.

Eukaryotic cells are also much more complex than prokaryotes, because they contain organelles such as mitochondria and chloroplasts (in plants) that contain their own DNA, which must also be replicated. Eukaryotic cells therefore have more than three DNA polymerases; there have been five DNA polymerases identified so far.

OBJECTIVE TYPE QUESTIONS

1. In the Meselson-Stahl experiment, which model of DNA replication was eliminated by the analysis of DNA isolated from bacteria one replication cycle after shifting from ^{15}N to ^{14}N medium?

   A. Conservative   B. Semiconservative

   C. Dispersive   D. Semidispersive

2. DNA replicates through a process called
   A. dispersive replication
   B. semidisperive replication
   C. conservative replication
   D. semiconservative replication

3. The scientists who determined the semiconservative nature of DNA replication.
   A. Schleiden and Schwann
   B. Griffith and Avery
   C. Meselson and Stahl
   D. Watson and Crick

4. When DNA replication occurs, the enzyme which separates the two sides of the helix is called
   A. DNA polymerase
   B. DNA amylase
   C. DNA ligase
   D. DNA helicase

5. The enzyme which bonds new nitrogenous bases to those existing on the original DNA strand is called
   A. DNA polymerase
   B. DNA amylase
   C. DNA ligase
   D. DNA helicase

6. Okazaki fragments are formed because
   A. DNA polymerase can only create a new strand of DNA from the 5' end to the 3' end
   B. DNA polymerase can only create a new strand of DNA from the 3' end to the 5' end
   C. newly formed DNA tends to break apart easily into fragments
   D. DNA helicase sometimes inadvertently breaks the DNA

7. One important difference between DNA replication in prokaryotes and eukaryotes is that
   A. prokaryotes do not use enzymes in the replication process
   B. there is only one replication origin in prokaryotes
   C. there are no Okazaki fragments in prokaryotes
   D. replication is conservative, not semiconservative, in prokaryotes

8. In the semiconservative replication of DNA, progeny DNA molecules consist of
   A. one-half of the molecules with two parental strands and one-half of the molecules with two new strands
   B. all molecules with interspersed parental and new segments
   C. all molecules with one parental and one new strand
   D. all molecules with two new strands

9. In the Meselson-Stahl DNA replication experiment, what percent of the DNA was composed of one light strand and one heavy strand after one generation of growth in ^{14}N containing growth media?
   A. 0
   B. 25
   C. 50
   D. 100

10. In the Meselson-Stahl DNA replication experiment, if the cells were first grown for many generations in ^{15}N containing media, and then switched to ^{14}N containing media, what percent of the DNA had one light strand and one heavy strand after two generations of growth in ^{15}N growth media?

    A. 25   B. 50

    C. 75   D. 100

Answers

1. B	2. D	3. C	4. D	5. A
6. A	7. B	8. C	9. D	10. B

SHORT ANSWER TYPE QUESTIONS

1. Differentiate between conservative, semiconservative and dispersive modes of replication.
2. Name the enzyme that carries out the following functions during DNA replication.
   (a) Unwinds the helical DNA by breaking the hydrogen bonds between complementary bases.
   (b) Synthesizes a short RNA primer at the beginning of each origin of replication.
   (c) Adds DNA nucleotides to the RNA primer.
   (d) Digests away the RNA primer and replaces the RNA nucleotides of the primer with the proper DNA nucleotides.
   (e) Links the DNA fragments of the lagging strand together.
3. Explain what do you mean by the termination of DNA replication.
4. Write a short note on the importance of DNA polymerases.
5. What happens to the RNA primers that are essential for initiation of DNA replication?

LONG ANSWER TYPE QUESTIONS

1. Explain the process of DNA replication.
2. Write a note on Meselson-Stahl's experiment.
3. Enlist the different enzymes of replication and discuss their respective functions.
4. Give an account of Frederick Griffith's Transforming Principle as an evidence for DNA as the genetic material.
5. Write an explanatory note on bacterial conjugation.

Protein Synthesis 22

INTRODUCTION

Protein synthesis is the process whereby DNA encodes for the production of amino acids and proteins. In order to make even one protein, the body must invoke the aid of messenger RNA, transfer RNA, DNA, amino acids, ribosomes, and multiple enzymes.

A protein is simply a long chain of amino acids linked together by bonds. There are only twenty amino acids consisting of carbon, hydrogen, nitrogen, oxygen, and two that contain sulfur. Ten of these amino acids have side groups that are attracted to water, while the other ten do not. Therefore, when a protein is in a water-based environment, the hydrophobic amino acids fold inwards while the hydrophilic remain on the outside. The backbone of amino acids form strong covalent bonds and the actual amino acids form temporary weak bonds. These weak bonds allow the amino acids to change shape, remain mobile, and attain flexibility. The important quality about proteins is that the position of the amino acids determines their function. The proteins serve many functions including support, structure, movement, transport, communication, and disease defense. Protein-containing structures include hair, nails, hooves, cartilage, muscle, hormones, antibodies, blood proteins, and enzymes.

This process of synthesis of proteins can be divided into two parts:

Transcription

It is the process by which the information contained in a section of DNA is transferred to messenger RNA by RNA transcription. One strand of the DNA double helix is used as a template by the RNA polymerase to synthesize a messenger RNA (mRNA). This mRNA migrates from the nucleus to the cytoplasm. During this step, mRNA goes through different types of maturation including one called splicing when the non-coding sequences are eliminated. The coding mRNA sequence can be described as a unit of three nucleotides called a codon.

Translation

The ribosome binds to the mRNA at the start codon (AUG) that is recognized only by the initiator tRNA. The ribosome proceeds to the elongation phase of protein synthesis. During this stage, complexes, composed of an amino acid linked to tRNA, sequentially bind to the appropriate codon in mRNA by forming complementary base pairs with the tRNA anticodon. The ribosome moves from codon to codon along the mRNA. Amino acids are added one by one, translated into polypeptidic sequences directed by DNA and represented by mRNA. At the end, a release factor binds to the stop codon, terminating translation and releasing the complete polypeptide from the ribosome.

CENTRAL DOGMA OF MOLECULAR BIOLOGY

The central dogma of molecular biology was first enunciated by Francis Crick in 1958.

It serves as a framework of transfer of information between different biopolymers; DNA, RNA and protein. The general transfers describe the normal flow of biological information: DNA can be copied to DNA by DNA replication, DNA information can be copied into mRNA by transcription, and proteins can be synthesized using the information in mRNA as a template by translation (Fig. 22.1). Proteins do not code for the production of protein, RNA or DNA. They are involved in almost all biological activities, structural or enzymatic.

In 1970, Temin suggested the existence of the process called reverse transcription or teminism which involves the transfer of information from RNA to DNA (the reverse of normal transcription) with the help of an enzyme called 'RNA dependent DNA polymerase'. This process is known to occur in the case of retroviruses, such as HIV, and in higher eukaryotes. It is not, however, the general case in most living organisms.

MECHANISM OF PROTEIN SYNTHESIS

Protein synthesis is the process by which proteins are created from the DNA blueprints. There are two major processes involved:

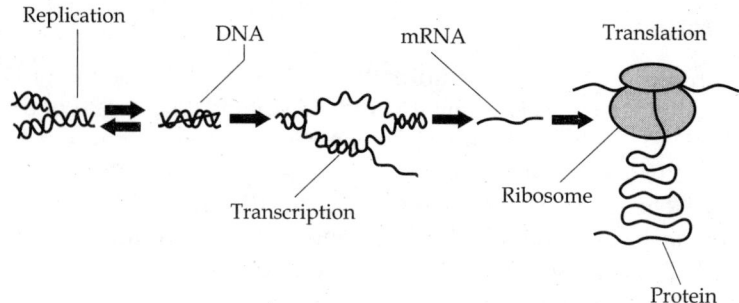

FIG. 22.1 Central dogma of protein synthesis

1. Transcription: DNA to RNA
2. Translation: RNA to Protein

Transcription

Transcription is the first step in protein synthesis. It is the synthesis of RNA under direction from DNA. In many ways transcription is quite similar to DNA replication, except, in this case, instead of new DNA nucleotides being added, RNA nucleotides are added to the DNA to form a RNA strand known as the primary transcript. The primary transcript eventually goes on to become a messenger RNA (mRNA) strand after some modification. It is mRNA that directs the following step of protein synthesis, translation.

Transcription begins when the enzyme, RNA polymerase breaks the hydrogen bonds between the two sides of a DNA strand and begins adding RNA nucleotides. RNA polymerase first binds to promoter regions, which include the initiation site and several other nucleotides, on the DNA strand. However, RNA polymerase can not recognize the promoter regions on its own. Thus, transcription factors search along the DNA strand for promoter regions and the RNA polymerase recognizes the transcription factors. Out of two strands of DNA, the one from which RNA is copied is called the antisense or template strand, and the other strand, to which it is identical, the sense or coding strand.

The DNA-dependent RNA polymerases that catalyze transcription are complex, multimeric proteins. The *E.coli* RNA polymerase consists of six polypeptides, viz., two α, one each of β, β^1, ω and σ. The complete RNA polymerase enzyme is called 'holoenzyme' (Fig. 22.2). In eukaryotes there are three different RNA polymerases I, II and III. RNA polymerase II transcribes majority of nuclear structural genes and is responsible for pre-mRNA synthesis.

When RNA polymerase successfully binds to a promoter region, it begins adding RNA nucleotides. Since RNA polymerase only works in a 5' to 3' direction only one side of the DNA strand is active in transcription. The addition of RNA nucleotides is directed by the DNA. Specific nucleotide

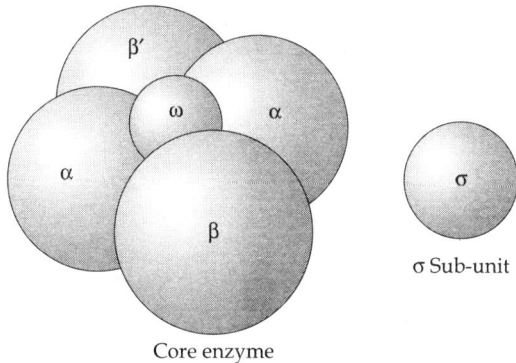

Core enzyme

σ Sub-unit

FIG. 22.2 Holoenzyme

sequences on the DNA strand mark initiation and termination sites, where the transcription begins and ends. The initiation sequence, termination sequence, and all the intervening nucleotides together are collectively known as a transcription unit.

As in DNA replication, bases that are complementary to those on the original DNA strand are added in a $5' \to 3'$ direction. However, RNA lacks the base thymine (thymine is specific to DNA) which is complementary to adenine. Instead the RNA has the base uracil. Thus, when adding complementary bases during transcription the complementary base pairings are cytosine for guanine, as in DNA replication, and adenine for uracil. These nucleotides are added to the DNA strand by RNA polymerase. RNA polymerase unwinds one turn of the DNA helix at a time and adds nucleotides. When one turn of the helix has been completed and the next is about to begin unwinding, the newly added RNA nucleotides separate from the DNA strand and the DNA strand rewinds. Nucleotide addition progresses at a rate of about 60 nucleotides per second.

Transcription continues until RNA polymerase reaches a termination site on the DNA strand. This site has a sequence of nucleotides, the most common of which being AATAAA in eukaryotic cells, that tell RNA polymerase to stop adding nucleotides. Thus the formation of the primary transcript has been completed (Fig. 22.3).

In eukaryotic cells, the primary transcript undergoes modification before it becomes messenger RNA (mRNA) and is ready to leave the nucleus and go into the cytoplasm. The 5' end of the molecule, where transcription began, is capped off with a modified form of guanine. This cap serves to protect the mRNA molecule from degradation in the cytoplasm. Also, it serves as a signal to ribosomal subunits in the cytoplasm. At the 3' end of the molecule a poly-A tail is added, which consists of 150-200 adenine nucleotides. It also protects the mRNA molecule from degradation and helps in the export of the mRNA from the nucleus to the cytoplasm. Also, the molecule is shortened in a process known as RNA splicing. The non-coding regions of the molecule, known as

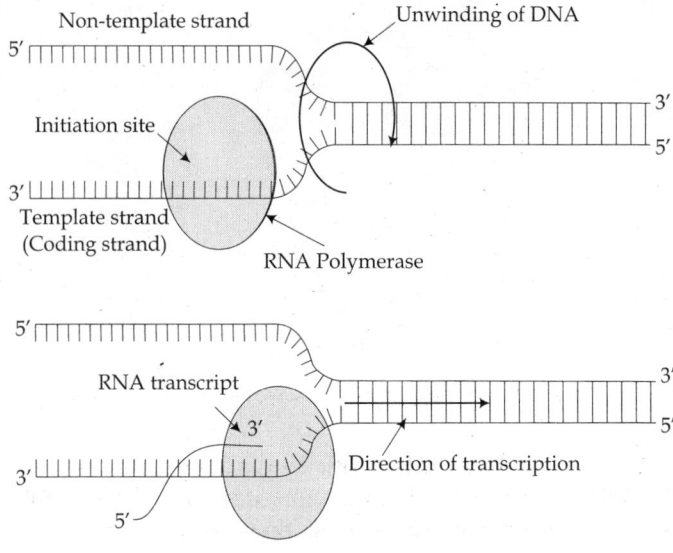

FIG. 22.3 Transcription

introns, are cut out by splicesomes. This leaves only the functional coding regions, called exons, in the mRNA molecule.

Thus transcription is completed and an mRNA molecule is formed. Messenger RNA is now able to leave to nucleus and go to the cytoplasm and perform its function in the next step in protein synthesis that is translation.

Translation

Translation is the process in which the information coded in the nucleotides of mRNA directs the aminoacid sequence of a polypeptide chain during protein synthesis. This process can be divided into following three steps:

1. Initiation
2. Elongation
3. Termination

The process of translation involves multiple molecules including mRNA, tRNA, the two ribosomal units and various enzymes and proteins.

Initiation

The cellular factory responsible for synthesizing proteins is the ribosome. The ribosome consists of structural RNA and about 80 different proteins. In its inactive state, it exists as two subunits; a large subunit and a small subunit. When the small subunit encounters an mRNA, the process of translation of the mRNA to protein begins.

Activation of amino acids: The tRNA acts as a translator between mRNA and protein and each tRNA molecule has a specific anticodon site and an acceptor site. Each tRNA also has a specific charger protein; this protein can only bind to that particular tRNA and attach the correct amino acid to the acceptor site. The energy to make this bond comes from ATP. These charger proteins are called aminoacyl tRNA synthetases.

Transfer of amino acids to tRNA: There are two sites in the large subunit, A-site and P-site, for subsequent amino acids to bind to, and thus, be close enough to each other for the formation of a peptide bond. The A-site accepts a new tRNA molecule bearing an amino acid, and the P-site bears the tRNA molecule attached to the growing chain (Fig. 22.4).

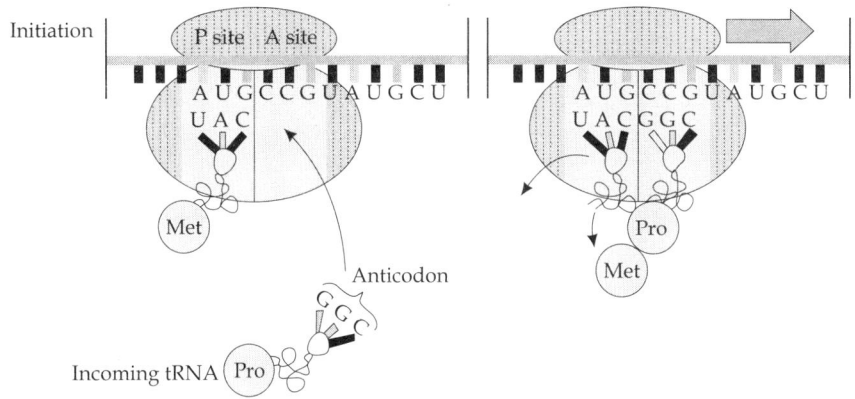

FIG. 22.4 Translation

Initiation of synthesis: The start signal for translation is the codon AUG, which codes for methionine, *'met'* in eukaryotes and *N-formylmethionine, 'fmet'* in prokaryotes. Not every protein necessarily starts with methionine, however. In *E. coli* the process of initiation takes place with the help of three initiation factors called IF1, IF2 and IF3. Often this first amino acid will be removed in post-translational processing of the protein. A tRNA charged with methionine binds to the translation start signal. The large subunit binds to the mRNA and the small subunit, and elongation begins.

Elongation

Peptide bond formation: During elongation, tRNA that matches the codon in the A-site comes in. With the help of the enzyme peptidyl transferase, a peptide bond is formed between the two amino acids, the energy coming from the bond between the first amino acid and its tRNA. Once the bond is made, the now uncharged tRNA drifts away, and the whole complex moves over three bases (to the right). Each of the three base mRNA sequences, codons, instructs the tRNA to bring a specific amino acid to the ribosome.

Elongation factors: This process occurs with the help of elongation factors like Ef-Tu, Ef-Ts and Ef-G in prokaryotes and Ef-1 and Ef-2 in

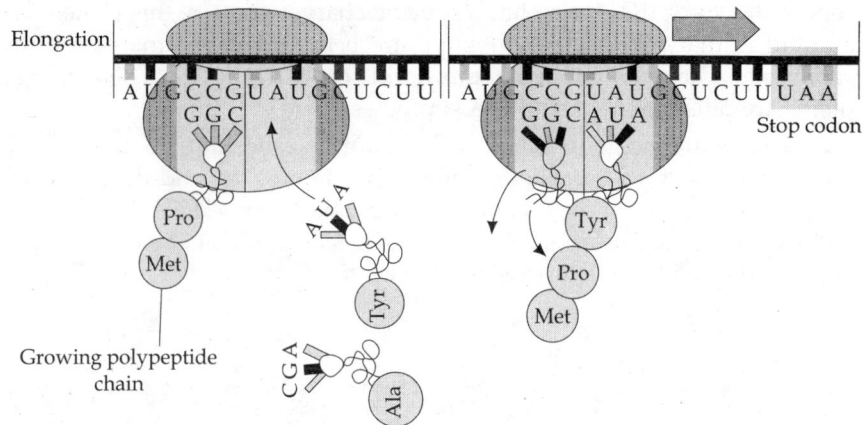

FIG. 22.5 Elongation

eukaryotes (Fig. 22.5). This step repeats until a specific codon called the stop codon (UAA, UAG or UGA) is reached. This leads to termination.

Termination

Since, there are no tRNAs that bind with the stop codons, useless proteins plug up the A-site. The bond between the last tRNA and the last amino acid on the long polypeptide chain is broken and the whole complex breaks up (Fig. 22.6). The process of chain termination requires release factors RF-1, RF-2 and RF-3 in prokaryotes and RF in eukaryotes.

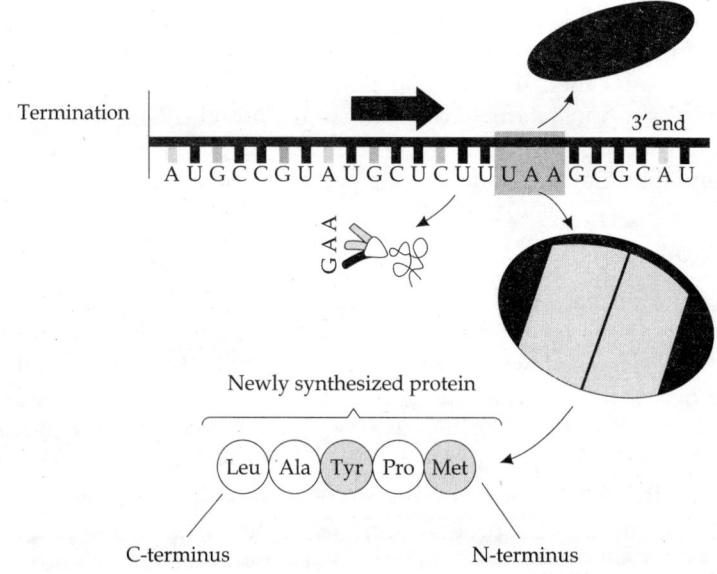

FIG. 22.6 Termination

It should be noted that mRNA can have many ribosome complexes translating it at the same time, called polysomes. The ribosomes are spaced at regular intervals and translate the mRNA, allowing for fast production of many proteins (Fig. 22.7).

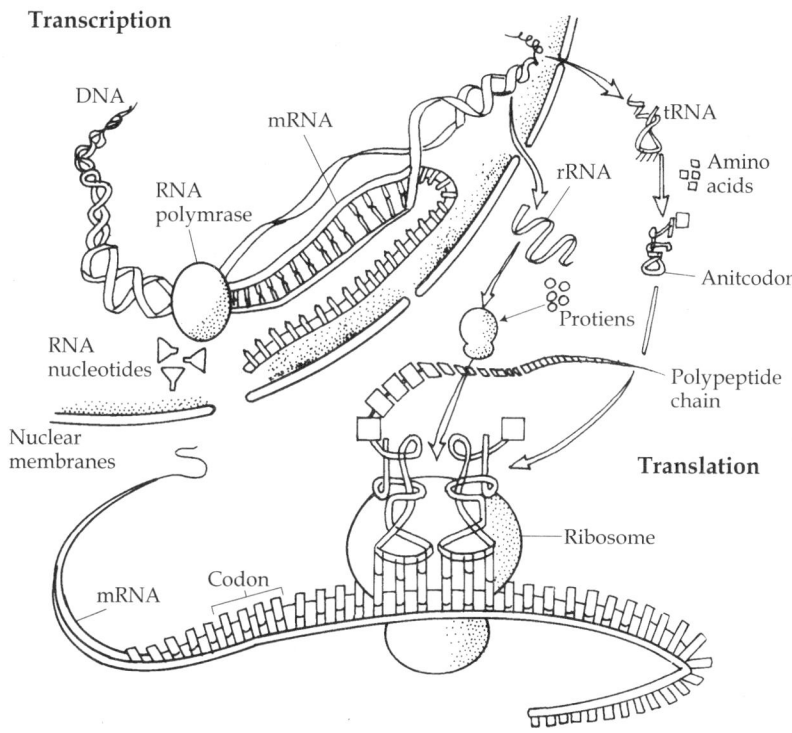

FIG. 22.7 Protein synthesis

OBJECTIVE TYPE QUESTIONS

1. In eukaryotes, gene expression is related to the coiling and uncoiling of _____ .

2. A DNA subunit composed of a phosphate group, a five-carbon sugar, and a nitrogen-containing base is called a(n) _____ .

3. The name of the five-carbon sugar that makes up a part of the backbone of molecules of DNA is _____ .

4. Knowing the order of the bases in a gene permits scientists to determine the exact order of the amino acids in the expressed _____ .

5. Due to the strict pairing of nitrogen base pairs in DNA molecules, the two strands are said to be _____ to each other.

6. According to base-pairing rules, Adenine pairs with Thymine _____ and Guanine pairs with _____ .

7. The enzyme that is responsible for replicating molecules of DNA by attaching complementary bases in the correct sequence is _____ .

8. Enzymes called _____ are responsible for unwinding the DNA double helix by breaking the hydrogen bonds that hold the complementary strands together.

9. The process by which DNA copies itself is called _____ .

10. Molecules of _____ carry instructions for protein synthesis from the nucleus to the cytoplasm.

11. The enzyme responsible for making RNA is called _____ .

12. The form of ribonucleic acid that carries genetic information from the DNA to the ribosomes is _____ .

13. A _____ is a sequence of DNA at the beginning of a gene that signals RNA polymerase to begin transcription.

14. Of the 64 codons of mRNA, 61 code for _____ , 3 are _____ signals, and one is a _____ signal.

15. Nucleotide sequences of tRNA that are complementary to codons on mRNA are called _____ .

Answers

1. DNA	2. Nucleotide	3. Deoxyribose
4. Protein	5. Complementary	6. Cytosine
7. DNA Polymerase	8. Helicases	9. Replication
10. RNA	11. RNA polymerase	12. mRNA
13. Promoter	14. Amino acids, Stop, Start	
15. Anticodons		

SHORT ANSWER TYPE QUESTIONS

1. Describe the aminoacylation (charging) of a tRNA molecule.

2. The DNA strand that is read to produce the mRNA (the template strand), assuming Phe = UUU and Tyr = UAU in mRNA. (Just give the sequence corresponding to a stretch of 6 amino acids starting with Phe.)

3. Write a note on central dogma of protein synthesis.

4. Write a short note on the structure of the holoenzyme.

5. Explain briefly the structure of mRNA.

LONG ANSWER TYPE QUESTIONS

1. Describe the process of translation in detail.

2. Write a descriptive account on tRNA molecule.

3. Explain the role of DNA polymerases.

4. Differentiate the protein synthesis in prokaryotes and eukaryotes.

5. Write a note on initiation complex.

Genetic Code

23

INTRODUCTION

The genetic code is a set of rules, which maps DNA sequences to proteins in the living cell, and is employed in the process of protein synthesis. Nearly all living things use the same genetic code, called the standard genetic code, although a few organisms use minor variations of the standard code. The sequence of nucleotides, coded in triplets (codons) along the mRNA determines the sequence of amino acids in protein synthesis. The DNA sequence of a gene can be used to predict the mRNA sequence, and the genetic code can in turn be used to predict the amino acid sequence.

HISTORY

When James Watson, Francis Crick and Rosalind Franklin deciphered the structure of DNA, efforts began to understand the nature of the encoding of proteins. George Gamov postulated that a three-letter code must be employed to encode for 20 different amino acids used by living cells to encode proteins. The first actual deciphering of a codon was done by Marshall Nirenberg and Johann Matthaei in 1961. They used a cell-free system to translate a polyuracil (poly-U) RNA sequence (UUUUU..., etc.), and thereby deduced that the codon UUU specified the amino-acid phenylalanine. Subsequent work by Har Gobind Khorana identified the rest of the code, and shortly thereafter Robert Holley identified tRNA as the adapter molecule that facilitated translation. In

1968, Khorana, Holley and Nirenberg shared the Nobel Prize in Physiology or Medicine for their work.

CHARACTERISTICS OF GENETIC CODE

The genetic code has the following general characteristics:

1. **The code is triplet:** The gene sequence inscribed in DNA and RNA is composed of trinucleotide units called codons, each coding for a single amino acid.

2. **The code is non-overlapping:** The genetic code is read in groups (or 'words') of three nucleotides. After reading one triplet, the reading frame shifts over three letters, not just one or two. In the following example, the code would not be read GAC, ACU, CUG, UGA...

Rather, the code would be read GAC, UGA, CUG, ACU...

GACUGACUGACU

3. **The code is degenerate:** There are 64 different triplet codons, and only 20 amino acids. Some amino acids are coded by more than one codon so that some redundancy is built into the system. In some cases, the redundant codons are related to each other by sequence; for example, leucine is coded by the codons CUU, CUA, CUC, and CUG. Note that the codons are same except for the third nucleotide position. This third position is known as the 'wobble' position of the codon. This is because in a number of cases, the identity of the base at the third position can wobble, and the same amino acid will still be coded. This property allows some protection against mutation—if a mutation occurs at the third position of a codon, there is a good chance that the amino acid specified in the encoded protein will not change. A codon is said to be four-fold degenerate if any nucleotide at its third position specifies the same amino acid; it is said to be two-fold degenerate if only two of four possible nucleotides at its third position specify the same amino acid.

   Degeneracy is mandatory in order to produce enough different codons to code for 20 amino acids and a stop codon. Because there are four bases, triplet codons are required to produce at least 21 different codes. For example, if there were two bases per codon, then only 16 amino acids could be coded for ($4^2 = 16$). Because at least 21 codes are required, then 4^3 gives 64 possible codons, meaning that some degeneracy must exist. For example, *phenylalanine, tyrosine, histidine, glutamine, lysine, asparagine, glutamic acid, aspartic acid* and *cysteine* have two codons each.

Isoleucine has three codons. *Valine, proline, threonine, alanine* and *glycine* have four codons each. *Leucine, arginine* and *serine* have six codons each.

4. **The code is commaless:** There are no spaces or commas separating neighboring codons. This is like having a sentence in English language consisting entirely of three letter words where there are no spaces between the words. This property is especially important in understanding the effects of mutations on proteins. For example, a code with commas would be represented as:

CUC, ACC, UUU

Leu　Thr　Phe

A mutation resulting in a deletion or addition of a base would affect only one amino acid, and thus there would be only a slight modification in the original genetic message.

CUC,　　_CC,　　UUU

Leu　　Changed-aa　Phe

On the other hand, a mutation in a commaless code would result in a drastic change in the entire genetic message.

UCACCUUU_

Ser　Pro

5. **The code is non-ambiguous:** The genetic code is unambiguous. Each codon specifies a particular amino acid, and only one amino acid. In other words, the codon ACG codes for the amino acid threonine, and not for any other amino acid.

6. **Initiation and stop codons:** The code contains signals for starting and stopping translation of the code. The start codon is AUG. AUG codes for the amino acid methionine and the first AUG encounters signals for translation to begin. The start codon sets the reading frame with AUG as the first triplet and subsequent triplets are read in the same reading frame. Translation continues until a stop codon is encountered. There are three stop codons: UAG, UGA and UAA. In classical genetics, the three stop codons were given names: UAG was *amber*, UGA was *opal* (sometimes also called *umber*), and UAA was *ochre*. To be recognized as a stop codon, the triplet must be in the same reading frame as the start codon.

7. **The code has polarity:** The code is always read in a 5' → 3' direction. The first AUG read in that direction sets the reading frame, and subsequent codons are read in frame, until the stop codon is encountered.

8. **The code is universal:** The code is almost universal. However, certain bacteria, mitochondria and protista have minor variations in their codes. The near universality of the code suggests that the code arose very early in the evolution of life.

Table: RNA codon table showing 64 codons and their respective coded amino acids.

		2nd base			
		U	**C**	**A**	**G**
1st base	**U**	UUU Phenylalanine	UCU Serine	UAU Tyrosine	UGU Cysteine
		UUC Phenylalanine	UCC Serine	UAC Tyrosine	UGC Cysteine
		UUA Leucine	UCA Serine	UAA Ochre (*Stop*)	UGA Opal (*Stop*)
		UUG Leucine	UCG Serine	UAG Amber (*Stop*)	UGG Tryptophan
	C	CUU Leucine	CCU Proline	CAU Histidine	CGU Arginine
		CUC Leucine	CCC Proline	CAC Histidine	CGC Arginine
		CUA Leucine	CCA Proline	CAA Glutamine	CGA Arginine
		CUG Leucine	CCG Proline	CAG Glutamine	CGG Arginine
	A	AUU Isoleucine	ACU Threonine	AAU Asparagine	AGU Serine
		AUC Isoleucine	ACC Threonine	AAC Asparagine	AGC Serine
		AUA Isoleucine	ACA Threonine	AAA Lysine	AGA Arginine
		AUG Methionine, *Start*	ACG Threonine	AAG Lysine	AGG Arginine
	G	GUU Valine	GCU Alanine	GAU Aspartic acid	GGU Glycine
		GUC Valine	GCC Alanine	GAC Aspartic acid	GGC Glycine
		GUA Valine	GCA Alanine	GAA Glutamic acid	GGA Glycine
		GUG Valine	GCG Alanine	GAG Glutamic acid	GGG Glycine

THE WOBBLE HYPOTHESIS

Holley determined the sequence of yeast tRNA^{Ala} in 1965. He observed that the sequence of the anticodon bases in this tRNA contained the nucleotide Inosine at the 5′ position. Inosine (I) is the nucleotide of the base hypoxanthine, a deaminated adenine. Inosine monophosphate (IMP) is actually a precursor in the biosynthesis of purine nucleotides. Francis Crick in 1966 proposed the Wobble hypothesis to generalize this observation. He suggested that while the interaction between the codon in the mRNA and the anticodon in the tRNA needed to be exact in two of the three nucleotide positions, this did not have to be so in the third position. He proposed that non-standard base-pairing might occur between the nucleotide base in the 5' position of the anticodon and the 3' position of the codon.

Crick recognised that the following base-pair schemes were possible:

5′ anticodon base	3′ codon base
A	U
C	G
G	C or U
U	A or G
I	A or C or U

This hypothesis not only accounts for the number of tRNAs that are observed, it also accounts for the degeneracy that is observed in the Genetic Code. The degenerate base is that in the wobble position.

OBJECTIVE TYPE QUESTIONS

1. A genetic code requires at least 20 different code words, one for each _____ .

   A. Amino acid
   B. Pair of bases
   C. Codon
   D. Intron

2. Genetic codon is non-ambiguous, i.e.

   A. Same codon may code for more than one amino acid
   B. Many codons may code for same amino acid
   C. Only one codon for a single amino acid
   D. None of the above

3. A triplet would code for a given amino acid as long as

   A. The three bases are present in a particular sequence
   B. The first two bases are the same and with same sequence
   C. The second and third bases are the same and with same sequence
   D. All are correct

4. Wobble's hypothesis assigns greatest importance to

   A. First base
   B. First two bases
   C. Last base of codon
   D. Last two bases

5. Which of the following is not a non-sense codon?

   A. UAA
   B. UAG
   C. UGA
   D. AAU

6. The language of the genetic code is written as a linear sequence of amino acids in the polypeptide. (**True or False**)

7. The genetic code has four letters, A, T, G, and C. (**True or False**)

8. There are 64 codons in the genetic code. (**True or False**)

9. The codon AUG is the start signal and is the first one to be translated. (**True or False**)

10. Stop codons are first translated. (**True or False**)

Answers

1. A	2. C	3. A	4. B	5. D
6. False	7. True	8. True	9. True	10. False

SHORT ANSWER TYPE QUESTIONS

1. Read and translate the codons on mRNA into the appropriate amino acids. GUACGAAAA
2. Write a short note on degeneracy of the genetic code.
3. Comment upon start and stop codons.

LONG ANSWER TYPE QUESTIONS

1. Explain the triplet nature of the genetic code.
2. Explain the characteristics of genetic code.
3. Write a note on wobble hypothesis.

Regulation of Gene Expression in Prokaryotes

24

INTRODUCTION

Most bacteria unlike plant and animal cells are exposed to a constantly changing physical and chemical environment. It is well known that these bacteria exhibit remarkable capability to adapt to such diverse environmental conditions. Within limits, bacteria can react to changes in their environment through changes in patterns of structural proteins, transport proteins, toxins, enzymes, etc., which adapt them to a particular ecological situation. This adaptability of bacteria depends on their ability to 'turn on' and 'turn off' the expression of specific sets of genes in response to the environmental conditions. The expression of particular genes is 'turned on' when the products of these genes are needed for the survival of bacteria in the particular environment. Whereas, the expression is 'turned off' when the products of certain genes are not needed in given environmental conditions. Therefore, the regulation of gene expression in this way helps in increasing the overall fitness of an organism by conserving the energy used in synthesizing the proteins or products which are not needed at a particular time. For example, *E. coli* does not produce fimbriae for colonization purposes when living in a planktonic (free-floating or swimming) environment. *Vibrio cholerae* does not produce the cholera toxin that causes diarrhea unless it is in the human intestinal tract. *Bacillus subtilis* does not make the enzymes for tryptophan biosynthesis if it can find pre-existing tryptophan in its environment. If *E. coli* is fed glucose and lactose together, it will use the glucose first because it needs too less enzymes to use glucose than it does to use lactose. In *Neisseria gonorrhoeae*, the bacterium will develop a sophisticated iron gathering and transport system if it senses that iron is in short supply in its environment.

Bacteria have developed sophisticated mechanisms for the regulation of both catabolic and anabolic pathways. Generally, bacteria do not synthesize degradative (catabolic) enzymes unless the substrates for these enzymes are present in their environment. For example, synthesis of enzymes that degrade lactose would be wasteful unless the substrate for these enzymes (lactose) was available in the environment. Similarly, bacteria have developed diverse mechanisms for the control of biosynthetic (anabolic) pathways. Bacterial cells shut down biosynthetic pathways when the end products of the pathway are not needed or are readily obtained by transport from the environment. For example, if a bacterium could find a preformed amino acid like tryptophan in its environment, it would make sense to shut down its own pathway of tryptophan biosynthesis, and thereby conserve energy. However, in real bacterial life, the control mechanisms for all these metabolic pathways must be reversible, since the environment can change quickly and drastically.

In this chapter, some of the common mechanisms are discussed by which bacterial cells can regulate and control their metabolic activities.

CONDITIONS AFFECTING GENE EXPRESSION

Gene expression is a multi-step process that can be regulated at different levels, viz., transcription of DNA, mRNA processing, post transcriptional modification and translation into a gene product, followed by folding, post-translational modification and targeting. The amount of protein that a cell expresses depends on the tissue, the developmental stage of the organism and the metabolic or physiologic state of the cell.

Numerous terms are used to describe patterns of gene expression, these include:

Constitutive gene: It is a gene that is transcribed continually and whose products are essential components of almost all living cells such as tRNA molecules, rRNA molecules, ribosomal proteins, RNA polymerase etc.

Housekeeping gene: It is typically a constitutive gene that is transcribed at a relatively constant level. The housekeeping gene's products are typically needed for maintenance of the cell. It is generally assumed that their expression is unaffected by experimental conditions.

Facultative gene: It is a gene which is only transcribed when needed compared to a constitutive gene. Their gene products are needed only under certain environmental conditions and continuous synthesis of such products would otherwise lead to wastage of energy.

Inducible gene: It is a gene whose expression is either responsive to environmental change or dependent on the position of the cell cycle. The process by which the expression of genes is turned on in response to a substance in the environment is called induction. The substances or molecules responsible for induction are called inducers.

Most bacteria are capable of growth using any one of several carbohydrates as an energy source such as glucose, sucrose, arabinose, lactose etc. However, if the environment contains glucose, the bacteria prefer to metabolize it than any other carbohydrate. In a medium containing lactose as the sole carbon source, the bacteria synthesize two enzymes, β-galactosidase and β-galactoside permease which are required for catabolism of lactose. In absence of lactose the synthesis of these two enzymes would be a waste. Therefore, the regulatory mechanism helps to turn on the synthesis of these two enzymes when they are needed and turn off their synthesis in a medium devoid of lactose.

Also, bacteria possess the metabolic capacity to synthesize many organic molecules that are required for growth. For example, *E. coli* has five genes coding for enzymes that are needed for the synthesis of tryptophan. These five genes must be expressed in *E. coli* cells growing in a medium not containing tryptophan. However, when the external medium contains tryptophan, these enzymes are not synthesized and the expression of the genes is turned off . The process of 'turning off' of the gene expression is called repression.

THE OPERON MODEL

An operon is a group of key nucleotide sequences including an operator, a common promoter, and one or more structural genes that are controlled as a unit to produce messenger RNA (mRNA). Operons occur primarily in prokaryotes and nematodes. They were first described by François Jacob and Jacques Monod in 1961.

An operon contains one or more structural genes which are transcribed into RNA. Upstream of the structural genes lies a promoter sequence which provides a site for RNA polymerase to bind and initiate transcription. Close to the promoter lies an operator sequence. The operon may also contain regulator genes such as a repressor gene which codes for a protein that binds to the operator and inhibits transcription. These regulator genes need not be the part of operon itself, they may be located on an entirely different gene. The repressor molecule will be transported to the promoter site to block the transcription of the structural genes.

A promoter is a DNA sequence that enables a gene to be transcribed. The promoter is recognized by RNA polymerase, which then initiates transcription. In RNA synthesis, promoters are the means to demarcate which genes should be used for messenger RNA formation and also control that which proteins are manufactured by the cell.

An operator is a segment of DNA that regulates the activity of the structural genes of an operon it is linked to, by interacting with a specific repressor or activator. It is a regulatory sequence for 'shutting down' a gene or turning it 'on'.

Control of operon genes is a type of gene regulation that enables the organisms to regulate the expression of various genes depending on environmental conditions. Operon regulation can be either negative or positive. Negative regulation involves the binding of a repressor to the operator to prevent transcription.

In negative inducible operons, a regulatory repressor protein is normally bound to the operator and it prevents the transcription of the genes on the operon. If an inducer molecule is present, it binds to repressor and changes its conformation so that it is unable to bind to the operator. This allows for the transcription of the genes on the operator.

In negative repressible operons, transcription of the genes on the operon normally takes place. Repressor proteins are produced by a regulator gene but they are unable to bind to the operator in their normal conformation. However, certain molecules called corepressors can bind to the repressor protein and change its conformation so that it can bind to the operator. The activated repressor protein binds to the operator and prevents transcription.

Operons can also be positively controlled. With positive control, an activator protein stimulates transcription by binding to DNA (usually at a site other than the operator).

In positive inducible operons, activator proteins are normally unable to bind to the pertinent DNA. However, certain susbstrate molecules can bind to the activator proteins and change their conformations so that they can bind to the DNA and enable transcription to take place.

In positive repressible operons, the activator proteins are normally bound to the pertinent DNA segment. However, certain molecules can bind to the activator and prevent it from binding to DNA. This prevents transcription.

THE *LAC* OPERON (Fig. 24.1)

The *lac* operon is an operon required for the transport and metabolism of lactose in *Escherichia coli* and some other enteric bacteria. It consists of three adjacent structural genes, a promoter, a terminator, and an operator. The *lac* operon is regulated by several factors including the availability of glucose and lactose.

The three structural genes are: *lacZ*, *lacY*, and *lacA*. *lacZ* encodes β-galactosidease (LacZ), an intracellular enzyme that cleaves the disaccharide lactose into glucose and galactose. *lacY* encodes β-galactoside permease (LacY), a membrane bound transport protein that pumps lactose into the cell. *lacA* encodes β-galactoside transacetylase (LacA), an enzyme that transfers an acetyl group from acetyl-CoA to β- galactoside. Only *lacZ* and *lacY* appear to be necessary for lactose catabolism.

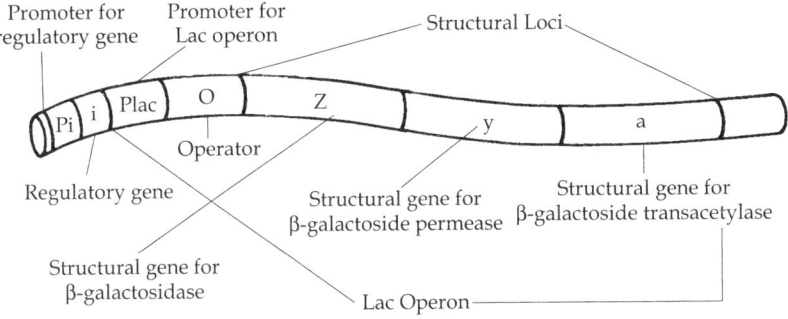

FIG. 24.1 Structure of *Lac* Operon

Specific control of the *lac* genes depends on the availability of the substrate lactose to the bacterium. The proteins are not produced by the bacterium when lactose is unavailable as a carbon source. The *lac* genes are organized into an operon, they are oriented in the same direction immediately adjacent on the chromosome and are co-transcribed into a single polycistronic mRNA molecule. Transcription of all genes starts with the binding of the enzyme RNA polymerase (RNAP), a DNA-binding protein, to a specific DNA binding site immediatelly upsteam of the genes, this site is called a *promoter*. From this position RNAP proceeds to transcribe all three genes *lacZ, lacY* and *lacA* into mRNA. The *lac* regulator gene, designated as *i* gene, codes for a repressor (Fig. 24.2 a, b).

The regulatory response to lactose requires an intracellular *regulatory protein* called the *lactose repressor*. The *lacI* gene encoding repressor lies nearby the *lac* operon and it is always expressed (referred to as constitutive

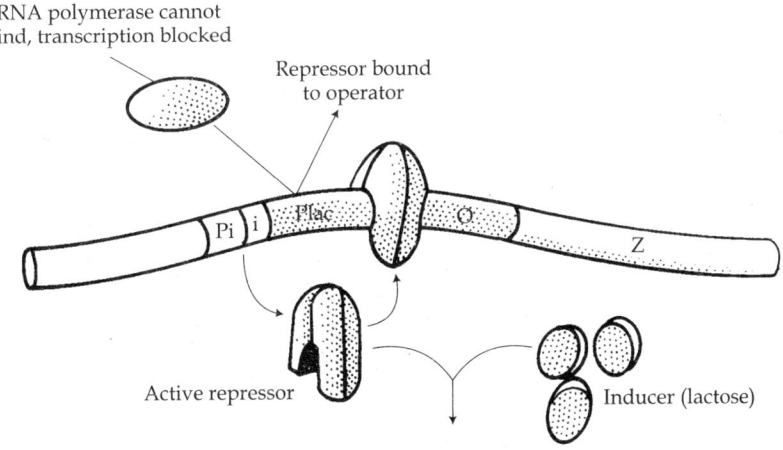

FIG. 24.2a

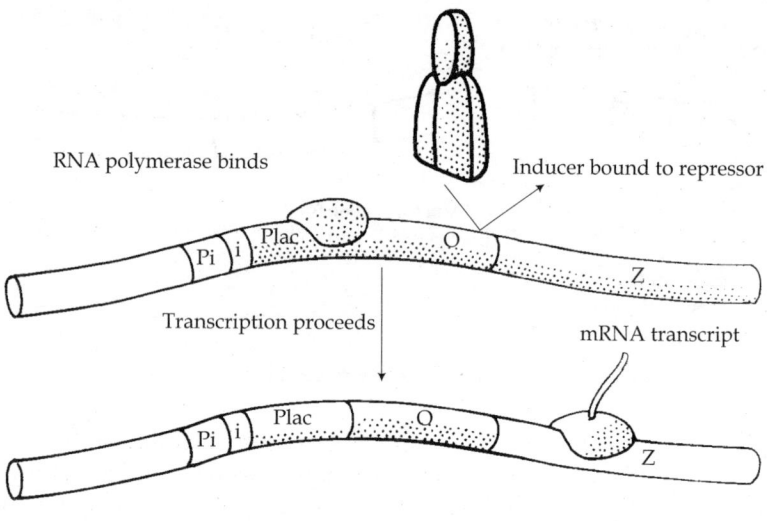

RNA polymerase binds

Inducer bound to repressor

Pi i Plac O Z

Transcription proceeds

mRNA transcript

Pi i Plac O Z

FIG. 24.2b

expression). If lactose is missing from the growth medium, the repressor binds very tightly to a short DNA sequence just downstream of the promoter near the beginning of *lacZ* called the *lac operator*. Repressor bound to the operator interferes with binding of RNAP to the promoter, and therefore mRNA encoding lacZ and lacY is only made at very low levels. When cells are grown in the presence of lactose, a lactose metabolite called allolactose binds to the repressor, causing a change in its shape. Thus altered, the repressor is unable to bind to the operator, allowing RNAP to transcribe the *lac* genes and thereby leading to high levels of the encoded proteins.

The *lac* operon is repressed, even in the presence of lactose, if glucose is also present. This repression is maintained until the glucose supply is exhausted. The repression of the *lac* operon under these conditions is termed catabolite repression and is a result of the low levels of cAMP that result from an adequate glucose supply. The repression of the *lac* operon is relieved in the presence of glucose if excess cAMP is added. As the level of glucose in the medium falls, the level of cAMP increases. Simultaneously there is an increase in inducer binding to the *lac* repressor. The net result is an increase in transcription from the operon. The ability of cAMP to activate expression from the *lac* operon results from an interaction of cAMP with a protein termed **CRP** (for cAMP receptor protein). The protein is also called **CAP** (for catabolite activator protein). The cAMP-CRP complex binds to a region of the *lac* operon just upstream of the region bound by RNA polymerase and that somewhat overlaps that of the repressor binding site of the operator region. The binding of the cAMP-CRP complex to the *lac* operon stimulates RNA polymerase activity 20-to-50-fold (Fig. 24.3 a, b).

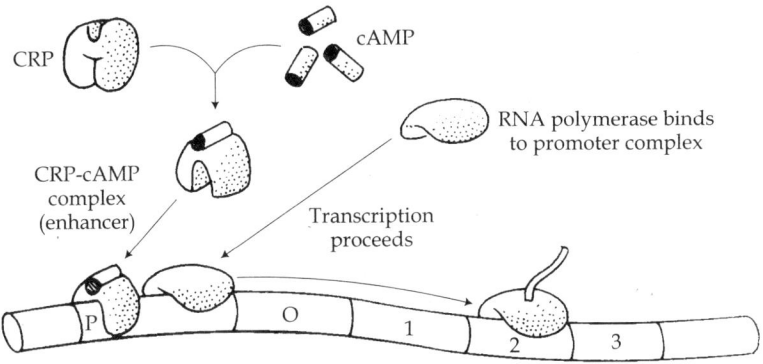

FIG. 24.3a Low glucose

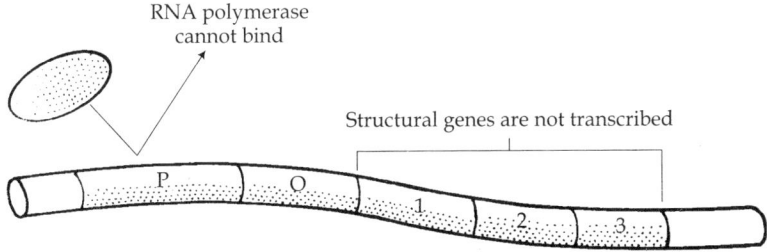

FIG. 24.3b High glucose

THE *TRP* OPERON (Fig. 24.4 a, b)

The trp operon encodes the genes for the synthesis of tryptophan. This cluster of genes, like the *lac* operon, is regulated by a repressor that binds to the operator sequences. The activity of the *trp* repressor for binding to the operator region is enhanced when it binds with tryptophan; in this capacity, tryptophan is known as a corepressor. Since the activity of the *trp* repressor is enhanced in

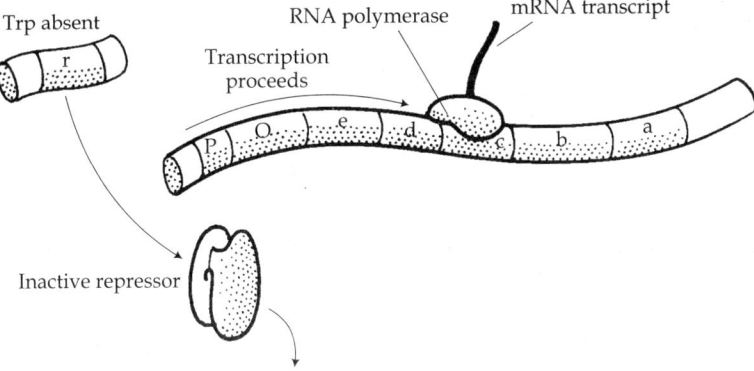

FIG. 24.4a The *trp* operon

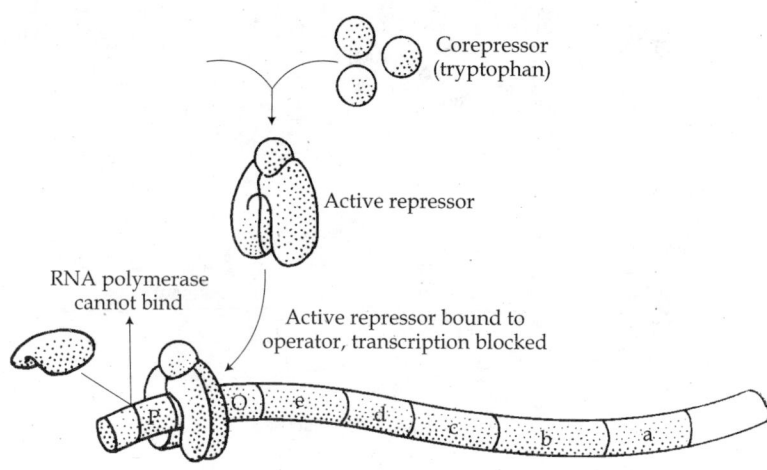

FIG. 24.4b   Trp present

the presence of tryptophan, the rate of expression of the *trp* operon is graded in response to the level of tryptophan in the cell.

Expression of the *trp* operon is also regulated by attenuation. The attenuator region, which is composed of sequences found within the transcribed RNA, is involved in controlling transcription from the operon after RNA polymerase has initiated synthesis. The attenuator sequences of the RNA are found near the 5' end of the RNA termed the leader region of the RNA. The leader sequences are located prior to the start of the coding region for the first gene of the operon (the *trpE* gene). The attenuator region contains codons for a small leader polypeptide that contains tandem tryptophan codons. This region of the RNA is also capable of forming several different stable stem-loop structures.

Depending on the level of tryptophan in the cell and hence the level of charged trp-tRNAs, the position of ribosomes on the leader polypeptide and the rate at which they are translating allows different stem-loops to form. If tryptophan is abundant, the ribosome prevents stem-loop 1-2 from forming and thereby favors stem-loop 3-4. The latter is found near a region rich in uracil and acts as the transcriptional terminator loop. Consequently, RNA polymerase is dislodged from the template.

The operons coding for genes necessary for the synthesis of a number of other amino acids are also regulated by this attenuation mechanism. However, it should be clear that this type of transcriptional regulation is not feasible for eukaryotic cells.

FEEDBACK INHIBITION

In bacterial cells, enzymatic reactions may be regulated by two unrelated modes: (1) control or regulation of enzyme activity (feedback inhibition or end product inhibition), which mainly operates to regulate biosynthetic pathways;

and (2) control or regulation of enzyme synthesis including, end-product repression, which functions in the regulation of biosynthetic pathways, and enzyme induction and catabolite repression, which regulate mainly degradative pathways. The process of feedback inhibition regulates the activity of pre-existing enzymes in the cells. The processes of end-product repression, enzyme induction and catabolite repression are involved in the control of enzyme synthesis. These latter processes which regulate the synthesis of enzymes may be either a form of positive control or negative control. End-product repression and enzyme induction are mechanisms of negative control which lead to a decrease in the transcription of proteins. Catabolite repression is considered a form of positive control because it affects an increase in transcription of proteins.

Feedback inhibition (or end product inhibition) is a mechanism for the inhibition of preformed enzymes that is seen primarily in the regulation of whole biosynthetic pathways, e.g. pathways involved in the synthesis of the amino acids. Such pathways usually involve many enzymatic steps, and the end product is able to feedback to the first step in the pathway and to regulate its own biosynthesis.

In feedback inhibition, the end product of a biosynthetic pathway inhibits the activity of the first enzyme that is unique to the pathway, thus controlling production of the end product. The first enzyme in the pathway is an allosteric enzyme. Its allosteric site will bind to the end product (e.g., amino acid) of the pathway which alters its active site so that it cannot mediate the enzymatic reaction which initiates the pathway. Other enzymes in the pathway remain active, but they do not see their substrates. The pathway is shut down as long as adequate amounts of the end product are present. If the end product is used up or disappears, the inhibition is relieved, the enzyme regains its activity, and the organism can resume synthesis of the end product. Thus, if an *E. coli* bacterium swims out of a glucose minimal medium into milk or some other medium rich in growth factors, the bacterium can stop synthesizing any of the essential metabolites that are made available directly from the new environment.

OBJECTIVE TYPE QUESTIONS

1. Which of the following is not part of the *lac* operon of *E. coli*?
   A. genes for inducible enzymes of lactose metabolism
   B. genes for the repressor, a regulatory protein
   C. genes for RNA polymerase
   D. a promoter, the RNA polymerase binding site

2. The inducer:
   A. combines with a repressor and prevents it from binding to the promoter
   B. combines with a repressor and prevents it from binding to the operator

    C. binds to the promoter and prevents the repressor from binding to the operator

    D. binds to the operator and prevents the repressor from binding at this site

3. An *E. coli* strain is *lac Z*. The structural gene for β-galactosidase is encoded at the *lac Z* locus. How would you describe the regulation of lactose metabolism in these cells?

    A. normal regulation of lactose metabolism

    B. constitutive expression of *lac Z*

    C. inability to synthesize the *lac Z⁻* gene product, β-galactosidase

    D. *lac Z* gene is inducible, but unable to be repressed by high glucose

4. Cells of an *E. coli* strain that are *trp-lac Z-met + bio +* were mixed with cells of an *E. coli* strain that are *trp + lac Z + met-bio-* and cultured for several hours. Then cells were removed, washed, and transferred to minimal media containing lactose as the only sugar source. A few cells were able to grow on minimal media with lactose, and formed colonies. How did these few cells become *trp + lac Z + met + bio +*?

    A. Transformation   B. Transduction

    C. Sexduction   D. Conjugation

5. *lacA* encodes _____ .

    A. β-galactoside transacetylase   B. β-galactoside permease

    C. β-galactosidease   D. All of the above

Answers

    1. C   2. B   3. C   4. D   5. A

SHORT ANSWER TYPE QUESTIONS

1. Write a note on components of *lac* operon.
2. Explain the structure of inducible gene.
3. Write a note on housekeeping genes.
4. Differentiate between constitutive and facultative genes.
5. Describe the functions of promoter, operator and structural genes.

LONG ANSWER TYPE QUESTIONS

1. Describe the model of operon.
2. Differentiate between inducible and repressible operon.
3. Explain in detail the working of *lac* operon.
4. Write a detailed account on feedback inhibition.
5. Explain the process of attenuation.

Regulation of Gene Expression in Eukaryotes

25

INTRODUCTION

The DNA of an organism encodes all of the RNA and protein molecules that compose its cells. However, a complete description of its DNA sequence tells us nothing about how the organism works and is able to maintain itself alive. We further need to know how the most essential elements in a genome—the genes—are used. Even the simplest single-celled bacterium can use its genes selectively that is by switching genes on and off to produce different metabolic enzymes depending on the food sources available in its environment. In eukaryotes, gene expression is under an even more elaborate control because its regulation is not only responsible for an organism's response to its environment, but also stands at the basis of cell differentiation.

Regulation can happen at various levels (before, during or after transcription or translation) and concerns many different elements (chromosome structure, various types of regulatory signals etc.). The expression of a gene is never a single affair but is part of a vast system involving many genes whose production is finely tuned, with genes often directly or indirectly regulating themselves inside complex networks.

More than thirty years ago, Francois Jacob and Jacques Monod described the first paradigm for differential gene expression. They proposed that regulation of the switching on and off of genes involved in sugar (lactose) metabolism in bacterial cells is accomplished by the binding of regulatory proteins to regulatory DNA sequences. Gene expression in eukaryotic cells occurs when: (i) genes encoded by the cell's

DNA are copied into mRNAs by transcription (much of the regulation of differential gene expression takes place at the level of transcription) and (ii) specific mRNAs are used as templates to make specific proteins in translation. But every gene is not expressed in every cell type (differential gene regulation). Thus, regulation of gene expression can be described as: A human cell contains about 100,000 genes, but each specific cell type expresses only about 10,000 genes. The production of a specific collection of proteins by a specific cell type is the result of differential gene expression.

PROMOTERS AND ENHANCERS

Promoters constitute binding sites for RNA polymerase and the general transcription factors. In eukaryotic cells, RNA Polymerase II (RNAPII; a large complex of proteins whose enzymatic activity does the actual copying of DNA into RNA during transcription) is responsible for the transcription of protein-encoding genes. RNAP's helper proteins are the so-called general transcription factors. Together, RNAPII and the general transcription factors are involved in the transcription of nearly every protein-encoding gene. Enhancers are DNA binding sites for gene-specific trancriptional regulatory proteins (proteins that regulate the transcription of one or a subset of genes). In bacteria, generally one protein binds to a single regulatory sequence in the DNA and turns transcription on or off. In eukaryotes, gene transcription is regulated by enhancer DNA sequences that contain not one but multiple regulatory DNA sequence elements that represent binding sites for multiple transcription factors. DNA-bound transcriptional regulatory proteins come together [through protein-protein interactions] to form gene-specific transcriptional regulatory complexes. So it is not a single factor and binding site, but, rather, the specific combination of DNA sequence elements and regulatory proteins that determines which genes will be transcribed in which cell types, a concept known as combinatorial regulation of gene expression. These gene-specific transcription complexes relay transcriptional regulatory information to RNAPII and help to determine whether a gene is transcriptionally active or silent under specific conditions (e.g., at a specific time in development; in a particular cell type).

Now that the idea of how transcriptional regulatory complexes are built is clear, let us return to the problem of cell-type specific gene expression. If every cell in the body contains the same DNA, why do only brain cells produce neurotransmitter proteins and only liver cells produce albumin? Through studies conducted over the past 10 to 15 years, transcriptional regulation has emerged as a major control point in cell-type specific gene expression. The specific transcription can be considered as an example of cell-type transcriptional regulatory processes that take place in the fat cell.

REGULATION OF GENE EXPRESSION

Gene expression in eukaryotes is controlled by a variety of mechanisms that range from those that prevent transcription to those that prevent expression after the protein has been produced. The various mechanisms can be placed into one of these four categories: transcriptional, post-transcriptional, translational, and post-translational.

Chromatin Structure: Unlike DNA of prokaryotes that is usually circular, smaller, and less elaborately folded and structured, eukaryotic DNA is complex with large amounts of protein to form chromatin, is highly extended and tangled during interphase but is condensed into short thick chromosomes during mitosis. Thus, eukaryotic chromosomes contain enormous amounts of DNA which must be packed correctly forming nucleosomes.

Interphase chromatin is much less condensed than mitotic chromatin. The two types of chromatin are: Heterochromatin, remains highly condensed and is not actively transcribed, and Euchromatin, less condensed and is actively transcribed.

In eukaryotes most of the DNA codes for protein or RNA, and coding sequences may be interrupted by long stretches of noncoding DNA called introns. The introns, contain promoter sequence at 5' end, the RNA polymerase attaches to promoter and transcribes both introns and exons. The primary RNA transcript (hnRNA) is processed by the removal of introns, addition of GTP cap at 5' end, addition of poly-A at 3' end. The eukaryotic genes may be regulated by other noncoding control sequences called enhancers which are located thousands of bases away from promoter and enhance transcription.

Gene amplification: It is the selective synthesis of DNA that results in multiple copies of a single gene. It permits the oocytes to make huge numbers of ribosomes for the synthesis of the needed protein.

DNA Methylation: DNA methylation is the addition of methane groups (-CH3) to bases of DNA. It is a common method of gene silencing. DNA is typically methylated by methyltransferase enzymes on cytosine nucleotides in a CpG dinucleotide sequence (also called " CpG island").

Rearrangements in the Genome

1. Transposons: These are the segments that move DNA from one location to another, either on the same chromosome or on a different one. If a transposon inserts itself within another gene, it can prevent the gene from expressing itself.

2. Yeast Mating Type Genes: There are two mating types in yeast and rearrangements can change from one to the other.

Thus, the physical structure of the DNA, as it exists compacted into chromatin, can affect the ability of transcriptional regulatory proteins (termed transcription factors) and RNA polymerases to find access to specific genes and to activate transcription from them. The presence of the histones and CpG methylation most affect accessibility of the chromatin to RNA polymerases and transcription factors.

Transcriptional initiation: This is the most important mode for control of eukaryotic gene expression. Specific factors that exert control include the strength of promoter elements within the DNA sequences of a given gene, the presence or absence of enhancer sequences (which enhance the activity of RNA polymerase at a given promoter by binding specific transcription factors), and the interaction between multiple activator proteins and inhibitor proteins.

Transcription of the different classes of RNAs in eukaryotes is carried out by three different polymerases. RNA pol I synthesizes the rRNAs, except for the 5S species. RNA pol II synthesizes the mRNAs and some small nuclear RNAs (snRNAs) involved in RNA splicing. RNA pol III synthesizes the 5S rRNA and the tRNAs. The vast majority of eukaryotic RNAs are subjected to post-transcriptional processing.

The most complex controls observed in eukaryotic genes are those that regulate the expression of RNA pol II-transcribed genes, the mRNA genes. Almost all eukaryotic mRNA genes contain a basic structure consisting of coding exons and non-coding introns and basal promoters of two types and any number of different transcriptional regulatory domains. The basal promoter elements are termed CCAAT-boxes (pronounced "cat") and TATA-boxes because of their sequence motifs. The TATA-box resides 20 to 30 bases upstream of the transcriptional start site and is similar in sequence to the prokaryotic Pribnow-box (consensus $TATA^T/_AA^T/_A$, where $^T/_A$ indicates that either base may be found at that position) (Fig. 25.1).

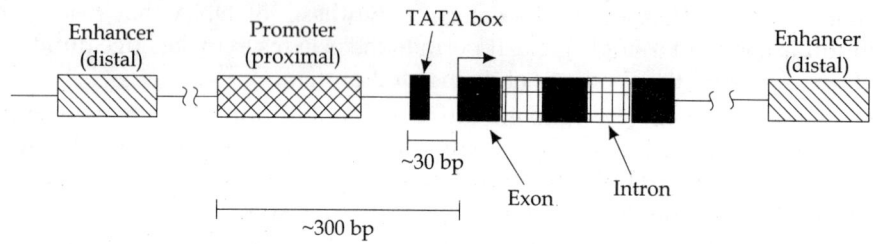

FIG. 25.1   Typical structure of a eukaryotic mRNA gene

Numerous proteins identified as TFIIA, B, C, etc. (for transcription factors regulating RNA pol II), have been observed to interact with the TATA-box. The CCAAT-box (consensus $GG^T/_CCAATCT$) resides 50 to 130 bases upstream of the transcriptional start site. The protein identified as C/EBP (for CCAAT-box/Enhancer Binding Protein) binds to the CCAAT-box element.

There are many other regulatory sequences in mRNA genes, as well, that bind various transcription factors. Theses regulatory sequences are predominantly located upstream (5′) of the transcription initiation site, although some elements occur downstream (3′) or even within the genes themselves. The number and type of regulatory elements to be found varies with each mRNA gene. Different combinations of transcription factors also can exert differential regulatory effects upon transcriptional initiation. The various cell types each express characteristic combinations of transcription

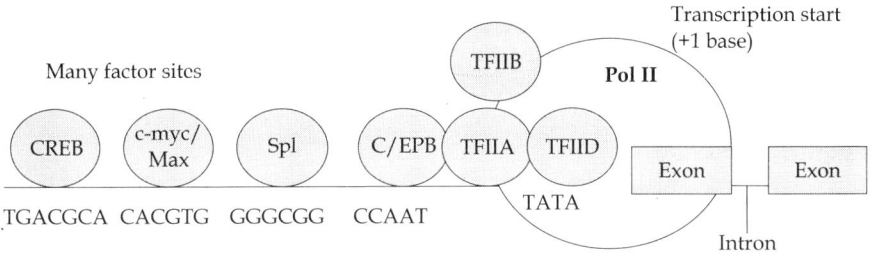

FIG. 25.2

factors; this is the major mechanism for cell-type specificity in the regulation of mRNA gene expression (Fig. 25.2).

Transcript Processing and Modification: After transcription, the RNA must be processed before it can be translated. RNA processing involves addition of a 5' cap, addition of a 3' poly (A) tail, and removal of introns. This processing represents another level of regulation of gene expression, particularly in regard to splicing out of introns. Regulation can be of two types: (a) whether RNA gets processed; and (b) which exons are retained in the mRNA.

The first type of regulation can determine whether or not an mRNA gets translated. If an RNA is not processed, it will not be transported out of the nucleus, and will not be translated.

The second type of regulation can affect the function of the protein produced. Some genes have exons that can be exchanged in a process known as **exon shuffling**. For example, a gene with four exons might be spliced differently in two different cell types. In cell 1, exons 1, 2, and 4 would be used in the mRNA:

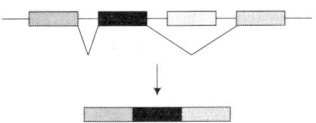

In cell 2, on the other hand, exons 1, 3, and 4 would be used:

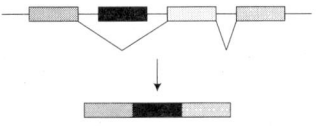

In each of these cases, the polypeptide produced could have a different function. In mammals, for example, the calcitonin gene produces a hormone in one cell type, and a neurotransmitter in another cell type, due to exon shuffling. In *Drosophila*, alternate splicing of the *sex-lethal* RNA can produce an mRNA encoding a functional polypeptide, or one with a premature stop codon that encodes a short, nonfunctional polypeptide.

Translational initiation: Since many mRNAs have multiple methionine codons, the ability of ribosomes to recognize and initiate synthesis from the correct AUG codon can affect the expression of a gene product. Several examples have emerged demonstrating that some eukaryotic proteins initiate at non-AUG codons. This phenomenon has been known to occur in *E. coli* for quite some time, but only recently it has been observed in eukaryotic mRNAs.

Post translational modification: These mechanisms starts after the protein has been produced.

Protein Activation: Some proteins are not active when they are first formed. They must undergo modification such as folding, enzymatic cleavage, or bond formation. Common modifications also include glycosylation, acetylation, fatty acylation, disulfide bond formations, etc. For example, Bovine proinsulin is a precursor to the hormone insulin. It must be cleaved into two polypeptide chains and about 30 amino acids must be removed to form insulin.

Feedback Control: Some enzymes in a metabolic pathway may be negatively inhibited by products of the pathway.

Hormonal Control of Gene Expression: Hormones are molecules that are produced in one cellular location in an organism, and whose effects are seen in another tissue or cell type. In mammals, hormones can be proteins or steroids. The protein hormones do not enter the cell, but bind to receptors in the cell membrane and mediate gene expression through intermediate molecules. Steroids though actually enter the cell and interact with steroid receptor proteins to control gene expression.

Glucocorticoid is one type of steroid whose method of controlling gene expression has been determined. The steroid interacts with a receptor protein, and this interaction serves two functions. First, binding stimulates the release of the protein Hsp90 that is bound to the receptor protein. When Hsp90 is bound to the receptor protein, gene expression is not activated. This would be expected, if the steroid is the signal required for the expression of specific genes in the tissue. When the steroid is bound and Hsp90 is released, the receptor protein forms a dimer (two proteins together) with another copy of the receptor protein. This complex then binds to specific enhancer sequences and gene expression is activated.

OBJECTIVE TYPE QUESTIONS

1. Prokaryotes and eukaryotes use several methods to regulate gene expression, but the most common method is
   A. Translational control
   B. Transcriptional control
   C. Post-transcriptional control
   D. Control of mRNA passage from the nucleus

2. A(n) _____ is a piece of DNA with a group of genes that are transcribed together as a unit.

    A. Promoter
    B. Repressor
    C. Operator
    D. Operon

3. What effect would the addition of lactose have on a repressed *lac* operon?

    A. The operator site on the operon would move
    B. It would reinforce the repression of that gene
    C. The *lac* operon would be transcribed
    D. It would have no effect whatsoever

4. A type of DNA sequence that is located far from a gene but can promote its expression is a(n)

    A. Promoter
    B. Activator
    C. Enhancer
    D. TATA box

5. Which of the following is *not* found in a eukaryotic transcription complex?

    A. Activator
    B. RNA
    C. Enhancer
    D. TATA-binding protein

Answers

1. B 2. D 3. C 4. C 5. B

SHORT ANSWER TYPE QUESTIONS

1. Write notes on:
    (i)   euchromatin
    (ii) heterochromatin
    (iii) promoters
    (iv) exon shuffling
    (v)   DNA methylation
    (vi) gene splicing
    (vii) gene amplification
2. List the 7 differences between eukaryotes and prokaryotes that are relevant to gene regulation.
3. Explain regulation of RNA processing.
4. Explain the effect of chromatin structure on gene expression.
5. Describe the post-transcriptional control in eukaryotes.

LONG ANSWER TYPE QUESTIONS

1. Explain the process of regulation of gene expression in eukaryotes.
2. Differentiate between promoters and enhancers.
3. Write a note on hormonal control of gene expression.

4. Suggest how environmental concerns and development into tissues results in a need for different types of transcriptional control in eukaryotes.

5. Describe the different classes of eukaryotic promoters (I, II and III).

Mutations

26

INTRODUCTION

Mutations are sudden, permanent, transmissible (if the change is in a germ cell) changes in the genetic material. An organism exhibiting a novel phenotype as a result of the presence of a mutation is called a mutant. The term 'mutation' was used by Hugo De Vries (1901) to describe phenotypic changes which were heritable. Thus, he differentiated between environmental and heritable variations and called them somatic and germinal mutations.

The heritable changes, in broad sense, refer to all those changes which result in altered pattern of heredity. These can be classified as:

Chromosomal mutations: these include the structural changes in chromosomes.

Gene or Point mutations: these include changes that alter the chemical structure of the gene at molecular level.

The term mutation is now used in a restricted sense to cover only the changes in individual genes called gene mutations or point mutations and excluding the changes in chromosome structure or number. Hence, a mutation is any change in the sequence of the DNA encoding a gene. Our studies on mutations have lead to a better understanding about the nature of genes. Originally, genes were thought to have beads-on-string structure where each bead was a single entity responsible for a phenotype. This theory led to the concept that only single mutation was possible for a specific gene. Detailed genetic experiments proved that the gene actually consists of many individual units and specific changes in these units can lead to several mutant phenotypes. We now know that these units are nucleotides. Therefore, understanding the nature of mutations is important to our understanding of a gene.

Gene mutations or point mutations involve changes in single base-pairs, the substitution of one base-pair for another, or the duplication or deletion of single base-pairs.

Without mutations, all genes would exist in only one form and the organisms would not be able to evolve and adapt to environmental changes. Hence, mutations are necessary to provide genetic variability to allow organisms to adapt to new environmental conditions.

HISTORY

The earliest recorded point mutation was observed by Seth Wright in 1791 on his farm located on the banks of Charles River in Dover, Massachusetts. Wright noticed a peculiar male lamb with unusually short legs in his flock of sheep. He thought that it would be an advantage to have a whole flock of these short-legged sheep which could not jump over the low stone fences in his neighborhood. He used this new short legged ram for breeding his 15 ewes. Two of the 15 lambs produced had short legs, the short legged sheep were then bred together and a line was developed in which the new trait was expressed in all individuals. The short legged breed of sheep was known as Ancon breed (Fig. 26.1).

FIG. 26.1 Short-legged sheep of ancon breed

H.J. Muller (1890-1967)   T.H. Morgan (1866-1945)

FIG. 26.2

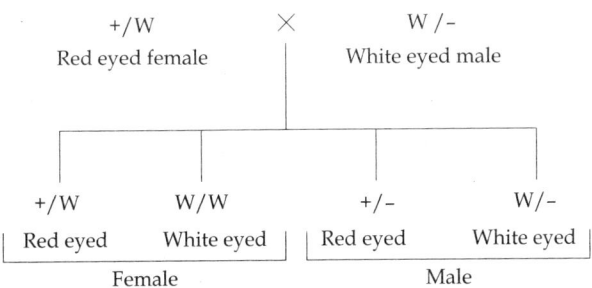

+/W	$\times$	W /−
Red eyed female		White eyed male

+/W	W/W	+/−	W/−
Red eyed	White eyed	Red eyed	White eyed
Female		Male	

FIG. 26.3 Appearance of a white-eyed male fly in a cross between red-eyed female and white eyed male

T.H. Morgan in 1910 (Fig. 26.2) began the scientific study on mutations. He worked on *Drosophila melanogaster* and reported white eyed male individuals among red eyed male individuals (Fig. 26.3). It was found that the gene for this character is located on sex chromosome (X-chromosome) and expresses itself in a male individual. When these rare white eyed males were crossed to their sister red eyed females, white eyed females could also be obtained in some cases. This signified that the females involved were heterozygous.

After the discovery of white eyed mutant, Morgan and his coworkers studied different mutations in *Drosophila* and about 500 different mutations were observed by different scientists all over the world. Moreover, mutations were also studied in other organisms like rodents, fowl, pea, maize, snapdragon, *E. coli*, *Neurospora*, bacteriophage, man, etc.

CLASSIFICATIONS

There are different classifications of mutations depending on different aspects.

Classification of mutations on the basis of occurrence:

(i) **Spontaneous mutations:** These occur randomly under the natural environmental conditions and have a low frequency.

(ii) **Induced mutations:** These arise from exposure to mutagenic agents such as radiations and/or chemicals.

(iii) **Gametic/germinal mutations:** These occur in the germ-line cells and are heritable.

(iv) **Somatic mutations:** These occur in the non-sex cells and are non-heritable.

Classification of mutations on the basis of their effect on organism:

(i) **Morphological mutations:** These involve alterations in external form.

(ii) **Nutritional or biochemical mutations:** These result due to the deficiency of a nutrient or a chemical compound and can be overcome by supplying the same.

(iii) **Behavioral mutations:** These alter the behavior patterns of an organism.

(iv) **Regulatory mutations:** Such mutations may disrupt the regulatory mechanisms and permanently activate or inactivate a gene.

(v) **Lethal mutations:** These involve genotypic changes leading to death of an individual.

(vi) **Conditional mutations:** These allow the mutant phenotype to be expressed only under certain conditions called restrictive conditions.

GERMINAL AND SOMATIC MUTATIONS

Eukaryotic organisms have two primary cell types, viz., germ cells and somatic cells. Mutations can occur in either cell type. If a gene is altered in a germ cell, the mutation is termed a germinal mutation. Because germ cells give rise to gametes, some gametes will carry the mutation and it will be passed on to the next generation when the individual successfully mates. Typically germinal mutations are not expressed in the individual containing the mutation. The only instance in which it would be expressed is if it affected gamete production.

Somatic cells give rise to all non-germline tissues. Mutations in somatic cells are called somatic mutations. Because they do not occur in cells that give rise to gametes, the mutation is not passed along to the next generation by sexual means. To maintain this mutation, the individual containing the mutation must be cloned. Two examples of somatic clones are navel oranges and red delicious apples. Horticulturists first observed the mutants. They then grafted mutant branches onto the stocks of 'normal' trees. After the graft was established, cuttings from that original graft were grafted onto tree stocks. In this way the mutation was maintained and proliferated.

Most tissues are derived from a cell or a few progenitor cells. If a mutation occurs in one of the progenitor cells, all of its daughter cells will also express the mutation. For this reason, somatic mutations generally appear as a sector on the mutated individual.

Cancer tumors are a unique class of somatic mutations. The tumor arises when a gene involved in cell division, a proto-oncogene, is mutated. All of the daughter cells contain this mutation. The phenotype of all cells containing the mutation is uncontrolled cell division. This results in a tumor that is a collection of undifferentiated cells called tumor cells.

SPONTANEOUS VERSUS INDUCED MUTATIONS

Some mutations arise as natural errors in DNA replication (or as a result of unknown chemical reactions); these are known as spontaneous mutations. The rates of such mutations have been determined for many species. *E. coli* has a spontaneous mutation rate of $1/10^8$ (one error in every 10^8 nucleotides replicated). Humans have a higher spontaneous mutation rate: between $1/10^6$ and $1/10^5$ (probably as a result of the higher complexity of human replication).

Mutations can also be caused by agents in the environment; these are induced mutations. Induced mutations increase the mutation rate over the spontaneous rate. Looking at a single mutation in an individual, one cannot say whether the mutation is spontaneous or induced. Induced mutations can only be discerned by looking at the mutation rate in a population, and comparing it to the spontaneous mutation rate for the species. If the observed mutation rate is higher, then induced mutations can be assumed. Agents in the environment that cause an increase in the mutation rate are called mutagens.

In general, the appearance of a new mutation is a rare event. Most mutations that were originally studied occurred spontaneously. This class of mutation is termed spontaneous mutations. Historically, geneticists recognized these in nature. The mutations were collected, and the inheritance of these mutations was analyzed. But these mutations clearly represent only a small number of all possible mutations. To genetically dissect a biological system further, new mutations were created by scientists by treating an organism with a mutagenizing agent. These mutations are called induced mutations.

Mutations can be induced by several methods. The three general approaches used to generate mutations are radiation, chemical and transposon insertion. The first induced mutations were created by treating *Drosophila* with X-rays. In addition to X-rays, other types of radiation treatments that have proven useful include gamma rays and fast neutron bombardment. These treatments can induce point mutations (changes in a single nucleotide) or deletions (loss of a chromosomal segment).

Chemical mutagens work mostly by inducing point mutations. Point mutations occur when a single base pair of a gene is changed. These changes are classified as transitions or transversions. Transitions occur when a purine is converted to a purine (adenine to guanine or guanine to adenine) or a pyrimide is converted to a pyrimidine (thymine to cytosine or cytosine to thymine). A transversion results when a purine is converted to a pyrimidine or a pyrimidine is converted to a purine.

Two major classes of chemical mutagens are routinely used. These are alkylating agents and base analogs. Each has a specific effect on DNA. Alkylating agents (such as ethyl methane sulphonate (EMS), ethyl ethane sulphonate (EES) and mustard gas) can mutate both replicating and non-replicating DNA. By contrast, a base analog (5-bromouracil and 2-aminopurine) only mutate DNA when the analog is incorporated into replicating DNA. Each class of chemical mutagen has specific effects that can lead to transitions, transversions or deletions.

Scientists are now using the power of transposable elements to create new mutations. Transposable elements are mobile pieces of DNA that can move from one location in a genome to another. Often when they move to a new location, the result is a new mutant. The mutant arises because the presence of a piece of DNA in a wild type gene disrupts the normal function of that gene. As more and more is being learned about genes and genomes, it is becoming apparent that transposable elements are a power source for creating insertional mutants.

The detailed knowledge of the structure and function of transposable elements is now being applied in the pursuit of new mutations. Stocks are created in which a specific type of element is present. This stock is then crossed to a genetic stock that does not contain the element. Once that element enters the virgin stock, it can begin to move around that genome. By carefully observing the offspring, new mutants can be discovered. This approach to developing mutants is termed insertional mutagenesis.

MUTATIONS OCCUR RANDOMLY IN NATURE

Most populations of pests are no longer affected by the anticoagulants and pesticides that have been traditionally used against them. Similarly, the pathogenic microorganisms are becoming resistant to antibiotics such as penicillin and streptomycin developed to control them. The development of these substances produced a new environment for the organisms involved. These organisms responded to the imposed environmental changes by evolving to forms that are resistant to these chemicals by the occurrence of mutations that produced resistance to these pesticides and antibiotics and the mutant organisms were at a large selective advantage in environments where these agents were present. The sensitive organisms were killed and the mutants multiplied to produce new resistant populations.

J. Lederburg and E.M. Lederberg developed a technique called replica plating experiment to demonstrate the random nature of the mutation. Replica plating is a technique in which one or more secondary Petri plates containing different solid (Agar-based) selective growth media (lacking nutrients or containing chemical growth inhibitors such as antibiotics) are inoculated with the same colonies of microorganisms from a primary plate (or master dish), reproducing the original spatial pattern of colonies. The technique involves pressing a velvet-covered disk to a primary plate, and then imprinting secondary plates with cells in colonies removed by the velvet from the original plate. Generally, large numbers of colonies (roughly 30-300) are replica plated due to the difficulty in streaking each out individually onto a separate plate (Fig. 26.4).

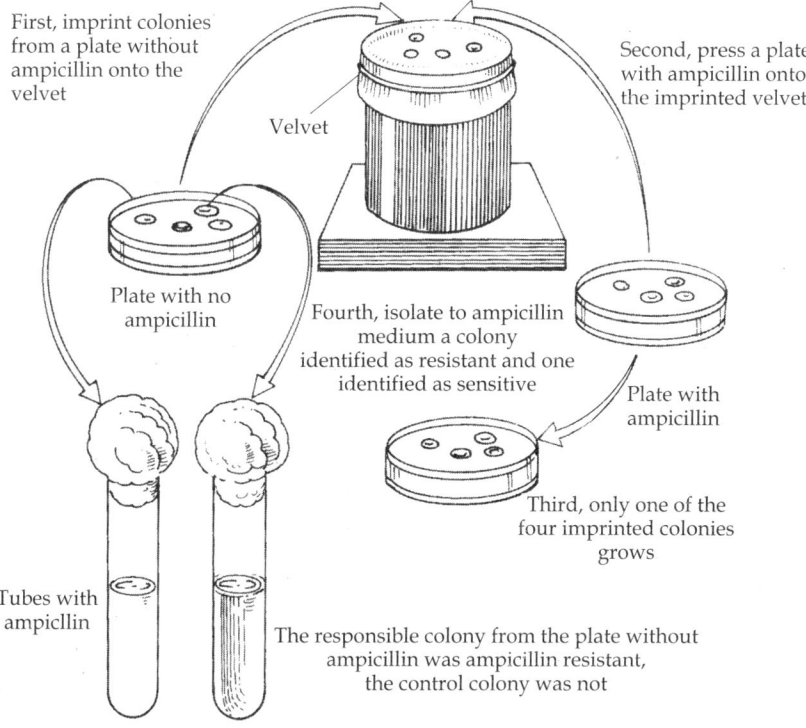

First, imprint colonies from a plate without ampicillin onto the velvet

Second, press a plate with ampicillin onto the imprinted velvet

Velvet

Plate with no ampicillin

Fourth, isolate to ampicillin medium a colony identified as resistant and one identified as sensitive

Plate with ampicillin

Third, only one of the four imprinted colonies grows

Tubes with ampicllin

The responsible colony from the plate without ampicillin was ampicillin resistant, the control colony was not

FIG. 26.4 Lederberg's replica plating technique

The purpose of replica plating is to be able to compare the master plate and any secondary plates to screen for a selectable phenotype. For example, a colony which appeared on the master plate but failed to appear at the same location on a secondary plate shows that the colony was sensitive to a substance on that particular secondary plate. Common screenable phenotypes include auxotrophy and antibiotic resistance.

Replica plating is especially useful for negetive selection. For example, if one wanted to select colonies that were sensitive to ampicillin, the primary plate could be replica plated on a secondary Amp$^+$ agar plate. The sensitive colonies on the secondary plate would die but the colonies could still be deduced from the primary plate since the two have the same spatial patterns from ampicillin resistant colonies. The sensitive colonies could then be picked off from the primary plate.

By increasing the variety of secondary plates with different selective growth media, it is possible to rapidly screen a large number of individual isolated colonies for as many phenotypes as there are secondary plates.

This experiment suggests that the environmental stress does not direct or cause genetic changes rather, it simply selects rare pre-existing mutations that result in phenotypes that are better adapted to the new environment.

THE MOLECULAR BASIS OF MUTATION

The double-helix structure of DNA proposed by Watson and Crick (1953), pointed out that the bases in DNA are not static. Hydrogen atoms can move from one position in a purine or pyrimidine to another position, for example, from an amino group to ring nitrogen. Such chemical fluctuations are called tautomeric shifts. In a DNA molecule, under normal conditions, the purine adenine (A) is linked to the pyrimidine thymine (T) by two hydrogen bonds, while the purine guanine (G) is linked to the pyrimidine cytosine (C) by three hydrogen bonds. Due to the tautomeric shifts, the more stable keto forms of thymine and guanine and amino forms of adenine and cytosine may shift to less stable enol and imino forms, respectively (Fig. 26.5).

When the bases are present in their rare imino or enol states, they can form adenine-cytosine and guanine-thymine base pairs and that results in two main types of point mutations: base substitutions and base additions or deletions (Fig. 26.6).

Base substitutions are those mutations in which one base pair is replaced by another. Base substitutions again can be divided into two subtypes: transitions and transversions. A transition is the replacement of a base by the other base of the same chemical category (purine replaced by purine: either A to G or G to A; pyrimidine replaced by pyrimidine: either C to T or T to C). A transversion is the opposite—the replacement of a base of one chemical category by a base of the other (pyrimidine replaced by purine: C to A, C to G, T to A, T to G; purine replaced by pyrimidine: A to C, A to T, G to C, G to T) (Fig. 26.7). In describing the same changes at the double-stranded level of DNA, we must state both members of a base pair: an example of a transition would be G-C $\rightarrow$ A-T; that of a transversion would be G-C $\rightarrow$ T-A as the complementary change will take place on the other strand.

Addition or deletion mutations are actually of *nucleotide* pairs; nevertheless, the convention is to call them *base*-pair additions or deletions. The simplest of these mutations are single-base-pair additions or single-base-pair deletions. There are examples in which mutations arise through simultaneous addition or deletion of multiple base pairs at once.

For single-base substitutions, there are several possible outcomes, which are direct consequences of two aspects of the genetic code: degeneracy of the code and the existence of translation termination codons.

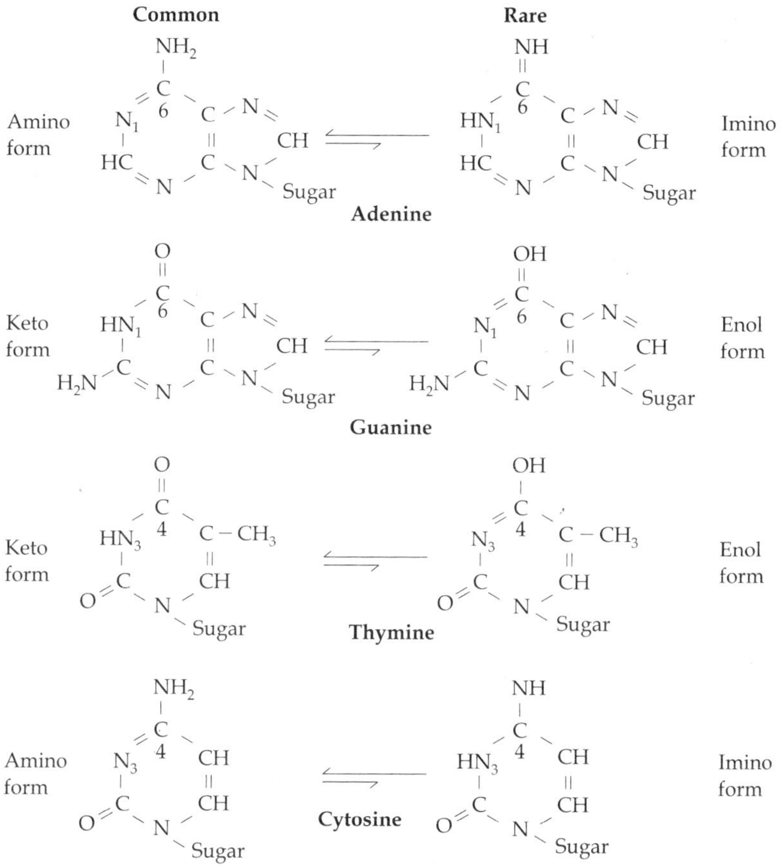

FIG. 26.5 Tautomeric forms of the four common bases of DNA

1. **Silent substitutions:** the mutation changes one codon into another codon which codes for the same amino acid and the amino acid sequence in polypeptide chain is unaltered.

2. **Missense mutations:** the codon for one amino acid is replaced by a codon for another amino acid.

3. **Nonsense mutations:** the codon for one amino acid is replaced by a translation termination (stop) codon.

Silent substitutions never alter the amino acid sequence of the polypeptide chain. The severity of the effect of missense and nonsense mutations on the polypeptide will differ on a case-by-case basis. For example, if a missense mutation causes the substitution of a chemically similar amino acid, referred to as a synonymous substitution then it is likely that the alteration will have a less-severe effect on the structure and function of protein. Alternatively, chemically different amino acid substitutions, called non-synonymous substitutions, are more likely to produce severe changes in protein structure

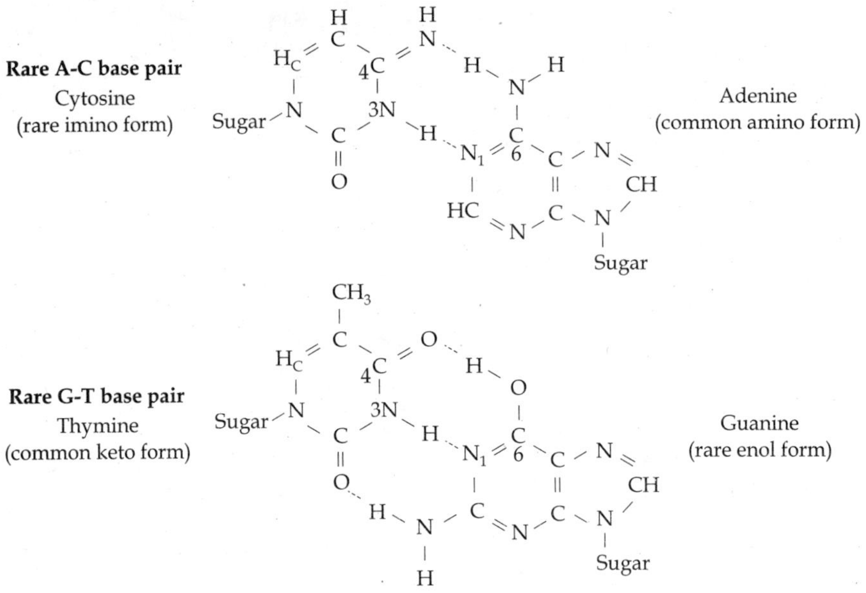

FIG. 26.6 Examples of mismatched A-C and G-T base pairs that form when one of the bases exists in rare tautomeric form

and function. Nonsense mutations will lead to the premature termination of translation. Thus, they have a considerable effect on protein function. Typically, unless they occur very close to the 3' end of the open reading frame, so that only a partly functional truncated polypeptide is produced, nonsense mutations will produce completely inactive protein products.

Like nonsense mutations, single-base additions or deletions have consequences on polypeptide sequence that extend far beyond the site of the mutation itself. Because the sequence of mRNA is "read" by the translational apparatus in groups of three base pairs (codons), the addition or deletion of a single base pair of DNA will change the reading frame starting from the location of the addition or deletion and extending through to the carboxyl-terminal of the protein. Hence, these lesions are called frameshift mutations. These mutations cause the entire amino acid sequence to bear no relation to the original amino acid sequence. Thus, frameshift mutations typically exhibit complete loss of normal protein structure and function.

To illustrate the effects of these mutations, consider the following phrase, read as a triplet code (just like the genetic code):

The fat cat ate the hot dog.

A base substitution might have an effect like this:

The fat car ate the hot dog.

or perhaps:

The fat cat ate the hot hog.

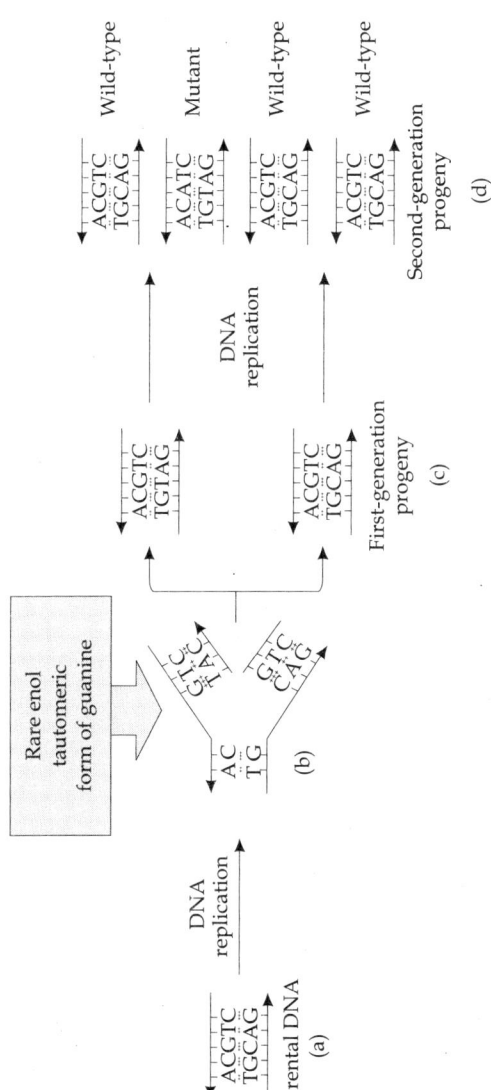

FIG. 26.7   Mutation via tautomeric shifts in the bases of DNA

In each case, the phrase still makes sense, but the meaning has been slightly changed.

An insertion, on the other hand, would have a more profound effect:

The fma tca tat eth eho tdo g.

Insertion of a single letter ('m' in this case) causes the phrase to become gibberish, because the reading frame has been changed. A deletion would have the same effect:

The atc ata tet heh otd og.

RADIATION INDUCED MUTATION

Radiations are capable of inducing mutations in DNA. Ionizing radiations, such as X-rays, gamma rays and cosmic rays are of high energy and can penetrate living tissues. In this process, these high energy rays collide with atoms and cause the release of electrons, forming ions. Ultraviolet rays dissipate their energy to the atoms they encounter, thereby raising the electrons in their outer orbitals to higher energy levels, a state referred to as excitation. Molecules containing atoms in either ionic forms or excited states are chemically more reactive and this forms the basis of mutagenic effects of ionizing radiation and ultraviolet light.

Ionizing radiation produces a range of effects on DNA both through free radical effects and direct action. It can cause:

- breaks in one or both strands, this can lead to rearrangements, deletions, chromosome loss and death (if unrepaired)
- damage and/or loss of bases
- crosslinking of DNA to itself or proteins

The genetic effects of radiation were reported by Muller in 1927 in *Drosophila* and by Stadler in 1928 in plants (barley); both showed that the frequency of induced mutations is a function of X-ray dose. Their experiments revealed that there was a linear relationship between X-ray dose and induced mutation level, that there was no threshold or 'safe' dose of radiation and that all doses are significant, and finally, that 'split dose' experiments showed that the genetic effects of radiation are cumulative (Fig. 26.8).

In 1927, H.J. Muller (Fig. 26.2) demonstrated that the X-ray treatment increased the frequency of sex-linked recessive lethal mutations in *Drosophila melanogaster.*

The different methods for the detection of sex-linked lethals are presented below:

The ClB method: This method involves the use of a ClB stock which carries an inversion in the heterozygous state to work as crossover suppressor (C), a recessive lethal (l) on X-chromosome in heterozygous state and a dominant marker, *Barred* (B) for the barred eye. Male flies irradiated for induction of mutations were crossed to ClB females. The ClB female flies thus obtained as

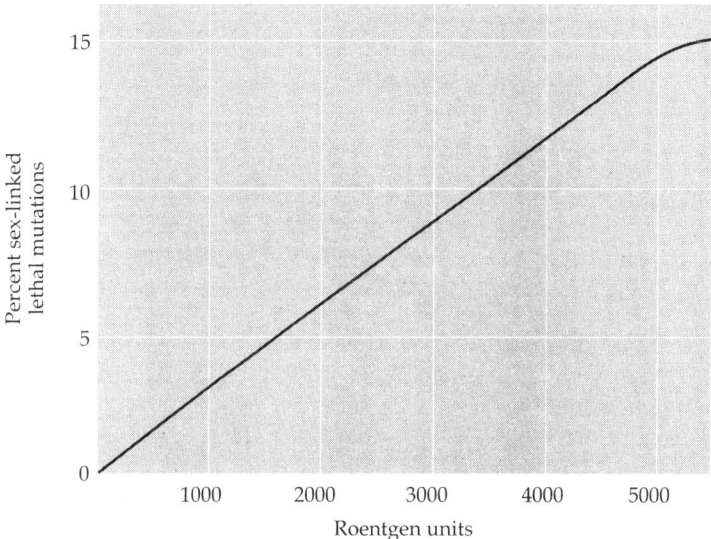

FIG. 26.8 Relationship between the frequency of sex-linked lethal mutations induced in *Drosophila* sperm and ionizing irradiation disage

progeny are crossed to normal males while the male progeny receiving ClB X-chromosome dies. In the next generation, in a cross between ClB females and normal males, 50% of males receiving ClB X-chromosome die and the other 50% receive the X-chromosome that may or may not carry the induced mutation. In case lethal mutation was induced no males will be observed and if no lethal mutation was induced, 50% males will survive (Fig. 26.9). This method for the detection of sex-linked lethal mutations has been improved and is replaced by Muller-5 method.

Muller-5 method: This method makes use of a *Muller-5 Drosophila* stock, which carries two marker genes, dominant 'Bar' (for barred eye) and recessive 'apricot' but does not carry a lethal as in ClB. In the F_2 generation, 50% males are Muller-5 in phenotype and remaining 50% are wild type. If a lethal mutation is induced in X-chromosome of irradiated male no wild type males would appear in F_2 generation. Hence, absence of wild type males in F_2 is an indication of an induced lethal mutation (Fig. 26.10).

Ultraviolet (non-ionizing) radiation can also cause mutations. UV radiation is less energetic, and therefore non-ionizing, but its wavelengths are preferentially absorbed by bases of DNA and by aromatic amino acids of proteins, so it, too, has important biological and genetic effects.

UV is normally classified in terms of its wavelength: **UV-C** (180-290 nm) — 'germicidal' — most energetic and lethal, it is not found in sunlight because it is absorbed by the ozone layer; **UV-B** (290-320 nm) — major lethal/mutagenic fraction of sunlight; **UV-A** (320 nm-visible) — 'near UV' — also has deleterious effects (primarily because it creates oxygen radicals) but it produces very few

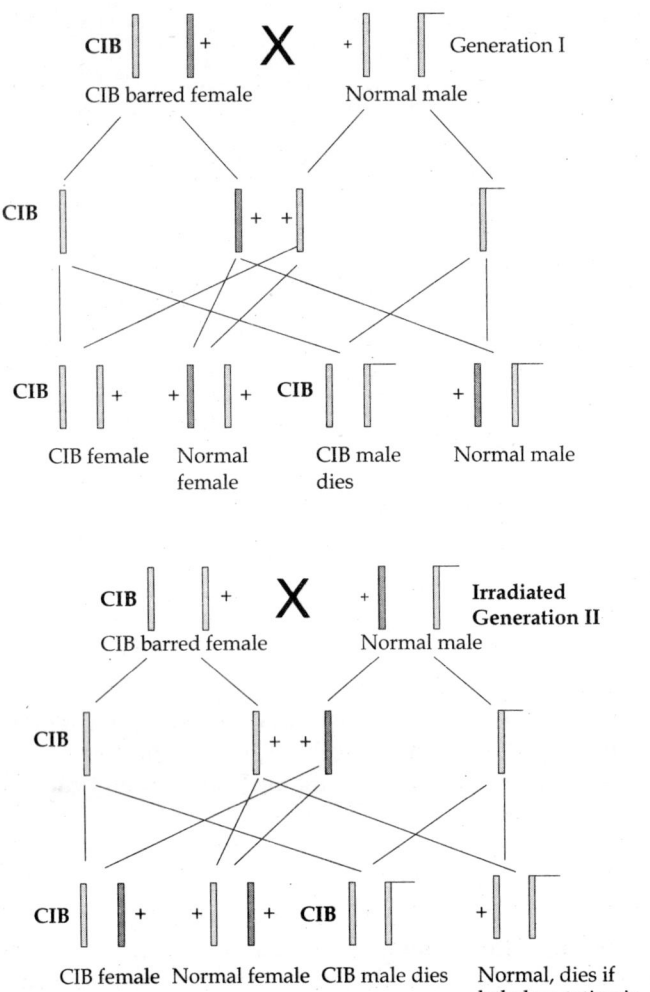

FIG. 26.9 Muller's CIB method for detection of sex-linked lethal mutations in *Drosophila*

pyrimidine dimers. The primary effect of UV on DNA is the creation of thymine dimers. Thymine dimers occur when two thymines are adjacent on a strand of DNA. UV radiation can cause the formation of a covalent bond between the two thymines, which prevents their participation in base pairing. Thymine dimers are very deleterious to a cell. They can completely interrupt replication, effectively causing a cell to die. These dimers, like bulky lesions from chemicals, block transcription and DNA replication and are lethal if unrepaired. They can stimulate mutation and chromosome rearrangement as well.

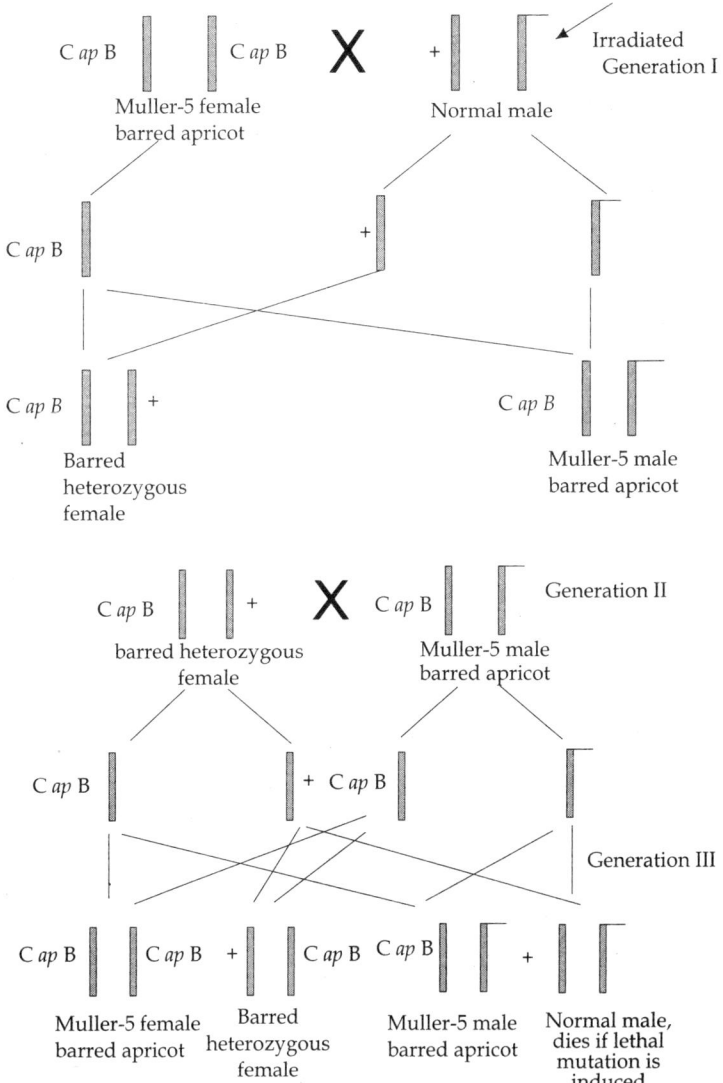

FIG. 26.10 Muller-5 method for detection of sex-linked mutations in *Drosophila*

REPAIR OF THYMINE DIMERS

Several mechanisms are available for the removal or correction of thymine dimers from DNA. The type of mechanism used is dependent upon the conditions of the cell.

- **Photoreactivation:** It has been observed that a brief exposure to blue light following UV exposure can reverse the effects of the UV radiation. In other words, the blue light can cause a thymine dimer to be corrected. This is due to the function of an enzyme called photolyase or photoreactivation enzyme (PRE), which cleaves the covalent bonds linking the thymine dimers using the energy from a photon of blue light. This is essentially a reversal of the reaction that produced the thymine dimer in the first place.

- **Excision Repair:** This is a repair system that does not require light. Instead of just breaking the bonds of the thymine dimer (as was done by photolyase), the excision repair system removes (excises) the region surrounding the offending nucleotides. Several proteins are involved in this process (in prokaryotes these are the products of the 'uvr' genes, for 'UV repair'). The steps of excision repair in prokaryotes are as follows (Fig. 26.11):

  (i) The distortion in the DNA (caused by the thymine dimer) is recognized by a protein complex. A pair of endonucleases makes nicks in the DNA strand on either side of the thymine dimer (generally the nicks are 12 nucleotides apart).

  (ii) The 12-nucleotide piece of DNA between the nicks is removed, and DNA polymerase I fills in the gap left behind.

  (iii) DNA ligase seals the final nick in the DNA.

- **Recombination Repair:** Sometimes, DNA replication begins before a thymine dimer can be repaired by one of the other mechanisms. When the replication machinery hits the dimer, replication stops. Occasionally, the replication will reinitiate just beyond the dimer, leaving a gap in the DNA. Now, if the cell tries to fix the dimer by excision repair, there is no template to use for resynthesis of the DNA. Then, how does the DNA get repaired? The answer is that the cell uses recombination to provide a template strand for repair synthesis.

First, the damaged region undergoes recombination with the complemenary strand on the other DNA molecule. One strand is exchanged between the two DNA molecules. This essentially transfers the gap to the DNA molecule that does not have the dimer. The gap can now be filled in by DNA polymerase I, and the dimer can be repaired by excision, since a template strand now exists (Fig. 26.12).

REPAIR OF MUTATIONS

Cells have mechanisms for minimizing the amount of mutation that takes place. As stated previously, these are not perfect, but they do reduce greatly the frequency of mutation. The two mechanisms are proofreading and mismatch repair dicussed here.

- **Proofreading:** This occurs during DNA replication. As DNA polymerase III adds nucleotides to the growing chain, it checks each one for correct base pairing. If the correct nucleotide has not been inserted, the polymerase uses its 3' to 5' exonuclease activity to remove the incorrect nucleotide. The polymerase can then carry on and insert the

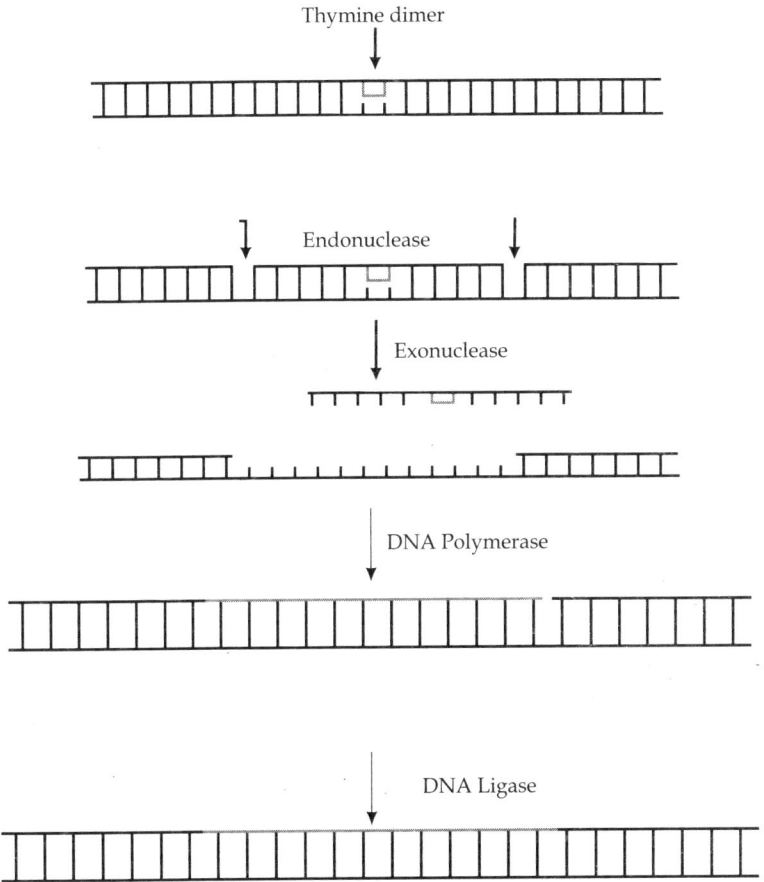

FIG. 26.11   Diagram of the excision repair pathway for the removal of thymine dimers from DNA

correct nucleotide. In bacteria, a wrong nucleotide gets inserted for every 10^5 nucleotides added during replication. Proofreading corrects most of these, so that the overall error rate in replication is one mistake for every 10^7 nucleotides added.

- **Mismatch Repair:** This mechanism is used soon after replication, to correct errors that escaped proofreading. Because mismatched bases do not hydrogen bond, they create a distortion in the double helix, which can be recognized and repaired by excision repair. DNA under normal circumstances is methylated; these methyl groups do not interfere with the function of the DNA in any way. Newly replicated DNA is not methylated however, the methyl groups are added enzymatically after replication. If mismatch repair is done immediately after replication (before methylation occurs), the original DNA strand will be methylated, and the newly-synthesized strand (the one containing

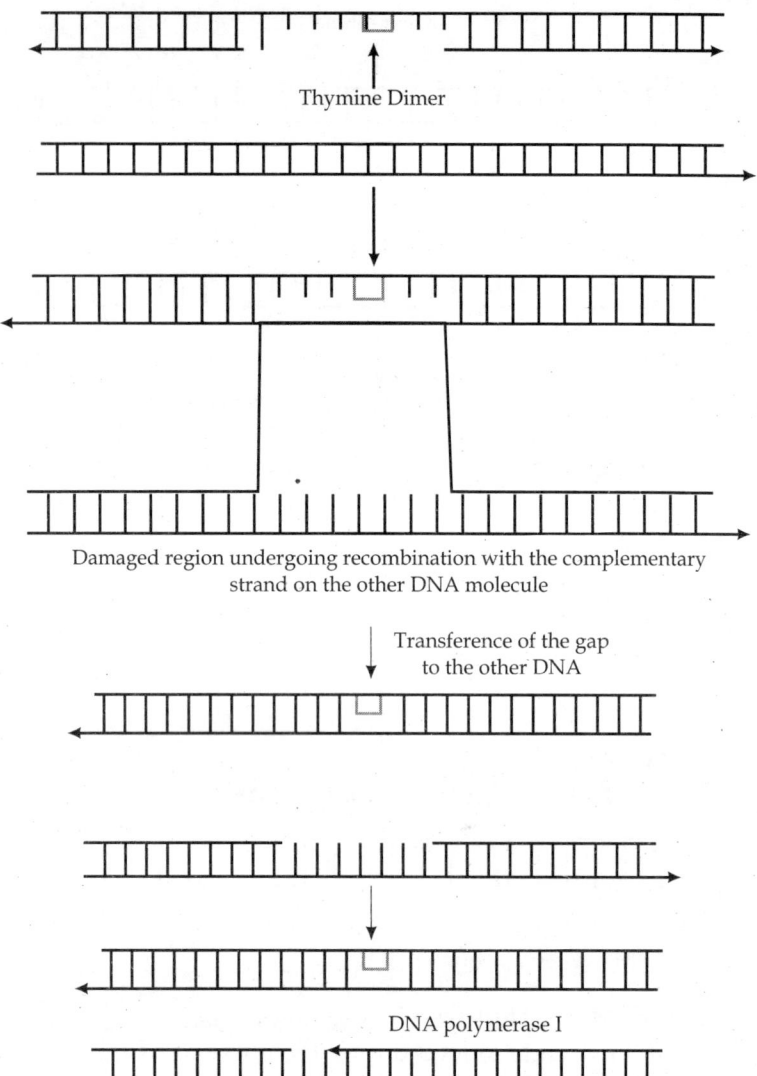

Thymine Dimer

Damaged region undergoing recombination with the complementary strand on the other DNA molecule

Transference of the gap to the other DNA

DNA polymerase I

FIG. 26.12 Diagrammatic representation of recombination repair

the error) will be unmethylated. The mismatch repair system therefore repairs the unmethylated strand.

Excision repair can be used to correct other problems as well. For example, if a deoxynucleotide containing uracil is ever inserted into a DNA molecule, the base is detected and removed by an enzyme called uracil DNA glycosylase. This enzyme removes the base, but leaves the sugar and phosphate in the DNA molecule. This base-less site is then recognized by specific endonucleases, which initiate excision repair.

REPAIR DEFECTS AND HUMAN DISEASES

Several recessive human genetic diseases are known or suspected to be caused by defective genes in repair systems. These defects often lead to an increased incidence of cancer because, as part of the general increased level of mutation, more mutations are produced in genes that can cause a cell to become cancerous.

One cancer-prone disease, xeroderma pigmentosum (XP), results from a defect in any one of eight genes involved in nucleotide excision repair. People suffering from this disorder are extremely prone to UV-induced skin cancers as a result of exposure to sunlight and have frequent neurological abnormalities.

Fanconi's anemia, ataxia-telangiectasia and Bloom's syndrome are three other inherited (autosomal recessive) diseases in humans in which the primary defect may be in a DNA repair pathway.

CHEMICALLY INDUCED MUTATION

The first report of mutagenic action of a chemical was in 1942 by Charlotte Auerbach, who showed that nitrogen mustard (component of poisonous mustard gas used in World Wars I and II) could cause mutations in cells. Since that time, many other mutagenic chemicals have been identified and there is a huge industry and several government agencies dedicated to finding them in food additives, industrial wastes, etc.

It is possible to distinguish chemical mutagens by their modes of action; some of these cause mutations by mechanisms similar to those which arise spontaneously while others are more like radiation in their effects.

Base Analogs

Some chemical compounds are sufficiently similar to the normal nitrogen bases of DNA that they are occasionally incorporated into DNA in place of normal bases; such compounds are called base analogs. Many of these analogs have pairing properties unlike those of the normal bases; thus they can produce mutations by causing incorrect nucleotides to be inserted during replication.

- **Bromouracil** (BU): artificially created compound extensively used in research. Resembles thymine (has Br atom instead of methyl group) and will be incorporated into DNA and pair with A like thymine. It has a higher likelihood for tautomerization to the enol form.
- **Aminopurine:** adenine analog which can pair with thymine or with cytosine; causes A:T to G:C or G:C to A:T transitions. Base analogs cause transitions, as do spontaneous tautomerization events.

The mutagen 5-bromouracil (5-BU) is an analog of thymine that has bromine at the carbon-5 position in place of the CH_3 group found in thymine.

Its mutagenic action is based on enolization and ionization. In 5-BU, the bromine atom is not in a position in which it can hydrogen-bond during base pairing, so the keto form of 5-BU pairs with adenine, as would thymine (Fig. 26.13). However, the presence of the bromine atom significantly alters the distribution of electrons in the base ring, so 5-BU can *frequently* change to either the enol form or an ionized form, the latter of which pairs with guanine (Fig. 26.13). 5-BU causes G-C → A-T or A-T → G-C transitions in the course of replication, depending on whether 5-BU has been enolized or ionized *in situ* or as an incoming base. Hence the action of 5-BU as a mutagen is due to the fact that the molecule spends more of its time in the enol or ion form.

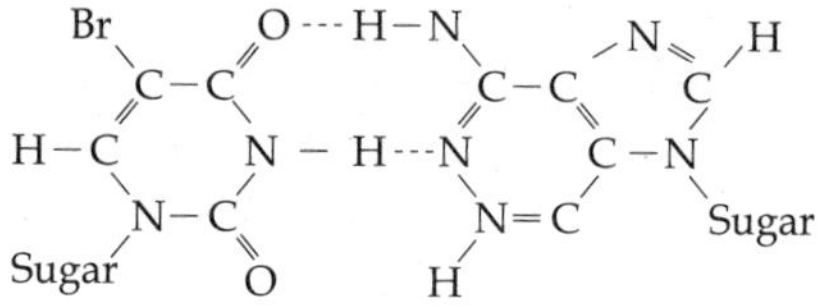

5- Bromouracil:
(Keto from)   **5- Bromouracil: Adenine base pair**

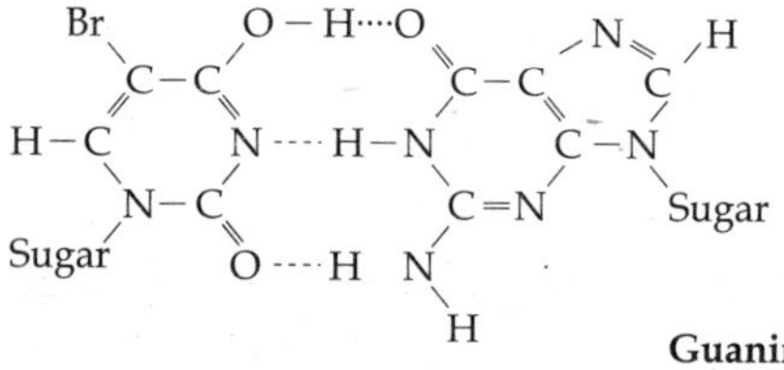

5-Bromouracil:
(enol from)   **5- Bromouracil: Guanine base pair**

FIG. 26.13

2-aminopurine (2-AP) is widely used in research. It is an analog of adenine that can pair with thymine but, when protonated, can also mispair with cytosine. Therefore, when 2-AP is incorporated into DNA by pairing with thymine, it can generate A-T → G-C transitions by mispairing with cytosine in subsequent replications. Or, if 2-AP is incorporated by mispairing with cytosine, then G-C → A-T transitions will result when it pairs with thymine. Genetic studies have shown that 2-AP, like 5-BU, is highly specific for transitions.

Other chemicals which alter structure and pairing properties of bases:

There are many such mutagens; some well-known examples are:

* **Nitrous acid** formed by digestion of nitrites (preservatives) in foods. It converts cytosine to uracil which base pairs with adenine instead of guanine, and adenine is deaminated to hypoxanthine which base pairs

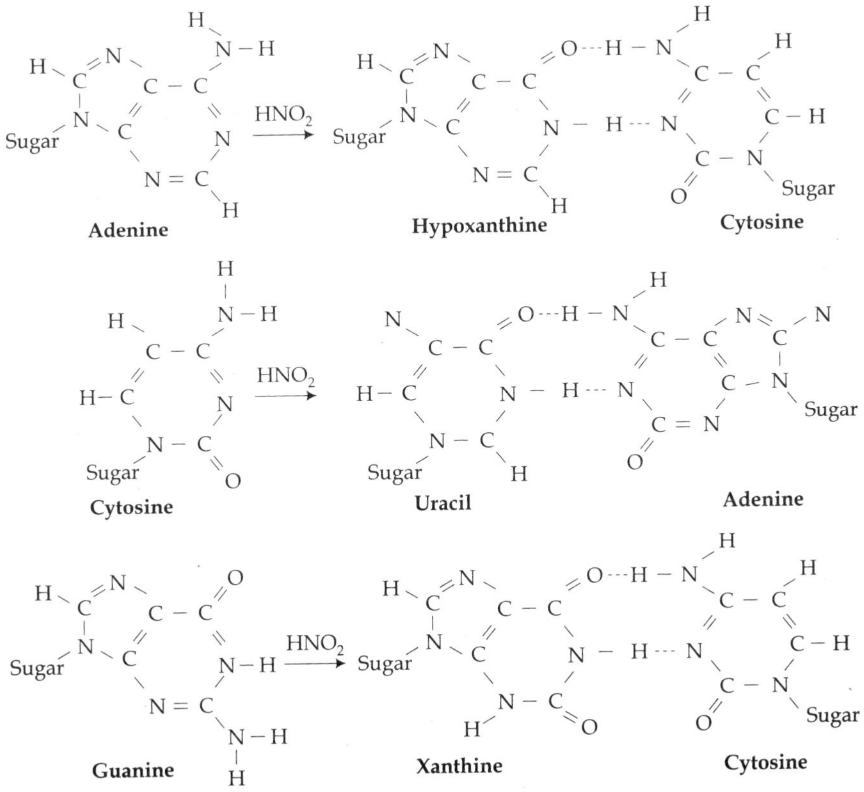

FIG. 26.14 Mutagenic action of nitrous acid

with cytosine instead of thymine. Thus, deamination by nitrous acid, like spontaneous deamination, causes transitions (Fig. 26.14).

- **Nitrosoguanidine, Methyl methanesulfonate, Ethylmethanesulfonate** chemical mutagens that react with bases and add methyl or ethyl groups. Depending on the affected atom, the alkylated base may then degrade to yield a baseless site, which is mutagenic and recombinogenic, or mispair to result in mutations upon DNA replication.

Intercalating Agents

Acridine orange, Proflavin, Ethidium bromide (used in labs as dyes and mutagens)

All are flat, multiple ring molecules which interact with bases of DNA and insert between them. This insertion causes a 'stretching' of the DNA duplex and the DNA polymerase is 'fooled' into inserting an extra base opposite an intercalated molecule. The result is that intercalating agents cause frameshifts.

FREQUENCY OF MUTATIONS

The spontaneous mutation rate varies. Large gene provides a large target and tends to mutate more frequently. A study of the five coat color loci in mice showed that the rate of mutation ranged from 2×10^{-6} to 40×10^{-6} mutations per gamete per gene. Data from several studies on eukaryotic organisms shows that in general the spontaneous mutation rate is $2\text{-}12 \times 10^{-6}$ mutations per gamete per gene. Given that the human genome contains 100,000 genes, we can conclude that 1-5 human gametes would contain a mutation in some gene.

PRACTICAL APPLICATIONS OF MUTATIONS

Most of the mutations are deleterious and make the organisms containing the mutations less efficient; yet, the possibility of developing new desirable traits through induced mutations has led to new avenues in plant breeding. Plant breeders through out the world have worked out such possibilities and have succeeded in developing mutant varieties of barley, wheat, oats, soyabeans, tomatoes and fruit trees etc. that provide increased yield, resistance to diseases, increased protein content, etc.

OBJECTIVE TYPE QUESTIONS

1. The ultimate source of new genetic variation in a species is
   A. Mutation
   B. The combination of parental chromosomes during sexual reproduction
   C. The Hardy-Weinberg equation
   D. All of the above

2. Which of the following statements is true about mutations?
   A. They can produce new alleles of existing genes.
   B. They can be inherited if they are in somatic cells.
   C. They are never as simple as an error in a single codon in a DNA molecule.
   D. A and B

3. Human mutations typically occur at a rate of about _____ per sex cell in normal, healthy people.
   A. 1   B. 1,00
   C. 1000   D. 10,000

4. Most known human mutations
   A. are caused by radiation   B. appear to be spontaneous
   C. are due to viruses   D. are caused by chemicals

5. In order for a mutation to be selected for or against by natural selection, ordinarily it must be
   A. A gross chromosomal rearrangement or an irregular number of chromosomes
   B. Occur in the genotype
   C. Expressed in the phenotype
   D. All of the above

6. A mutation is defined as
   A. A change in an organism's DNA
   B. The growth of an abnormal cell structure
   C. The changing of a cell from one type to another
   D. A way of changing mRNA to proteins

7. Which of the following illustrates a single point mutation on a segment of DNA which reads TACACGGTG?
   A. TACACGTGTG   B. TACACGCTG
   C. GTGGCACAT   D. TTCACGGAG

8. A frame shift mutation has more serious consequences than a point mutation because a frame shift mutation
   A. Cannot ever be transcribed onto an mRNA molecule
   B. Prevents uracil from being used in the mRNA molecule
   C. Only affects a single codon
   D. Causes all of the codons after it to be different

9. Which of the following illustrates a frame shift mutation on a segment of DNA which reads TACACGCTG?
   A. TACACGTGTG   B. TACACGCTG
   C. GTGGCACAT   D. TTCACGGAG

10. Why are mutations important?
    A. They are always passed on to future generations.
    B. They are often random events.
    C. They only occur in sex cells.
    D. Variation that results from mutations is fundamental to the evolution of a species.

Answers

1. A	2. A	3. A	4. B	5. C
6. A	7. B	8. D	9. A	10. D

SHORT ANSWER TYPE QUESTIONS

1. Define the terms
   (i) Genotype
   (ii) Phenotype
   (iii) Allele
   (iv) Mutation
   (v) Spontaneous mutation
   (vi) Induced mutation
2. Classify mutations on the basis of their occurrence.
3. What exactly are mutations? Are they ever beneficial?
4. What role does the environment play in mutation?
5. How does mutation take place (at the genetic level) and what impact it has on evolution?

LONG ANSWER TYPE QUESTIONS

1. Describe two different mechanisms of spontaneous mutation.
2. Explain the following:
   (i) Sense mutation
   (ii) Nonsense mutation
   (iii) Frameshift mutation
   (iv) Missense mutation
3. Describe three ways in which chemical mutagens work.
4. Compare ultraviolet radiation and gamma radiation in terms of how they induce mutation.
5. Write an account on repair of mutations.

Cancer

27

INTRODUCTION

Cancer is a class of diseases characterized by uncontrolled cell division and the ability of these cells to invade other tissues, either by direct growth into adjacent tissue by invasion or by migration of cells to distant sites, a process known as metastasis. This unregulated growth is caused by a series of acquired or inherited mutations to DNA within cells thereby, damaging genetic information that defines the cell functions and removing normal control of cell division. Cancer is responsible for more than 100 life-threatening diseases.

Cancer can originate almost anywhere in the body and generally falls into four main categories:

Carcinomas: This is the most common type of cancer; it arises from the cells that cover external and internal body surfaces. Lungs, breast, and colon are the most frequent cancers of this type.

Sarcomas: These are cancers arising from cells found in the supporting tissues of the body such as bone, cartilage, fat, connective tissue, and muscle.

Lymphomas: These cancers arise in the lymph nodes and tissues of the body's immune system.

Leukemias: These are cancers of the immature blood cells that grow in the bone marrow and tend to accumulate in large numbers in the bloodstream.

LIST OF VARIOUS TYPES OF CANCERS

- **Bone cancer** (Ewing's Sarcoma, Osteosarcoma, Chondrosarcoma)
- **Brain and CNS tumors** (Acoustic Neuroma, Spinal Cord Tumours, Brain Tumour)
- **Breast cancer**
- **Colo-Rectal cancers** (Anal Cancer)
- **Endocrine cancers** (Adrenocortical Carcinoma, Pancreatic Cancer, Pituitary Cancer, Thyroid Cancer, Parathyroid Cancer, Thymus Cancer, Multiple Endocrine Neoplasia)
- **Gastrointestinal cancers** (Stomach or Gastric Cancer, Oesophageal Cancer, Small Intestine Cancer, Gall Bladder Cancer, Liver Cancer, Extra-Hepatic Bile Duct Cancer, Gastrointestinal Carcinoid Tumour)
- **Genitourinary cancers** (Testicular Cancer, Penile Cancer, Prostate Cancer)
- **Gynaecological cancers** (Cervical Cancer, Ovarian Cancer, Vaginal Cancer, Uterus/Endometrium Cancer, Vulva Cancer, Gestational Trophoblastic Cancer, Fallopian Tube Cancer, Uterine Sarcoma)
- **Head and neck cancer** (Oral Cavity, Lip, Salivary Gland Cancers, Larynx, Hypopharynx, Oropharynx Cancer, Nasal, Paranasal, Nasopharynx Cancer)
- **Leukemia** (Childhood Leukaemia, Acute Lymphocytic Leukaemia, Acute Myeloid Leukaemia, Chronic Lymphocytic Leukaemia, Chronic Myeloid Leukaemia, Hairy Cell Leukaemia)
- **Multiple myeloma**
- **Lung cancer**
- **Lymphoma** (Hodgkin's Disease, Non-Hodgkin's Lymphoma, Aids Related Lymphoma)
- **Eye cancer** (Retinoblastoma, Intra-Ocular Melanoma)
- **Skin cancer** (Melanoma, Non-Melanoma Skin Cancer)
- **Soft tissue sarcoma** (Childhood Soft Tissue Sarcoma, Adult Soft Tissue Sarcoma, Kaposi's Sarcoma)
- **Urinary system cancer** (Kidney Cancer/Wilm's Tumour, Bladder Cancer, Urethral Cancer, Transitional Cell Cancer,)
- **Other related disorders** (Metastatic Cancer, Carcinoid Tumours, Neurofibromatosis, Germ Cell Tumours, Desmoplasic Small Round Cell Tumour,
- **Other hematological disorders** (Myelodysplastic Syndromes, Myeloproliferative Disorders, Aplastic Anaemia, Fanconi Anaemia, Waldenstrom's Macroglobulinemia)

CARCINOGENESIS

Carcinogenesis (the creation of cancer) is the process by which normal cells are transformed into cancer cells. Normal, healthy cells in the body grow in a very orderly and well-controlled way, living for a set period of time and then dying on schedule. When a normal cell dies, the body replaces it with another normal cell. Cancer cells grow in an uncontrolled manner, and they lose the ability to die and therefore the diseased cells accumulate and a tumor is created. Tumors remain small until they are able to attract their own blood supply, which allows them to obtain the oxygen and nutrients they need to grow larger. Cancer can also spread (metastasize) and invade healthy tissue in other areas of your body.

Cell division is a physiological process that occurs in almost all tissues and under many circumstances. Normally homeostasis, the balance between proliferation (cell division) and programmed cell death, usually in the form of apoptosis or cell suicide, is maintained by tightly regulating both processes to ensure the integrity of organs and tissues. Mutations in DNA that lead to cancer disrupt these orderly processes by disrupting the programming regulating the processes.

This uncontrolled and often rapid proliferation of cells that leads to tumors can be classified as being either benign or malignant. Benign tumors are tumors that cannot spread by invasion or metastasis; hence, they only grow locally. Malignant tumors are tumors that are capable of spreading by invasion and metastasis. By definition, the term 'cancer' applies only to malignant tumors. Since, benign tumors do not spread to other parts of the body or invade other tissues, and they are rarely a threat to life unless they compress vital structures or are physiologically active (for instance, producing a hormone). Malignant tumors can invade other organs, spread to distant locations (metastasize) and become life threatening.

More than one mutation is necessary for carcinogenesis. In fact, a series of several mutations to certain classes of genes is usually required before a normal cell will transform into a cancer cell. Only mutations in those certain types of genes which play vital roles in cell division, cell death, and DNA repair will cause a cell to lose control of its proliferation.

PROPERTIES OF MALIGNANT CELLS

Cells capable of forming malignant tumors exhibit many properties which distinguish them from the cells of healthy tissue.

- They are resistant to apoptosis (programmed cell death).
- They have an uncontrolled ability to divide (i.e., they are immortal), and they often divide at an increased rate.
- These cells are self-sufficient with respect to growth factors.

- They are insensitive to antigrowth factors, and contact inhibition is suppressed.
- These cells may exhibit altered differentiation.

More aggressive malignant cells may also show additional characteristics.

- They have the ability to invade neighboring tissues, usually through the secretion of metalloproteinase that can digest extracellular matrix material.
- They can form new tumors (Metastasis) at distant sites.
- They secrete chemical signals that stimulate the growth of new blood vessels (Angiogenesis).

CAUSES OF CANCER

Cancer is often perceived as a disease that strikes for no apparent reason. While scientists do not yet know all the reasons, many of the causes of cancer have already been identified. Besides intrinsic factors such as heredity, diet, and hormones, scientific studies point to key extrinsic factors that contribute to the development of cancer. These extrinsic factors are chemicals (e.g., smoking), radiation, and viruses or bacteria. The population based studies have indicated that cancers arise with different frequencies in different areas of the world. For example, stomach cancer is especially frequent in Japan, colon cancer is prominent in the United States, and skin cancer is common in Australia. What is the reason for the high rates of specific kinds of cancer in certain countries? In theory, differences in heredity or environmental risk factors might be responsible for the different cancer rates observed in different countries.

Some atoms give off radiation, which is energy that travels through space. Prolonged or repeated exposure to certain types of radiation can cause cancer. Cancer caused by the ultraviolet radiation is most common in the people who spend long hours in strong sunlight. Ultraviolet radiation from sunlight is a low-strength type of radiation. Increased rates of cancer also have been detected in people exposed to high-strength forms of radiation such as X-rays or radiation emitted from unstable atoms called radioisotopes. Because these two types of radiations are stronger than ultraviolet radiation, they can penetrate through clothing and skin into the body. Therefore, high-strength radiation can cause cancers of internal body tissues.

In addition to chemicals and radiation, a few viruses can also trigger the development of cancer. In general, viruses are small infectious agents that cannot reproduce on their own, but instead enter into living cells and cause the infected cell to produce more copies of the virus. Like cells, viruses store their genetic instructions in large molecules called nucleic acids. In the case of cancer viruses, some of the viral genetic information carried in these nucleic acids is inserted into the chromosomes of the infected cell, and this causes the cell to become malignant. Only a few viruses that infect human cells actually cause cancer. Included in this category are viruses implicated in cervical cancer, liver cancer, and certain lymphomas, leukemias, and sarcomas.

People who develop AIDS after being infected with the human immunodeficiency virus (HIV) are at high risk for developing a specific type of cancer called Kaposi's sarcoma. Kaposi's sarcoma is a malignant tumor of blood vessels located in the skin. This type of cancer is not directly caused by HIV infection. Instead, HIV causes an immune deficiency that makes people more susceptible to viral infection. Infection by a virus called KSHV (Kaposi's sarcoma-associated herpesvirus) then appears to stimulate the development of Kaposi's sarcoma.

Viruses are not the only infectious agents that have been implicated in human cancer. The bacterium *Helicobacter pylori*, which can cause stomach ulcers, has been associated with the development of cancer, so people infected with *H. pylori* are at increased risk for stomach cancer.

ONCOGENES

Oncogene is the genetic material that carries the ability to induce cancer. An oncogene is a sequence of deoxyribonucleic acid (DNA) that when altered or mutated from its original form, the proto-oncogene causes the proliferation of normal cells. A variety of proto-oncogenes are involved in different crucial steps of cell growth, and a change in the sequence of proto-oncogene or in the amount of protein it produces can interfere with its normal role in cellular regulation. Uncontrolled cell growth, or neoplastic transformation ultimately result into the formation of a cancerous tumor (Fig. 27.1).

Oncogenes were first discovered in certain retroviruses (viruses composed of RNA instead of DNA and that contain reverse transcriptase) and were identified as cancer-causing agents in many animals. In the mid-1970s, the American microbiologists John Michael Bishop and Harold Varmus tested the theory that healthy body cells contain dormant viral oncogenes that, when triggered, cause cancer. They showed that oncogenes are actually derived from normal genes (proto-oncogenes) present in the body cells of their host.

The similarity between viral and cellular oncogenes can be explained by the life strategy of the retrovirus. Retroviruses are very small RNA viruses, i.e. they store genetic information in the form of RNA rather than DNA. On invasion of a cell, the RNA is released into the cytoplasm, where a DNA copy is made by the process of reverse transcription. The enzyme, reverse transcriptase, that catalyzes this process, is coded by the viral genome. The process is similar to transcription, in which RNA polymerase uses a DNA template to make an RNA copy. In this case, the reverse transcriptase uses an RNA template to make a DNA copy. The DNA copy has the ability to insert itself into the host DNA, where transcription generates many more copies of the viral RNA. On rare occasions, the virus may pick up a copy of a host gene, probably by inserting next to it. The host gene then becomes part of the viral

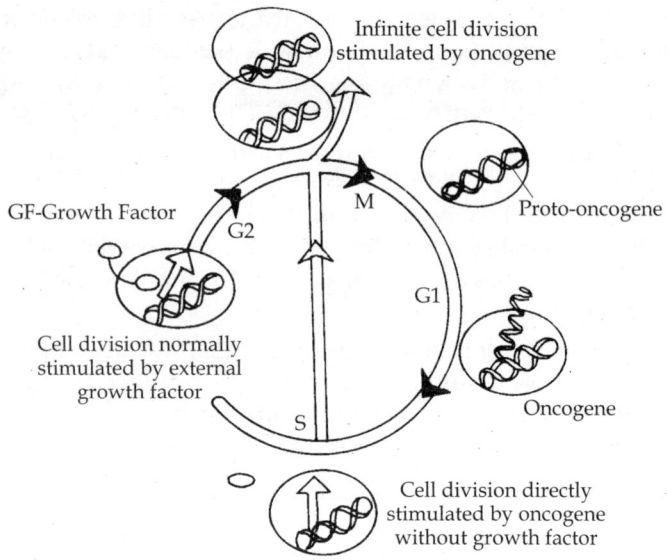

FIG. 27.1 Oncogenes are mutated forms of normal cellular genes involved in growth signalling pathways (proto-oncogenes). When these genes become mutated the cell does not require the presence of pro-growth signals (e.g., growht factors) in order to undergo cell division.

genome. It now is transcribed in infected cells under the viral promoter rather than the normal host promoter that is tightly regulated by a network of transcription factors. The excess transcription of the oncogene causes infected cells to escape growth regulation. This can be a major factor in tumor development.

Oncogenes are designated by three-letter abbreviations, such as *myc, ras,* and *mas.* The origin or location of the gene is indicated by the prefix of 'v' for virus or 'c' for cell or chromosome; additional prefixes, suffixes, and superscripts provide further delineation. About 60 human oncogenes have been identified. Breast cancer has been linked to the c-*neu* oncogene and lung cancer to the c-L-*myc* gene. Oncogenes arising in members of the *ras* gene family are found in 20 % of all human cancers, including lung, colon, and pancreatic cancer.

In humans, proto-oncogenes can be transformed into oncogenes in three ways, all of which result in a loss of or reduction in cell regulation. An alteration of a single nucleotide base pair, called a point mutation, can arise spontaneously or as a result of environmental influences such as chemical carcinogens or ultraviolet radiation. This seemingly minor event can lead to the production of an altered protein that cannot be properly regulated. Point mutations are responsible for converting certain c-*ras* proto-oncogenes to oncogenes. A second method of oncogenesis occurs by the process of translocation, in which a segment of the chromosome breaks off and attaches to another chromosome. If the dislocated chromosome contains a proto-oncogene, it may be removed from its regulatory controls and be continuously produced. Too many protein molecules can swamp the cellular process normally under their control, disrupting the delicate balance of the mechanisms of cell growth. Many leukemias and lymphomas are caused by translocations of proto-oncogenes. The third method of transformation involves amplification in the number of copies of the proto-oncogene, which also can result in overproduction of the protein and its concomitant effects. Amplified proto-oncogenes have been found in tumors from patients with breast cancer and neuroblastoma.

TUMOR SUPPRESSOR GENES

A second group of genes implicated in cancer are the 'tumor suppressor genes'. Tumor suppressor genes are normal genes whose absence can lead to cancer. In other words, if a pair of tumor suppressor genes are either lost from a cell or inactivated by mutation, their functional absence might allow cancer to develop. Individuals who inherit an increased risk of developing cancer often are born with one defective copy of a tumor suppressor gene. Because genes come in pairs (one inherited from each parent), an inherited defect in one copy will not lead to cancer because the other normal copy is still functional. But if the second copy undergoes mutation, the person may then develop cancer because there is no longer any functional copy of the gene. Tumor suppressor genes are a family of normal genes that instruct cells to produce proteins that restrain cell growth and division. Since tumor suppressor genes code for proteins that slow down cell growth and division, the loss of such proteins allows a cell to grow and divide in an uncontrolled fashion (Fig. 27.2).

Retinoblastoma is a juvenile eye cancer that is caused by a mutation in the *Rb* gene located on chromosome 13 of humans. This gene suppresses the development of cancer like its dominant phenotype. Therefore both alleles must be mutant for the disease to develop. The *Rb* gene product interacts with a protein called E2F, a nuclear transcription factor involved in cellular replication functions during the S phase of the cell cycle. By this interaction it prevents this function of E2F. The *Rb* gene product is only active when it is not phosphorylated by a kinase. It can not interact with E2F when it is phosphorylated. The mutant *Rb* gene product is always phosphorylated and can not regulate E2F, control of cell division at the S phase does not occur, and normal cells become cancerous.

One particular tumor suppressor gene codes for a protein called 'p53' that can trigger cell suicide (apoptosis). In cells that have undergone DNA damage, the p53 protein acts like a brake pedal to halt cell growth and division. If the damage cannot be repaired, the p53 protein eventually initiates cell suicide, thereby preventing the genetically damaged cell from growing out of control.

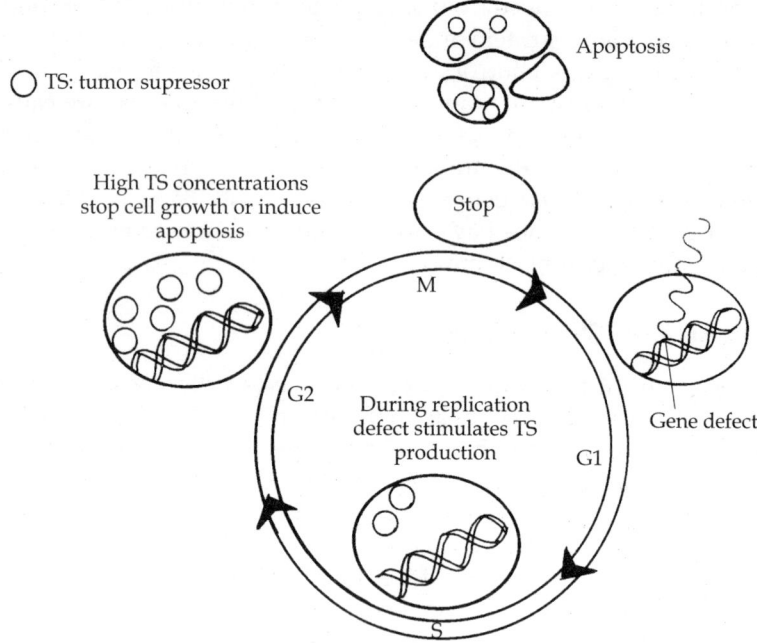

If the gene defect occurs on the TS gene itself, it is
"turned off" and the negative regulatory mechanisms fails

FIG. 27.2 Tumour suppressor genes are genes often involved in the apoptotic pathway. Normally tumour suppressors detect breaks or defects in the DNA—if present in low concentrations these proteins will pause the cell cycle and active DNA repair mechanisms. If present in high concentrations, tumour suppressors shut down the cell cycle or cause apoptosis. When these genes are mutated to be dysfunctional then the cell does not undergo either of these events.

DNA REPAIR GENES

There is a type of genes implicated in cancer called 'DNA repair genes'. DNA repair genes code for proteins whose normal function is to correct errors that arise when cells duplicate their DNA prior to cell division. Mutations in DNA repair genes can lead to a failure in repair, which in turn allows subsequent mutations to accumulate. People with a condition called *xeroderma pigmentosum* have an inherited defect in a DNA repair gene. As a result, they cannot effectively repair the DNA damage that normally occurs when skin cells are exposed to sunlight, and so they exhibit an abnormally high incidence of skin cancer. Certain forms of hereditary colon cancer also involve defects in DNA repair.

OBJECTIVE TYPE QUESTIONS

1. Cancer refers to a single disease characterized by uncontrolled growth and spread of abnormal cells. (True or False)
2. While healthy, normal cells carry out specific functions to sustain life; cancer cells serve no useful purpose in the body. (True or False)
3. Cells reproduce themselves only when they get injured and worn out.
   (True or False)
4. Benign tumors may grow and interfere with bodily function, but they rarely leave the body part where they originate. (True or False)
5. Neoplasia is a synonym for cancer, which is often used among medical professionals. (True or False)
6. Hyperplasia refers to an abnormal increase in the size of each cell.
   (True or False)
7. A cell may become cancerous if anything goes wrong in the process of cell reproduction. (True or False)
8. Abnormal cell division can only be caused by gene mutations. (True or False)
9. Cigarette smoking is solely responsible for lung cancer, which leads to a great number of deaths in both men and women. (True or False)
10. According to statistics, the chances of getting skin cancer from exposure to excessive sunlight are equal for people of different races. (True or False)

Answers

1. False	2. True	3. False	4. True	5. True
6. False	7. True	8. False	9. False	10. False

SHORT ANSWER TYPE QUESTIONS

1. Write a short note on oncogenes - their protein products and functions.
2. What are viral oncogenes?
3. Write a note on normal cells vs. cancer cells.
4. Enlist some properties of malignant cells.
5. State the ways in which cancer cells interfere with cell functions.

LONG ANSWER TYPE QUESTIONS

1. What is the difference between a carcinoma and a sarcoma?
2. What steps occur in the progression of cancer between initiation and metastasis?

3. What is the difference between a tumor suppressor gene and an oncogene?
4. What are genes, oncogenes and proto-oncogenes?
5. Discuss the mechanisms of carcinogenesis.

Genetic Disorders 28

INTRODUCTION

A genetic disorder is a disease caused by abnormal expression of one or more genes in a person causing a clinical phenotype. Abnormalities can range from a small mutation in a single gene or involving multiple genes to the addition or subtraction of an entire chromosomes or set of chromosomes.

CHROMOSOMAL DISORDERS

These disorders result from the changes that affect entire chromosome or segments of chromosomes and can cause problems with growth, development, and function of the body systems. Some commonly known chromosomal disorders have been described below:

Down Syndrome

Down syndrome was originally discovered in 1866 by John Langdon Down. However, it was not until 1959 that a French doctor, named Jerome Lejeune, studied and described that it was caused by the inheritance of an extra chromosome 21 (also known as 'trisomy 21') (Fig. 28.1). It is caused by the process of non-disjunction, that is, a pair of number 21 chromosome fails to separate during the formation of a gamete (egg or sperm). When the egg containing a pair of chromosome 21 unites with a normal sperm to form an embryo, that embryo ends up with three copies of chromosome 21 instead of the normal two. The extra

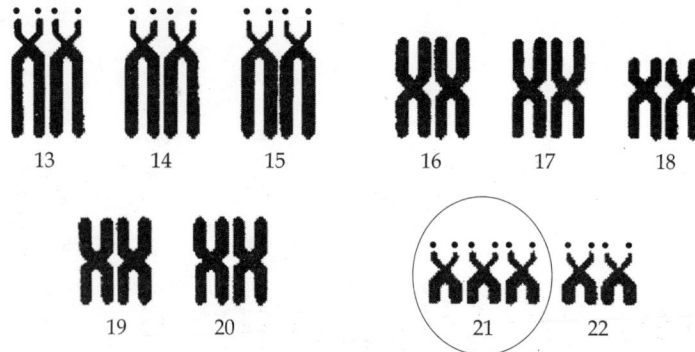

FIG. 28.1 Trisomy 21

chromosome is then copied in every cell of the body, and as a result of this each gene may be producing the protein product and too much protein can have serious consequences on the individual. It is the most common genetic disorder caused by a chromosomal abnormality. It affects 1 out of every 800 to 1,000 children.

People with Down syndrome have very distinct facial features: a flat face, a small broad nose, abnormally shaped ears, a large tongue, and upward slanting eyes with small folds of skin in the corners (Fig. 28.2). A characteristic feature is the transverse crease on the palm and clinodactyly of the fifth finger (Fig. 28.3). They also have an increased risk of developing a number of medically significant problems like respiratory infections, gastrointestinal tract obstruction, leukemia, heart defects, hearing loss, hypothyroidism, and various eye abnormalities. They exhibit moderate to severe mental retardation; children with Down syndrome usually develop more slowly than

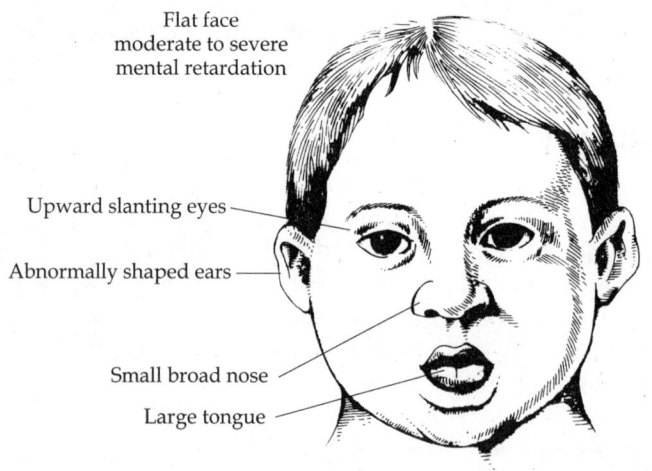

FIG. 28.2 Clinical features of Down syndrome

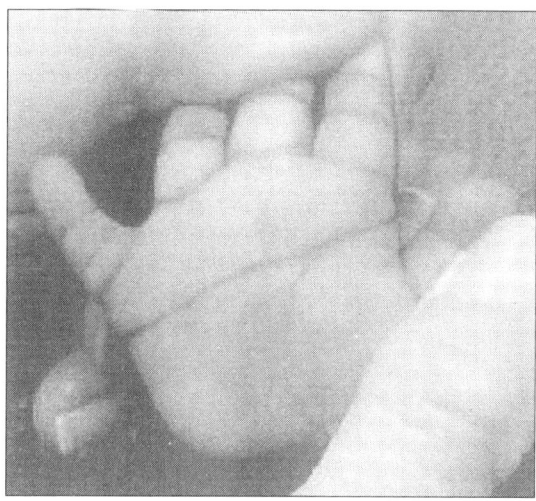

FIG. 28.3 Hand of an infant with Down syndrome. Note the transverse palmar crease and clinodactyly of the 5th finger.

their peers, and have trouble learning to walk, talk, and taking care of themselves. Due to these medical problems most people with Down syndrome have a less life expectancy and about half live upto 50 years of age.

Interestingly, non-disjunction events seem to occur more frequently in older women. This may explain why the risk of having a child with Down syndrome is greater among mothers over age 35.

In rare cases Down syndrome is caused by a Robertsonian translocation, which occurs when the long arm of chromosome 21 breaks off and attaches to another chromosome at the centromere. The carrier of such a translocation will not have Down syndrome, but can produce children with Down syndrome.

Till date no cure exists for Down syndrome, but physical therapy and/or speech therapy can help people with the disorder to some extent. Screening for common medical problems associated with the disorder, followed by corrective surgery, can often improve quality of life. Moreover, enriched environments significantly increase their capacity to learn and lead a meaningful life.

However, screening and diagnostic tests of the women during pregnancy can check the occurrence of this disorder. Screening tests identify a mother who is likely carrying a foetus with Down syndrome. The most common screening tests are the Triple Screen and the Alpha-Fetoprotein Plus. These tests measure levels of certain substances in the blood. Alternatively, ultrasounds (which use sound waves to look inside the mother's uterus) allow the doctor to examine the foetus in the womb for the physical signs of Down syndrome. The diagnostic tests like chorionic villus sampling (CVS), amniocentesis, and precutaneous umbilical blood sampling (PUBS) help to examine the chromosomes of foetus and determine the presence of an extra chromosome 21. These procedures are then followed by genetic counseling to the parents by experts.

Klinefelter Syndrome

The disorder is named after Harry Klinefelter, who first reported its symptoms in 1942. Klinefelter syndrome is a disorder that affects only males. Males normally have an X chromosome and a Y chromosome (XY). But males who have Klinefelter syndrome have an extra X chromosome (XXY), giving them a total of 47 instead of the normal 46 chromosomes (Fig. 28.4). It is typically caused by non-disjunction of chromosomes. When an egg containing two X chromosomes instead of normal one unites with a normal sperm to form an embryo, that embryo may end up with three copies of the sex chromosomes (XXY) instead of the normal two (XY). The extra chromosome is then copied in every cell of the foetus.

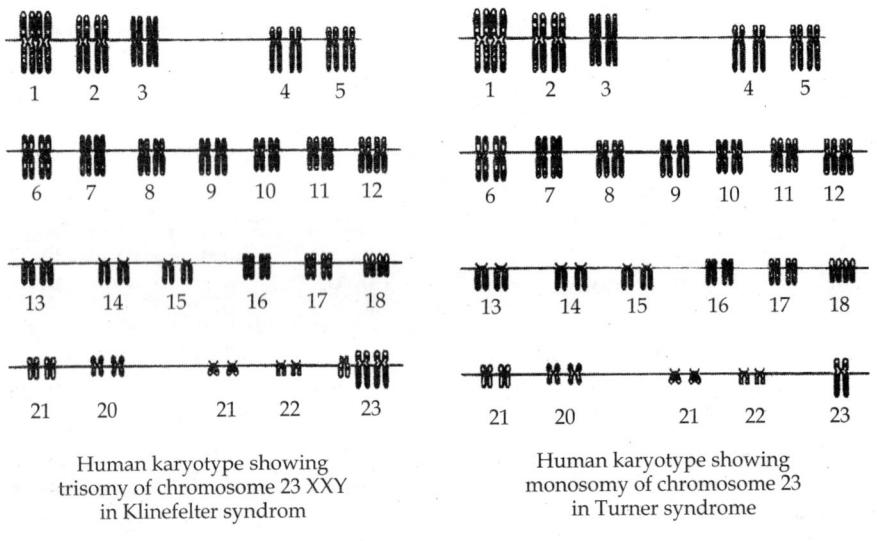

| Human karyotype showing trisomy of chromosome 23 XXY in Klinefelter syndrom | Human karyotype showing monosomy of chromosome 23 in Turner syndrome |

FIG. 28.4 Human karyotype showing trisomy and monosomy

People with this disorder develop as males with subtle characteristics that become apparent during puberty. They are often tall and usually do not develop secondary sex characteristics, such as facial hair or underarm and pubic hair. The extra X chromosome primarily affects the testes, which produce sperm and the male hormone testosterone. More often people with this disorder lead a normal life until they hit puberty or try to have children. At puberty, men with this syndrome often develop breast tissue, have a less muscular body, and grow very little facial or body hair (Fig. 28.5). Men with Klinefelter syndrome are sterile because they cannot produce sperm. Learning disabilities (not categorized as mental retardation) are also a common problem for them.

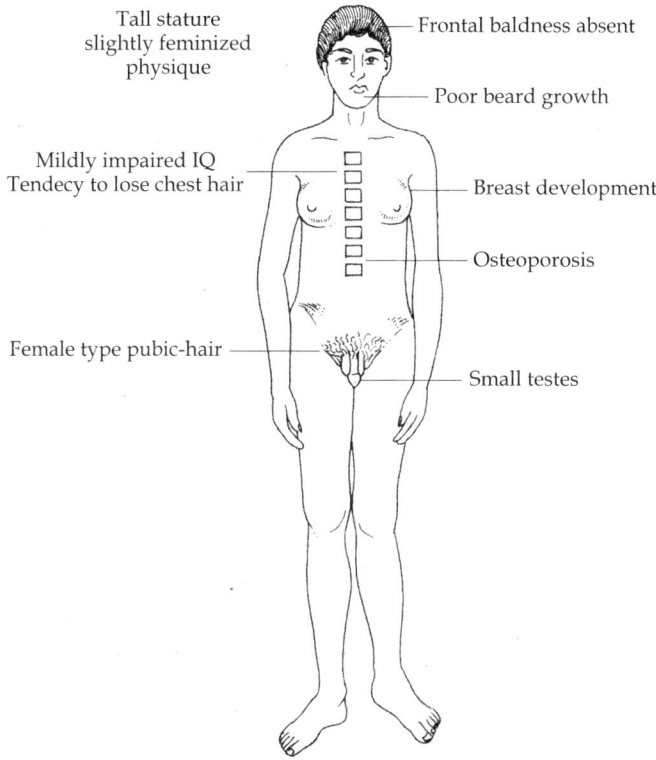

Tall stature
slightly feminized
physique

Frontal baldness absent

Poor beard growth

Mildly impaired IQ
Tendecy to lose chest hair

Breast development

Osteoporosis

Female type pubic-hair

Small testes

FIG. 28.5 Clinical features of Klinefelter syndrome

Klinefelter syndrome may also be diagnosed during a period of pregnancy. Doctors can look for the chromosome abnormality in cells taken from the amniotic fluid that surrounds the foetus (amniocentesis), or from the placenta (chorionic villus sampling (CVS)). Hormone replacement therapy is the best way to treat this disorder. Teenagers are typically given testosterone injections to replace the hormone that would normally be produced by the testes. Synthetic testosterone works like natural testosterone — it builds muscle and increases hair growth.

Turner Syndrome

Turner syndrome is named after Dr. Henry Turner, who in 1938 published a report describing the disorder. Turner syndrome is caused by a missing or incomplete X chromosome. People who have Turner syndrome develop as females. The genes affected are normally involved in growth and sexual development, that is why girls with the disorder are shorter than normal and have abnormal sexual characteristics. Normally, females inherit one X chromosome from their mother and one X chromosome from their father. But females who have Turner syndrome have one of their X chromosomes missing (Fig. 28.4). Turner syndrome is also caused by non-disjunction of sex chromosomes during the formation of an egg. When an abnormal egg with no X-chromosome unites with a normal sperm to form an embryo, that embryo may end up missing one of the sex chromosomes (X rather than XX or Y rather than XY). As the embryo grows and the cells divide, every cell of the foetus body will be missing one of the X chromosomes. However, a female foetus (normally XX) can survive with only one X chromosome, but a male fetus (normally XY) could not survive with only one Y chromosome because the Y chromosome carries only a few genes and X-chromosome genes are needed for cells to function normally. The turner abnormality is not inherited from an affected parent (not passed down from parent to child) because women with Turner syndrome are usually sterile and cannot have children.

About 30% of girls with the disorder are only missing the X chromosome in some of their cells. This mixed chromosome pattern is known as mosaicism. Girls with this pattern may have fewer symptoms because they still have some normal (XX) cells.

Turner syndrome affects growth and sexual development. Girls with this disorder are shorter than normal, and may fail to start puberty when they should. This is because the ovaries (which produce eggs, as well as the sex hormones estrogen and progesterone) fail to develop properly. Women with Turner syndrome appear to have a stocky appearance, arms that turn out slightly at the elbow, a receding lower jaw, a short webbed neck, and low hairline at the back of the neck (Fig. 28.6). Other medical symptoms include lymphedema (swelling of hands and feet), heart and/or kidney defects, high blood pressure, and infertility.

About half of the cases are diagnosed within the first few months of a girl's life by the characteristic physical symptoms (swelling of the hands and feet, or a heart defect). Other patients are diagnosed in adolescence when they fail to grow normally or go through puberty. Hormone replacement therapy is the best way to treat this disorder. Teenagers are treated with growth hormone to help them reach a normal height. They may also be given low doses of androgens (male hormones which females also produce in small quantities) to increase height and encourage normal hair and muscle growth. Some patients may take the female hormone estrogen to promote normal sexual development.

SINGLE GENE DISORDERS

These disorders result when a mutation causes the protein product of a single gene to be altered or missing. Some commonly known single gene disorders have been described below:

Galactosemia

Galactosemia was first discovered in 1908 by the physician Von Ruess. It is a rare disorder that affects the body's ability to break down a food sugar called

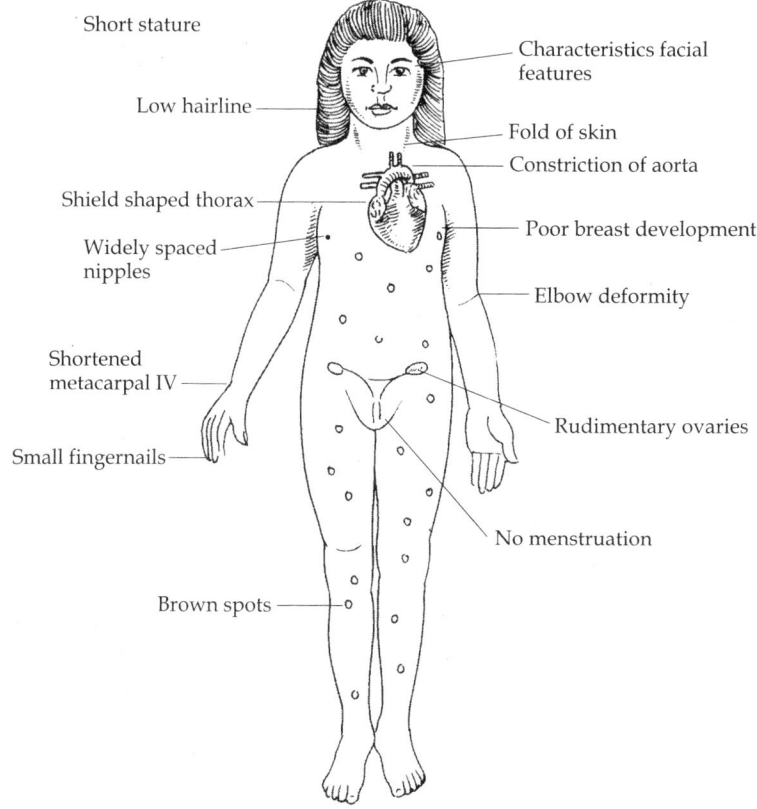

Short stature

Low hairline

Shield shaped thorax

Widely spaced
nipples

Shortened
metacarpal IV

Small fingernails

Brown spots

Characteristics facial
features

Fold of skin

Constriction of aorta

Poor breast development

Elbow deformity

Rudimentary ovaries

No menstruation

FIG. 28.6 Clinical features of Turner syndrome

galactose (found in milk and other dairy products). Normally, the body breaks down lactose into galactose and then into glucose (a sugar used for energy). People with galactosemia are missing an enzyme called GALT (galactose-1-phosphate uridyl transferase), which normally converts galactose into glucose. Without this enzyme, harmful amounts of galactose build up in the blood.

The people with this disorder inherit one defective gene from each parent. A person with one normal gene and one mutated gene is a carrier. A carrier produces less of the GALT enzyme than normal, but is still able to break down glucose and avoid having symptoms of galactosemia. However, carriers can still pass on the mutated gene to their children.

The build-up of galactose in the body can cause several severe symptoms like kidney failure, an enlarged liver, cataracts (clouding of the eye lens), poor growth, and mental retardation.

For those families with a history of the disorder, a doctor can determine during a woman's pregnancy whether the foetus has galactosemia or not either by taking a sample of fluid from around the foetus, i.e., amniocentesis, or by taking a sample of foetal cells from the placenta i.e., chorionic villus sampling or CVS. However, the only way to treat galactosemia is through dietary restrictions. People with the disorder must stay away from foods and drinks containing galactose, including milk, cheese, and legumes.

Phenylketonuria

Norwegian doctor Asbjørn Følling discovered phenylketonuria (PKU) in 1934. This disease is also known as Folling's disease. It is a rare metabolic disorder that affects the metabolism of protein. If not treated shortly after birth, PKU can be destructive to the nervous system, causing mental retardation. It is caused by a mutation in a gene on chromosome 12 which is responsible for coding an enzyme, phenylalanine hydroxylase (PAH), in the liver. The enzyme PAH breaks down the amino acid phenylalanine into other products requied by the body. When this gene is mutated, the shape of the PAH enzyme changes and it is unable to properly break down phenylalanine. Phenylalanine builds up in the blood and poisons nerve cells (neurons) in the brain.

Babies born with PKU usually have no symptoms at first. But if the disease is left untreated, babies experience severe brain damage. This damage can cause epilepsy, behavioral problems, and stunt the growth of the baby. Other symptoms include eczema (skin rash), a musty body odor (from too much phenylalanine), a small head (microcephaly), and excessively fair skin (because phenylalanine is necessary for skin pigmentation).

PKU is an autosomal recessive disorder, meaning that you need to inherit mutations in both copies of the gene to develop the symptoms of the disorder. A carrier does not have symptoms of the disease, but can pass on the defective gene to his or her children. If both parents carry one copy of the faulty gene, each of their children has a 25% chance of being born with the disease.

People suffering from the disease must eat a protein-free diet, because nearly all proteins contain phenylalanine. Infants are given a special formula without phenylalanine. Older children and adults have to avoid protein-rich foods such as meat, eggs, cheese, and nuts. They must also avoid artificial sweeteners with aspartame, which contains phenylalanine.

MULTIFACTORIAL DISORDERS

These disorders result from mutations in multiple genes often coupled with environmental causes.

Hypothyroidism

The thyroid is the largest endocrine gland in the body. It is present just below the larynx (voice box) and wraps around the trachea (windpipe). The thyroid gland produces thyroid hormone, which helps the body to grow and develop. It also plays an important role in the metabolism. Hypothyroidism (or underactive thyroid) is a common condition in which the thyroid gland produces too little thyroid hormone. About 1 in 5,000 children is born with congenital hypothyroidism, in which the thyroid fails to grow normally and cannot produce enough hormone. There is no known cause for most cases of congenital hypothyroidism. But about 10 to 20% of the time, the condition is caused by an inherited defect that alters the production of thyroid hormone.

The most common inherited form of hypothyroidism is a defect of the TPO (thyroid peroxidase) gene on chromosome 2. This gene plays an important role in thyroid hormone production.

Hypothyroidism may be caused by an autoimmune disease that attacks the thyroid gland, surgery or radiation to treat thyroid cancer and other conditions, or rare and random genetic events in which a mutation is acquired during early development.

In babies with the inherited form of hypothyroidism, the condition affects growth and cognitive development. It may cause mental retardation, delayed puberty, stunted growth, and ataxia (inability to coordinate muscle movements). In adults, hypothyroidism slows the body's metabolism, making the patient feel mentally and physically sluggish. Symptoms may include weakness, fatigue, muscle aches, mood swings, hair loss, memory loss, or slow speech. The symptoms of patients will depend upon how little thyroid hormone they produce, and for how long they have had the disorder.

When the body is deprived of thyroid hormone, the pituitary gland works overtime, producing extra thyroid-stimulating hormone (TSH). This glut of TSH may enlarge the thyroid into a condition called a goiter.

Cri-du-chat Syndrome (Cat cry syndrome)

The name of this syndrome is derived from the French word meaning 'cry of the cat', referring to the distinctive cat like cry of children with this disorder. The geneticist Jerome Lejeune identified cri-du-chat syndrome in 1963. Cri-du-chat is caused by a deletion (the length of which may vary) on the short arm of chromosome 5 (Fig. 28.7). Multiple genes are missing as a result of this deletion, and each may contribute to the symptoms of the disorder. One of the deleted genes known to be involved is TERT (telomerase reverse transcriptase). This gene is important during cell division because it helps to keep the tips of chromosomes (telomeres) intact.

Babies with cri-du-chat are usually small at birth, and may have respiratory problems. Often, the larynx does not develop correctly, which causes the cat-like cry. People who have cri-du-chat have very distinctive features. They may have a small head (microcephaly), an unusually round face, a small chin, widely set eyes, folds of skin over their eyes, and a small bridge of the nose (Fig. 28.7). Several problems occur inside the body, as well. A small number of children have heart defects, muscular or skeletal problems, hearing or sight problems, or poor muscle tone. With growth, they usually have difficulty walking and talking correctly. They may have behavioral

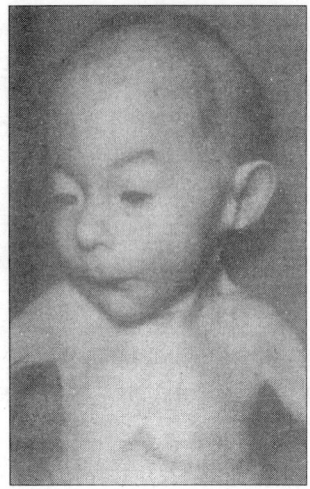

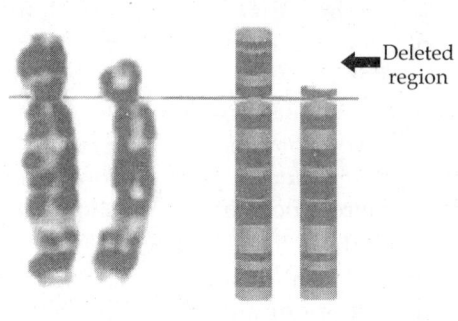

Deleted region

cri-du-chat chromosome 5 pair

A child suffering from
cri-du-chat syndrome

FIG. 28.7 Cri-du-chat syndrome

problems (such as hyperactivity or aggression), and severe mental retardation. Unfortunately, most people with this disorder do not survive to adulthood.

It is also possible to test for cri-du-chat (and other chromosomal abnormalitites) while the child is still in the mother's womb. They can either test a tiny sample of tissue from outside the sac where the baby develops (chorionic villus sampling (CVS)), or they can test a sample of the amniotic fluid (amniocentesis).

William's Syndrome

William's syndrome is a rare genetic disorder that affects the child's growth, physical appearance, and cognitive development. People with William's syndrome have some genes missing from chromosome 7, including the gene elastin. This gene is important as it codes for the protein product that gives blood vessels the stretchiness and strength required to withstand a lifetime of use. The elastin protein is made only during embryonic development and childhood, when blood vessels are formed.

The most common symptoms of William's syndrome are mental retardation, heart defects, and unusual facial features like small upturned nose, wide mouth, full lips, small chin, widely spaced teeth. Other symptoms include low birth weight, kidney abnormalities, and low muscle tone. People with this syndrome also exhibit characteristic behaviors, such as hypersensitivity to loud noises and an overly outgoing personality.

This disorder can be diagnosed and confirmed by using a special technique called FISH (fluorescent *in situ* hybridization) since the deletion is so small that it cannot be seen in a karyotype. However, the FISH technique allows DNA sequences to be labeled with a fluorescent chemical (called a probe) that lights up when exposed to ultraviolet (UV) light.

GENETIC COUNSELING

The chromosomes and the genes are made up of DNA molecules, which are the simplest building blocks of heredity. The DNA forms the unique 'blueprint' for every physical and biological characteristic of that person. Humans have 23 pairs of chromosomes in every living cell of our bodies. When the egg and sperm join for the formation of zygote, half of each chromosomal pair is inherited from each parent. This newly formed combination of chromosomes then copies itself again and again during foetal growth and development, passing identical genetic information to each new cell in the growing foetus. An error in just one gene (and in some instances, even the alteration of a single piece of DNA) can sometimes be the cause for a serious medical condition.

Most disorders cannot occur unless both the mother and father transmit the gene. Other genetic conditions, such as Down syndrome, are not inherited. In general, they result from an error (mutation) in the cell division process during conception or foetal development. Still others, such as achondroplasia (the most common form of dwarfism), may either be inherited or develop as a result of a genetic mutation.

Genetic tests are done by analyzing small samples of blood or body tissues. They determine whether the person concerned, or his/her partner, or the baby carry genes for certain inherited disorders. However, the genetic tests do not yield easy-to-understand results. They can reveal the presence, absence, or malformation of genes or chromosomes. Deciphering what these complex tests mean is where a genetic counselor comes in.

Genetic counseling is the process of:
- evaluating family history and medical records
- ordering genetic tests
- evaluating the results of this investigation
- helping parents understand and reach decisions about what to do next

The best time to seek genetic counseling is before becoming pregnant, when a counselor can help assess the risk factors. But even after one becomes pregnant, a meeting with a genetic counselor can still be helpful. For example, several babies have been diagnosed with spina bifida before birth. Recent research suggests that delivering a baby with spina bifida via cesarean section can minimize damage to the baby's spine and perhaps reduce the likelihood that the child will need a wheelchair.

One should consider genetic counseling if any of the following risk factors apply:

- if a standard prenatal screening test (such as the alpha fetoprotein test) yields an abnormal result
- if an amniocentesis yields an unexpected result (such as a chromosomal defect in the unborn baby)
- if either parent or a close relative has an inherited disease or birth defect
- if either parent already has children with birth defects or genetic disorders
- if the mother-to-be has had two or more miscarriages or babies that died in infancy
- if the mother-to-be will be 35 years of age or older at the time of child birth (chances of having a child with Down syndrome increase with the mother's age; a 35-year-old woman has a one in 350 chance of conceiving a child with Down syndrome. This chance increases to one in 110 by age 40 and one in 30 by age 45.)
- if parents are concerned about genetic defects that occur frequently in their ethnic or racial group (for example, couples of African descent are most at risk for having a child with sickle cell anemia; Irish descent may be carriers of Tay-Sachs disease; and couples of Italian, Greek, or Middle Eastern descent may carry the gene for thalassemia, a red blood cell disorder).

OBJECTIVE TYPE QUESTIONS

1. The likelihood of a child being born with a major genetic defect, such as mental retardation, can often be detected by sampling
   A. The mother's uterus cells
   B. Cells from the embryo or foetus
   C. The father's blood cells
   D. Blood cells of siblings

2. Which of the following can be detected now by examining a karyotype?
   A. Over 3,000 genetic defects
   B. An unborn child's gender or sex
   C. Both of the above
   D. None of the above

3. If there is a family history of genetic disorders, knowing the gender of an unborn child can be important because
   A. Male children are more likely to have autosomal defects show up in their phenotypes
   B. Female children are more likely to have autosomal defects show up in their phenotypes

    C. Male children are more likely to have X-linked traits show up in their phenotype

    D. Both A and C

4. Most genetic disorders are due to

    A. Gross chromosomal abnormalities such as irregular shapes or numbers of chromosomes

    B. The gender of an individual

    C. Neither of the above

    D. Both of the above

5. Which of the following statements is true regarding karyotype analysis?

    A. It is rarely done on the cells of unborn children because it cannot detect most genetic disorders.

    B. It is now an important medical tool used in predicting the likelihood that an unborn child will be normal.

    C. It is not done any more because human pregnancy has only a small risk of birth defects.

    D. It is done in case of communicable diseases

6. Which of the following statements is true for sex chromosome abnormalities in humans?

    A. They usually have mild effects and rarely are fatal.

    B. Most are not gender specific

    C. They cannot be diagnosed before birth.

    D. They are always fatal

7. Male sex chromosome abnormalities can be due to abnormal numbers of _____ chromosome.

    A. The X   B. The Y

    C. Either the X or the Y   D. Autosomal

8. If someone only has one X chromosome and no Y chromosomes in their somatic cells, they

    A. Are metafemales   B. Have Turner syndrome

    C. Have Klinefelter syndrome   D. Are super females

9. A chromosomal abnormality that causes a woman to be unusually short in stature (average 4' 7"), to have a webbed neck, and to generally lack feminine secondary sexual characteristics is

    A. Triple-X syndrome   B. Turner syndrome

    C. XYY syndrome   D. Klinefelter syndrome

10. A chromosomal abnormality that causes a man to have asexual to feminine body contours with large breasts; small penis, testes, and prostate gland; relatively little body hair; and sterility is

    A. Klinefelter syndrome   B. XYY syndrome

    C. Richard Speck syndrome   D. Turner syndrome

Answers

1. B	2. B	3. C	4. C	5. B
6. A	7. C	8. B	9. B	10. A

SHORT ANSWER TYPE QUESTIONS

1. What genetic disorders does prenatal genetic testing detect?
2. Are chromosomal abnormalities the same as genetic mutations? Give reasons.
3. What is a karyotype?
4. Enlist eight symptoms of Down syndrome.
5. What are the symptoms of cri-du-chat and how is it diagnosed?

LONG ANSWER TYPE QUESTIONS

1. Do some genetic disorders affect certain populations more than others? If so why?
2. What is the etiology of infertility in women with Turner syndrome?
3. What causes gynecomastia in males with Klinefelter syndrome?
4. Comment upon galactosemia.
5. What is mongolism and how does it occur?

Trends in Genetics 29

IMPORTANCE OF GENETICS

The word 'Genetics' is derived from the Greek word '*genno*' which means 'to give birth', it was first suggested to describe the study of inheritance and the science of variation by the prominent British scientist William Bateson in 1906. Genetics is the science which deals with genes, heredity, and the variation of organisms.

Genetic information is encoded and transmitted from generation to generation in deoxyribonucleic acid (DNA). DNA is a coiled molecule organized into structures called chromosomes within cells. Segments along the length of a DNA molecule form genes. Genes direct the synthesis of proteins that carry out all life-supporting activities in the cell. Although all humans share the same set of genes, individuals can inherit different forms of a given gene, making each person genetically unique.

During the past few decades the science of genetics has grown explosively thus generating a huge amount of new and important information. The modern science of genetics influences many aspects of daily life, from the food that we take to how we identify criminals or treat diseases. The genetic advances in different fields enable scientists to manipulate genes and alter a plant or animal to make it more useful. In agriculture, for instance, some food crops, such as oranges, potatoes, wheat, and rice, have been genetically altered to withstand insect pests, resulting in a higher crop yield. Similarly, tomatoes and apples have been modified so that they resist discoloration on their way to market thereby enhancing their appeal on supermarket shelves. The genetic makeup of cows has been modified to increase their milk production, and cattle raised for beef have been altered so that they grow faster. Genetic technologies have also helped to convict criminals via DNA fingerprinting. DNA recovered from semen, blood, skin cells, or hair found at a crime scene can be analyzed in a laboratory and compared with the DNA of a suspect. An individual's DNA is as unique as a set of fingerprints, and a DNA match can be used in a courtroom as evidence connecting a person to a crime. Moreover, the way industries produce certain substances proves costly and arduous due to elaborate manufacturing methods. In medicine, scientists can genetically alter bacteria so that they mass-produce specific proteins, such as insulin used by people with diabetes mellitus or human growth hormone used by children who suffer from growth disorders.

'Gene therapy' is another form of medical application that is still in its experimental form. It is a procedure with which scientists try to cure disease by replacing malfunctioning genes with healthy ones. Gene therapy has shown promise in treating some devastating conditions, including some forms of cancer and cystic fibrosis besides inherited disorders.

Advances in genetic technologies and related fields have allowed scientists to take a glimpse into the genetic makeup of every person. The information derived from this testing can serve many valuable purposes like it can save lives in case of major life threatening diseases, it can assist couples trying to decide whether or not to have children in case of genetic incompatibilities in parents or if the foetus is screened to be with congenital anomalies, and help law-enforcement officials in solving a crime. Despite of these breakthroughs, the genetic testing also raises some troubling social concerns about privacy and discrimination. For example, if an individual's genetic information becomes widely available, it could give health insurers cause to deny coverage to people with certain risk factors or encourage employers to reject certain high-risk job applicants. Furthermore, many genetically linked problems are more common among certain racial and ethnic groups—for example, the BRCA1 breast cancer allele is more common in Ashkenazi Jews, and the blood disorder called sickle-cell anemia is more prevalent among blacks of African ancestry. Many minority groups fear that the expansion of genetic testing could create whole new avenues of discrimination.

Moreover, certain genetic tests that shed light on traits such as personality, intelligence, and mental health or potential abilities are issues of grave concern. Genetic tests that indicate a person is unlikely to get along with other people could be used to limit a professional advancement of that person. In other cases, tests that identify a genetic risk of heart failure could discourage a person from competing in sports or other such activities. Certain new technologies like Gene therapy that allow the manipulation of genes have raised even more disturbing possibilities, that is, besides giving new hope for healthy lives to people with typically fatal diseases it can be used to alter or encourage traits now viewed as part of normal human variability, such as shortness or baldness. This concept, known as Eugenics, typically involves encouraging people with 'positive' genes to reproduce and discouraging those with 'inferior' genes from having offspring. It would further raise inferiority complex within people who for some reason are not able to afford such expensive therapies.

HUMAN GENOME PROJECT (Fig. 29.1)

The field of genetics has gained momentum recently by the Human Genome Project that has helped scientists to develop detailed maps that identify the chromosomal locations of different genes. The project was initiated in 1990 in the United States with government funding, and it rapidly grew into an international consortium of academic centers and drug companies in China, France, Germany, Japan, the United Kingdom, and the United States.

The completed human genome has provided scientists with a detailed blueprint of our complex genetic code. Large computer databases of genetic information enable scientists to look for patterns and relationships among the actions of different genes. The goals of the project were to identify all the approximately 30,000 genes in human DNA, to determine the sequences of the 3 billion chemical base pairs that make up human DNA, store this information in databases, improve tools for data analysis, transfer related technologies to the private sector, and address the ethical, legal, and social issues (ELSI) that may arise from the project.

The cellular and genomic principles that now enlighten us have greatly advanced our thinking and prepared the field to accept the notion of totally novel and unanticipated mechanisms. Moreover, new information in the field

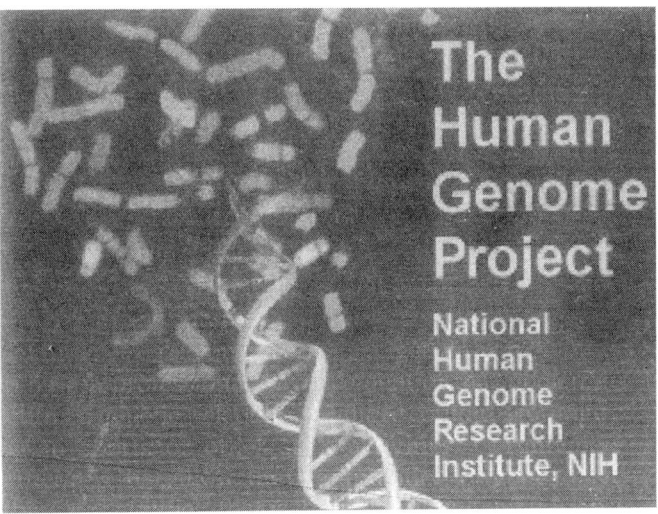

FIG. 29.1 The knowledge of human genome provides better understanding of basic biological processes

is accumulating at an unprecedented rate. A brief insight into the new aspects and mechanisms has been presented in the following sections.

DNA LIBRARIES

A DNA library is a storehouse of genetic information maintained in bacteria. These bacteria are clones created by recombinant DNA, and the foreign DNA they hold is the library's store of information. DNA libraries are helpful to scientists who require a plentiful supply of particular DNA segments to do their work (Fig. 29.2).

GENE CHIP

The gene chip, also known as a DNA chip or DNA microarray, is a thumbnail-sized chip of glass or silicon that carries DNA instead of electronic circuits

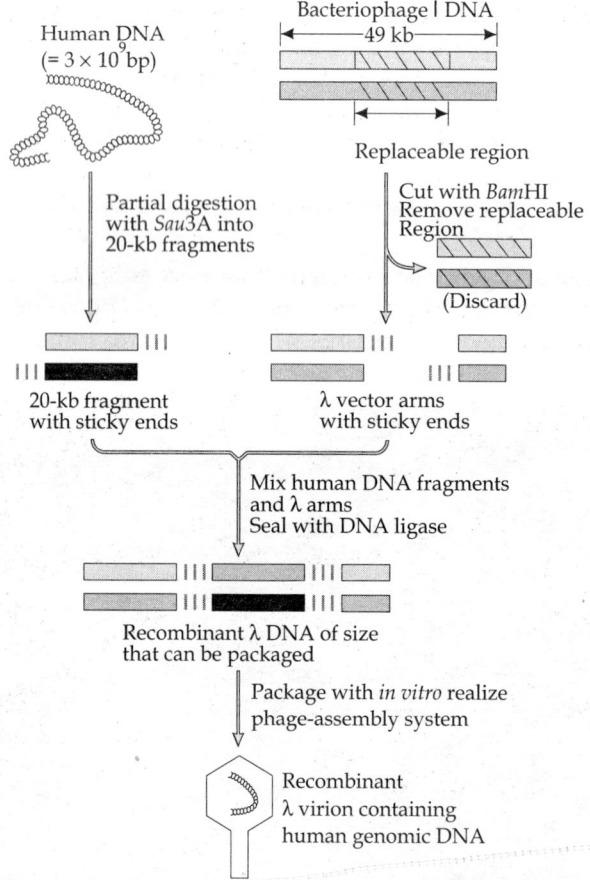

FIG. 29.2 Making a genomic library in bacteriophage lambda vector

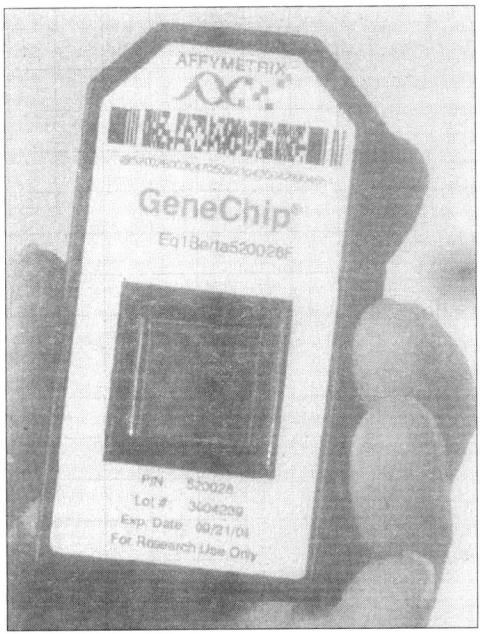

FIG. 29.3   Gene chip

(Fig. 29.3). Gene chips can identify the genes that are active within a cell and help identify mutated genes. Gene chips are a way of automating experiments that previously could only be done one at a time. They are a picture of what the RNA looks like at one particular moment. Thousands of DNA pieces in genes are spotted on a small surface enabling us to look at changes in gene expression profile of thousands of genes in one experiment.

RECOMBINANT DNA

The DNA molecules of all life forms, from plants to animals, have the same structure and the same four bases. Scientists have made use of these similarities in a technology called recombinant DNA. In this laboratory method, one or more genes of an organism are introduced into a second organism. The new genes, sometimes known as foreign DNA, become functional in the second organism and produce a desired protein. In this way, scientists can create changes in the genetic makeup of an organism that would be unlikely to occur through natural processes.

OBJECTIVE TYPE QUESTIONS

1. The word 'Genetics' was suggested for the first time by _____ .

   A. Mendel   B. Kornberg

   C. Bateson   D. Wilmut

2. What is the concept called wherein people with 'positive' genes are encouraged to reproduce and those with 'inferior' genes are dicouraged from having offspring.
   A. Euphenics   B. Eugenics
   C. Euthenics   D. None of the above

3. The human genome project was initiated in the year _____ .
   A. 1950   B. 1980
   C. 1990   D. 2000

4. _____ is the technique that can be used to convict criminals.
   A. Karyotyping   B. Chorionic sampling
   C. DNA fingerprinting   D. Gene therapy

5. _____ is a thumbnail-sized chip of glass or silicon that carries DNA and can identify the genes that are active within a cell and help identify mutated genes.
   A. DNA library   B. DNA microarray
   C. DNA fingerprint   D. None of the above

Answers

1. C   2. B   3. C   4. C   5. B

SHORT ANSWER TYPE QUESTIONS

1. Write a short note on DNA library.
2. Describe in brief the uses of recombinant DNA technology.
3. What is a gene chip? Describe its importance.

LONG ANSWER TYPE QUESTIONS

1. Write a note on importance of genetics.
2. Expain what is gene therapy. Describe its benefits and limitations.
3. Write an account on 'Human Genome Project'.

Bibliography

Adams MD, Celniker SE and Holt RA. *et al.* (2000). The genome sequence of *Drosophila melanogaster. Science,* 287: 2185-2195.

Alberts B, Bray D and Johnson A. *et al.* (1997). Essential Cell Biology. London: Garland Publishing.

Alberts B, Bray D, Lewis J, Raff M, Roberts K and Watson JD. (1997). Molecular Biology of Cell, 2nd ed. Garland publishing, New York.

Avery OT, MacLeod CM and McCarty M. (1944). Studies on the chemical nature of the substance inducing transformation of pneumococcal types. *J. Exp. Med.,* 79: 137.

Bainton D. (1981). The discovery of lysosomes. *J. Cell Biol.,* 91: 66s-76s.

Branden C and Tooze J. (1999). Introduction to Protein Structure, 2nd edn. New York: Garland Publishing.

Bretscher, MS. (1985). The molecules of the cell membrane *Sci. Amer.,* 253: (4) 100-109.

Brenner S, Jacob F and Meselson M. (1961). An unstable intermediate carrying information from genes to ribosomes for protein synthesis. *Nature,* 190: 576-581.

Brent R. (2000). Genomic Biology. *Cell,* 100: 169-182.

Cairns J. (1978). Cancer: Science and Society. San Francisco: WH Freeman.

Callan HG. (1982). Lampbrush chromosomes. *Proc. R. Soc. Lond. B Sci. Biol.,* 214: 417-448.

Carpita NC and McCann M. (2000). The Cell Wall. In Biochemistry and Molecular Biology of Plants (Buchanan BB, Gruissem W & Jones RL eds), pp 52–108. Rockville, MD: ASPB.

Carroll SB, Grenier JK and Weatherbee SD. (2001). From DNA to Diversity: Molecular Genetics and the Evolution of Animal Design. Maldon, MA: Blackwell Science.

Celis JE. (ed) (1998). Cell Biology: A Laboratory Handbook, 2nd edn. San Diego: Academic Press.

Compton DA. (2000). Spindle assembly in animal cells. *Annu. Rev. Biochem.,* 69: 95-114.

Cramer WA and Knaff DB. (1990). Energy Transduction in Biological Membranes: A Textbook of Bioenergetics. New York: Springer-Verlag.

Crick FHC. (1966). The genetic code: III *Sci. Amer.*, 215: (4) 55-62.

De Robertis EDP and De Robertis EMF. Jr. (1980). Cell and Molecular Biology, 7th ed. Saunders College/Holt, Rinehart and Winson, Philadelphia.

Dickerson RE and Geis I. (1983). Hemoglobin: Structure, Function, Evolution and Pathology. Menlo Park, CA: Benjamin Cummings.

Dressler D and Potter H. (1991). Discovering Enzymes. New York: Scientific American Library.

Du Praw EJ. (1968). Cell and Molecular Biology. Academic Press, New York.

Dyon RD. (1975). Essentials of Cell Biology. Allyn and Bacon, Boston.

Farquhar MG and Palade GE. (1998). The Golgi apparatus: 100 years of progress and controversy. (Part of a special issue on the Golgi). *Trends Cell Biol.*, 8: 2-10.

Fawcett DW and Bedford JM. (ed) (1979). The Spermatozoon. Baltimore: Urban and Schwarzenberg.

Gardner EJ, Simmons MJ and Snustad DP. (1991). Principles of Genetics, 8th ed. John Wiley, New York.

Garrett RH and Grisham CM. (1998). Biochemistry, 2nd edn. Orlando: Saunders.

Graur D and W-H. (1999) Fundamentals of Molecular Evolution, 2nd edn. Sunderland, MA: Sinauer Associates.

Gray MW, Burger G and Lang BF. (1999). Mitochondrial evolution. *Science*, 283: 1476-1481.

Gupta PK. (1995). Genetics. Rastogi Publications, India.

Harrison SC. (1992). Viruses. *Curr. Opin. Struct. Biol.*, 2: 293-299.

Hartwell L, Hood L and Goldberg ML *et al.* (2000). Genetics: from Genes to Genomes. Boston: McGraw Hill.

Hatch MD. (1987). C4 photosynthesis: a unique blend of modifed biochemistry, anatomy and ultrastructure. *Biochim. Biophys. Acta*, 895: 81-106.

Horton HR, Moran LA and Ochs RS *et al.* (2001). Principles of Biochemistry, 3rd edn. Upper Saddle River, NJ: Prentice Hall.

Hyams JS and Brinkely BR. (eds) (1989). Mitosis: Molecules and Mechanisms. San Diego: Academic Press.

Ingraham JL, Maaløe O and Neidhardt FC. (1983). Growth of the Bacterial Cell. Sunderland, MA: Sinauer.

International Human Genome Sequencing Consortium. (2001). Initial sequencing and analysis of the human genome. *Nature*, 409: 860-921.

Jacob F and Monod J. (1961). Genetic regulatory mechanisms in the synthesis of proteins. *J. Mol. Biol.*, 3: 318-356.

Jacobs C and Shapiro L. (1999). Bacterial cell division: a moveable feast. *Proc. Natl. Acad. Sci. USA,* 96: 5891-5893.

Kornberg A and Baker TA. (1992). DNA Replication, 2nd edn. New York: WH Freeman.

Kornberg A. (1960). Biological synthesis of DNA. *Science,* 131: 1503-1508.

Kornberg HL. (1987). Tricarboxylic acid cycles. *Bioessays,* 7: 236-238.

Lacey AJ. (ed) (1999). Light Microscopy in Biology: A Practical Approach, 2nd edn. Oxford: Oxford University Press.

Lamond AI and Earnshaw WC. (1998). Structure and function in the nucleus. *Science,* 280: 547-553.

Lawlor DW. (2001). Photosynthesis, 3rd edn. Oxford: BIOS.

Lehninger AL, Nelson DL and Cox MM. (1993). Principles of Biochemistry, 2nd edn. New York: Worth.

Lewin B. (2000) Genes VII. Oxford: Oxford University Press.

Lodish H, Berk A and Zipursky SL. *et al.* (2000). Molecular Cell Biology, 4th edn. New York: WH Freeman.

Lockhart DJ and Winzeler EA. (2000). Genomics, gene expression and DNA arrays. *Nature,* 405: 827-836.

Mathews CK, Van Holde KE and Ahern K-G. (2000). Biochemistry, 3rd edn. San Francisco: Benjamin Cummings.

Meselson M and Stahl FW. (1958). The replication of DNA in *E. coli. Proc. Natl Acad. Sci. USA,* 44: 671-682.

Mitchison TJ and Salmon ED. (2001). Mitosis: a history of division. *Nat. Cell Biol.,* 3: E17-E21.

Nirenberg MW. (1963). 'The genetic code: II'. *Sci. Amer.,* 208 (3): 80-94.

Nogales E, Wolf SG and Downing KH. (1998). Structure of the αβ-tubulin dimer by electron crystallography. *Nature,* 391: 199-203.

Pauling L and Corey RB. (1951). Configurations of polypeptide chains with favored orientations around single bonds: two new pleated sheets. *Proc. Natl. Acad. Sci. USA,* 37: 729-740.

Pauling L. (1960). The Nature of the Chemical Bond, 3rd edn. Ithaca, NY: Cornell University Press.

Pluta AF, Mackay AM and Ainsztein AM. *et al.* (1995). The centromere: hub of chromosomal activities. *Science,* 270: 1591-1594.

Purdue PE and Lazarow PB. (2001). Peroxisome biogenesis. *Annu. Rev. Cell Dev. Biol.,* 17: 701-752.

Rao K and Hall DO. (1999). Photosynthesis. Cambridge: Cambridge University Press.

Saenger W. (1984). Principles of Nucleic Acid Structure. New York: Springer.

Sharon N. (1980). Carbohydrates. *Sci. Amer.,* 243: (5) 90-116.

Schafer G, Purschke W and Schmidt CL. (1996). On the origin of respiration: electron transport proteins from archaea to man. *FEMS Microbiol. Rev.,* 18: 173-188.

Scheffler IE. (1999). Mitochondria. New York/Chichester: Wiley-Liss.

Singer SJ and Nicolson GL. (1972). The fluid mosaic model of the structure of cell membranes. *Science,* 175: 720-731.

Slayter EM and Slayter HS. (1992). Light and Electron Microscopy. Cambridge: Cambridge University Press.

Stent GS. (1971). Molecular Genetics: An Introductory Narrative. San Francisco: WH Freeman.

Strickberger MW. (1977). Genetics, 2nd ed. Macmillan, New York.

Sturtevant AH and Beadle GW. (1939). An Introduction to Genetics. Philadelphia: Saunders. Reprinted 1988 New York: Garland.

Swanson CP. (1971). The Cell, 3rd ed. Prentice Hall, New Jersey.

Swanson CP and Webster P. (1977). The Cell. Prentice-Hall, Englewood Cliffs, NJ.

Swanson CP, Mertz T and Young WJ. (1981). Cytogenetics- The chromosomes in division, Inheritence and Evolution, 2nd ed. Englewood Cliffs, Prentice Hall, New Jersey.

Warren G and Malhotra V. (1998). The organization of the Golgi apparatus. *Curr. Opin. Cell Biol.,* 10: 493-498.

Watson JD and Crick FHC. (1953). Molecular structure of nucleic acids. A structure for deoxyribose nucleic acid. *Nature,* 171: 737-738.

Watson JD, Gilman M, Witkowski J and Zoller M. (1992). Recombinant DNA, 2nd edn. New York: WH Freeman.

Watson JD, Hopkins NH and Roberts JW. *et al.* (1987). Molecular Biology of the Gene, 4th edn. Menlo Park, CA: Benjamin-Cummings.

Wolfe A. (1999). Chromatin: Structure and Function, 3rd edn. New York: Academic Press.

Zhimulev IF. (1998). Morphology and structure of polytene chromosomes. *Adv. Genet.,* 37: 1-566.

Zubay GL. (1998). Biochemistry, 4th edn. Dubuque, IO: William C Brown.

Glossary

Active transport- Transport in which a cell must expend energy. Active transport occurs against a concentration gradient.

Acentric- A chromosome or, more commonly, a chromosome fragment that lacks a centromere.

Acrocentric- [Greek *akron*-extremity + centric] A chromosome in which the centromere is located near one end, creating very unequal arms.

Adenine- A purine that occurs in the nucleotides of DNA and RNA. Commonly abbreviated A in DNA and RNA sequences.

Adenosine triphosphate- Commonly called **ATP**, this substance stores energy in the form of chemical bonds. ATP is produced in mitochondria (also in chloroplasts in plants) and is used to drive the vast number of chemical reactions in a cell that require energy. If more ATP is produced than is required, the excess is stored in the form of fat and glycogen (or starch in plants).

Amino acid- An organic compound that has both an amino group and a carboxyl group as part of its structure. The type formula is H_2–N–CH(R)–COOH, where the hydrogen (H) and the R group are attached covalently to the carbon (C) drawn to their left. The R group can be any of dozens of atomic structures, the simplest being a single H. Amino acids can have a variety of functions in metabolism, but they are especially interesting because they are the building blocks of proteins. They can be joined together in any order by peptide bonds, and there can exist a vast number of polypeptide chains that differ in their amino acid sequences.

Anaphase- Phase of mitosis in which the chromosomes begin to pull to opposite poles of the cell.

Anticodon- The three nucleotides on transfer RNA (tRNA) that are complementary to a codon in messenger RNA (mRNA). This complementarity binds the tRNA to the mRNA, and the attached amino acid is then transferred to the nascent polypeptide chain. This specificity of binding of the anticodons to the mRNA codons accounts for the fidelity with which a gene is translated into protein.

Antiparallel- Going in opposite directions, as in two arrows lying parallel but with arrow heads pointing in opposite directions. The two polynucleotide strands of DNA have direction, and pairing occurs only if the directions are opposite.

Antisense strand- The DNA strand that serves as the template in transcription. Also called the template strand.

Asexual- A type of reproduction that does not require the union of female and male gametes.

Autosome- A chromosome other than the chromosomes involved in sex determination. Humans have 22 pairs of autosomes + one pair of sex chromosomes.

Bacteria- [Sing. *bacterium*; Greek *bakterion*-little stick] One-cell organisms that lack nuclei, i.e. they are prokaryotes. Many of the "germs" that cause disease are bacteria. However, most are benign. Indeed, we depend on them for many natural processes. Several have been important experimentally in understanding the structure of genes and how they function.

Bacteriophage- [bacterio- + *phagein* to eat] A virus that attacks and destroys bacteria. Studies of bacteriophage were important in early development of the field of molecular biology.

Basal body- A structure found at the connection of cilia and flagella with the cell membrane. It is composed of microtubules in a circular configuration of nine triplets.

Base- A molecule that has the ability to bind hydrogen ions, H^+, thereby neutralizing their acidity. *Basic*, in this sense, is synonymous with *alkaline*. While there are many basic molecules that are important in biology, those that occur in nucleic acids merit special note, as they are often referred to collectively as bases. These are the purines adenine and guanine and the pyrimidines cytosine, thymine, and uracil.

Base pair- A common designation of the pairs of nucleotides that are complementary in a DNA double helix. As a matter of convenience, a DNA sequence is often represented by the sequence of bases on one strand. It is understood, but frequently not stated, that each such base represents the nucleotide pair. Thus, in writing a sequence such as -G-A-C-, the complementary sequence -C-T-G- is understood. The context of the discussion should make clear whether a single strand of DNA is inferred or whether the double-stranded structure is the subject.

Biogenesis- The concept that all life arises from living matter.

cAMP (cyclic adenosine monophosphate)- A molecule that plays a key role in the regulation of various processes within the cell.

Cancer- A malignant tumor. Such tumors can spread to other parts of the body and can invade other tissues.

Capsid- The protein shell of a virus.

Capsomere- Protein clusters making up discrete subunits of a viral protein shell.

Carcinogen- A substance that causes cancer. Use of the term is usually restricted to environmental agents, such as radiation and chemicals, although some of the products of normal metabolism may also be carcinogenic. In most cases, carcinogens are also mutagens, and it is the increase in mutation rates that account for the increase in cancer following exposure.

cDNA- DNA that has been generated by reverse transcriptase acting on RNA templates, especially messenger RNA. This procedure allows mRNA to be isolated from cells or tissues and copied into DNA, thereby revealing which genes are actively transcribed in those cells or tissues. Since the template is mRNA, only the exons are represented in the cDNA. [Note: the c of cDNA stands for complementary, meaning complementary to RNA].

Cell- The smallest unit of life that carries out its own processes.

Cellulose- A carbohydrate that is found in cell walls.

Cell membrane- The structure which surrounds the cell and regulates the movement of materials into and out of the cell. It is composed mostly of phospholipids.

Cell wall- A structure found in most prokaryotes and some eukaryotes which gives the cell greater structure. In prokaryotes, it is composed of peptidoglycans, and in eukaryotes, it consists or polysaccharides, pectins, and lignin.

Cell cycle- A term that refers to the sequence of stages in the life of a cell, including its replication. Following cell division, a cell is said to be in the G_1 phase. For most mammalian cells, this is the predominant period in the life of a cell. A cell in G_1 that is not scheduled to divide again may go into a more permanent phase called G_0. Cells in G_1 that are ready to enter into division must first make a copy of the DNA. This period of DNA synthesis is called the S phase. When synthesis is completed, the cell is in the G_2 phase for a short period. Division of the nuclear contents occurs in mitosis or meiosis, known as the M phase. At the completion of M phase, cytokinesis usually results in division of the cytoplasm, and two daughter cells are formed. Tthe daughter cells are again in G_1.

Central dogma- The original postulate that genetic information can be transferred only from nucleic acid to nucleic acid and from nucleic acid to protein, that is from DNA to DNA from DNA to RNA and from RNA to protein (although information transfer from RNA to DNA was not excluded and is now known to occur [reverse transcription]). Transfer of genetic information from protein to nucleic acid never occurs.

Centrioles- Essential tubular organelles found near the nucleus in pairs that aid in cellular division.

Centromere- [Latin *centrum*-center + Greek *meros*-part.] The position on a chromosome at which the spindle fibres attach in cell division. The centromere is the last part of the chromosome to divide.

Cilia- Tiny hairs along the outside of the cell membrane which are used to move the cell and capture food particles.

Chlorophyll- The green material found in chloroplast that is active in photosynthesis.

Chloroplast- The organelle in which photosynthesis takes place. It contains chlorophyll.

Chromoplast- An organelle in which photosynthesis take place. It contains pigments other than chlorophyll, resulting in a color other than green.

Chromatid- One of the two replicated copies of a chromosome prior to division of the centromere. The two chromatids that originate from the same chromosomes are *sister* chromatids. Those that originate from two homologous chromosomes are *nonsister* chromatids. In mitosis, sister chromatids should be identical. Nonsister chromatids will differ in specific allele combinations, as do any pair of homologous chromosomes.

Chromatin- [Greek *chroma* color.] The materials in nuclei that are readily stained by certain dyes. Chromatin is now known to be equivalent to chromosomes in a dispersed state. Chromosomal regions that stain very readily are designated as *heterochromatin*, and those that stain less are called *euchromatin*. Most genes appear to be within the euchromatic regions.

Chromosomes- Condensed form of chromatin visible during cellular division.

Chromosome library- A DNA library generated from a preparation of a specific chromosome. It is possible to sort metaphase chromosomes mechanically with high speed automated sorters. Such a preparation of chromosomes would then be digested with a restriction enzyme and the restriction fragments produced would be inserted into vectors, which in turn would be inserted into bacteria.

Chromosome rearrangement- The physical breakage of one or more chromosomes followed by rejoining in incorrect configurations. Common rearrangements are deletions, duplications, inversions, and translocations. Other, more complex rearrangements are also observed.

Codon- A sequence of three nucleotides that code for one amino acid in a polypeptide. Of the 64 possible codons, 61 code for amino acids and three for chain termination.

Coding region- The segments of a gene that code for amino acid sequences in proteins. Introns and untranslated segments at the 5' and 3' ends of the messenger RNA are not included in the coding regions.

Contractile vacuole- An organelle which pumps excess water of a cell to prevent it from bursting.

Cot value- (cot1/2); The product of Co (the original concentration of denatured DNA) and t (time in seconds), giving a useful index of DNA renaturation. Cot1/2 is the value when 50% renaturation has occurred which can be used to estimate the length of unique DNA in a sample.

Cross- A term, used especially in experimental genetics, referring to specific types of matings.

Crossing over- Exchange of chromosomal segments between non-sister chromatids of homologous chromosomes that occurs during prophase I of meiosis. This generates new chromosomal combinations of alleles.

Cyclic AMP (cAMP)- A form of AMP (adenosine monophosphate) used frequently as a second messenger in eukaryotics and in catabolite repression in prokaryotes.

Cytokinesis- [cyto- + Greek *kinesis*-motion] The process of division of the cytoplasm in the formation of two new cells.

Cytoplasm- Collective term for cytosol and all the organelles contained in it (outside the nucleus and within the plasma membrane).

Cytoskeleton- Network of microtubules that support and give structure to cell while aiding in intracellular transport.

Cytosol- Jelly-like material that contains the organelles between the nucleus and the plasma membrane.

Degenerate code- Any code in which more than one symbol or combination equates to the same item. The genetic code is degenerate because the same amino acid may be coded by more than one codon. This is not surprising, because only 20 amino acids are coded by 61 codons.

Denaturation- The separation of the two strands of a DNA double helix, or the severe disruption of the hydrogen bonded structure of any complex molecule without breaking the covalent bonds of its chains.

Density-gradient centrifugation- A method of separating macromolecules by their

(1) differential rate of sedimentation in a centrifugal gradient

(2) differential bouyancy in a density gradient.

Diploid- Having two sets of chromosomes. In humans, diploid cells have 46 chromosomes (two sets of 22 autosomes plus a pair of sex chromosomes, either two X chromosomes (females) or one X and one Y (males). All somatic cells are diploid. Germ cells are diploid prior to the first division of meiosis but are haploid following the first meiotic division.

DNA- *Deoxyribonucleic acid* is the double-helix molecule holding the genetic information of organisms that, along with protein, composes the chromatin.

DNA fingerprint- A term used originally to describe a genetic profile based on certain minisatellite genetic markers. If a sufficient number of markers is used, a particular combination of markers would be as characteristic of an individual as a fingerprint (except that monozygotic twins would have identical genetic profiles, though not quite identical fingerprints).

DNA ligase- An enzyme that catalyzes formation or rejoining of a covalent bond between adjacent nucleotides when one of a pair of complementary strands has a gap.

DNA polymerase- An enzyme that catalyzes the assembly of DNA polynucleotides from the four deoxyribonucleotides. In DNA replication, the full name of the enzyme is DNA-dependent DNA polymerase, reflecting the fact that the template is DNA.

DNA primer- A short segment of DNA that is complementary to one end of a DNA strand to be copied and that provides a 3' end to initiate assembly of the new strand by DNA polymerase. The primer would thus form the 5' end of the new strand.

DNA sequence- The order in which nucleotide pairs occur in a segment of DNA.

Double helix- A helix formed by two strands intertwined to form a spiral. The common form of DNA is a double helix.

Electrophoresis- The separation of charged molecules (ions) in an electric field. Under the alkaline conditions ordinarily used, DNA and most proteins have a negative electrical charge (anions). They move toward the positively-charged electrical pole and away from the negative pole. Molecules with different charge densities separate from each other because of their different rates of movement. *Gel electrophoresis* is a variation that is especially useful in separating DNA or protein molecules. The gel forms a matrix that acts somewhat like a sieve, separating molecules on the basis of size. In the case of DNA, the charge density is uniform, and size alone separates different polynucleotides, even those that differ by a single nucleotide. The mobility of proteins in gel electrophoresis is dependent both on charge and size.

Endoplasmic reticulum- A network of tunnels which extend away from the nucleus, used for the transport of proteins.

Enzyme- A protein utilized in chemical reactions.

Eukaryotes- Advanced cell type with a nuclear membrane surrounding genetic material and numerous membrane-bound organelles dispersed in a complex cellular structure.

Excision repair- A process whereby cells remove part of a damaged DNA strand and replace it through DNA synthesis using the undamaged strand as a template. The repair of a DNA lesion by removal of the faulty DNA segment and its replacement with a new segment.

Exon- A region of a gene that is present in the final functional transcript (mRNA) from that gene.

Facilitated diffusion- A method of transport across the cell membrane by which carrier proteins bond to a molecule on one side of the membrane, move through the membrane, and then release it on the other side.

Flagella- Large hairs which can whip back and forth to propel a cell.

Gamete- [Greek *gamete*-wife; *gametes*-husband] A haploid germ cell, either maternal (ovum) or paternal (sperm).

Gene- A unit of heredity. Originally, *gene* was used to describe a unit of phenotypic variation. It is also now used to describe a unit of function, i.e., a segment of DNA from which a single RNA molecule is transcribed. Ultimately, the two definitions merge as more knowledge is gained of the molecular basis of hereditary variations.

Genetic code- The sequence of DNA nucleotides that are translated into amino acid sequences in proteins. The code is a triplet code, i.e. a sequence of three nucleotides codes for one amino acid. This unit of three nucleotides is called a codon. Of the 64 possible sequences of three nucleotides, 61 code for amino acids. The remaining three are stop codons, that is, when a stop codon is encountered during translation, synthesis of that polypeptide chain is terminated. The genetic code is universal, in the sense that the same code is used by all forms of life, from viruses and prokaryotes to plants and animals.

Genetic engineering- A general term that refers to the assembly of DNA from different sources into single pieces of DNA, which then are cloned in bacteria or other vectors or inserted into the genome of an organism by some means.

Genome- A haploid complement of genes, used particularly in the abstract. Common usage also refers to the diploid complement of diploid organisms.

Golgi body- Stacks of membranous pouches which act as a transport station, packaging proteins from the endoplasmic reticulum and placing them into tiny vesicles.

Helicase- A protein that unwinds DNA at replication forks.

Heterochromatin- Densely staining condensed chromosomal regions, believed to be for the most part genetically inert.

Heteroduplex- A DNA double helix formed by annealing single strands from different sources; if there is a sequence difference between the strands, the heteroduplex may show single strand loops or bubbles (unpaired regions).

Histone- A type of basic protein that forms the unit around which DNA is coiled in the nucleosomes of eukaryotic chromosomes.

Homeobox- A short stretch of *nucleotides* whose *base sequence* is virtually identical in all the *genes* that contain it. It has been found in many organisms from fruit flies to human beings. In the fruit fly, a homeobox appears to determine when particular groups of genes are expressed during development.

Homologous- As commonly used in genetics, *homologous* describes the two members of a pair of chromosomes in a diploid organism. Members of a homologous pair have the same array of genes, although the gene pair may vary slightly in structure.

Hydrogen bond- A weak bond involving the sharing of an electron with a hydrogen atom; hydrogen bonds are important in the specificity of base pairing in nucleic acids and in the determination of protein shape.

Idiogram- A photograph or diagram of the chromosomes of a cell arranged in an orderly fashion.

Initiation codon- The codon that is used to initiate all translation. This codon, AUG in RNA codes and ATG in DNA codes, codes for the amino acid methionine.

Intermediate filament- A part of the cytoskeleton with a strong, ropelike structure which gives the cell strength and helps it to maintain its shape.

Interphase- Time period between cellular divisions in which cellular processes such as protein synthesis are carried out.

In vitro- In an experimental situation outside the organism. Biological or chemical work done in the test tube (literally in glass) rather than in living systems.

Inducer- An environmental agent that triggers transcription from an operon.

Initiation factors (IF1, IF2, IF3)- Proteins (prokaryotic with eukaryotic analogues) required for the proper initiation of translation.

Karyo-, -karyon- [Greek *karyon*, kernel.] Prefix and suffix referring to the cell nucleus. Examples: *karyotype, heterokaryon*.

Karyotype- A standardized array of chromosomes. Typically, chromosomes are cut out from a photograph of a metaphase spread and are then pasted into the appropriate position, starting with the two chromosomes 1, followed by chromosomes 2, etc. Karyotypes are very useful in evaluating the chromosome status of a person or of tissue samples and cell cultures.

Lac operon- An inducible operon including three loci involved in the uptake and breakdown of lactose in *Escherichia coli*.

Lagging strand- In DNA replication, the strand that is synthesized apparently in the 3′ to 5′ direction, but actually in the 5′ to 3′ direction by ligating short fragments synthesized individually. Strand of DNA being replicated discontinuously.

Lampbrush chromosomes- Chromosomes of amphibian oocytes having loops suggestive of a lampbrush.

Leading strand- Strand of DNA being replicated continuously. In DNA replication, the strand that is made in the 5′ to 3′ direction by continuous polymerization at the 3′ growing tip.

Leucoplast- Colorless plastids in autotrophs which store starch, proteins, and lipids.

Lyon hypothesis- The hypothesis, first clearly stated by Mary F. Lyon, that only one X chromosome is active in mammalian cells with two or more X chromosomes.

Lysosome- A sac similar to a vacuole which contains powerful digestive enzymes used to break down large food particles.

Meiosis- Cellular division that yields four gametes through two cellular divisions.

Metaphase- Phase of mitosis in which the chromosome pairs line up at the equator of the cell.

Metaphase plate- The array of chromosomes in a plane that characterizes metaphase. The metaphase plate forms between the poles of the spindle and perpendicular to the spindle axis.

Mitosis- Cellular division that yields two identical cells from one cell through a five-step process.

Microfilament- A part of the cytoskeleton which consists of actin and aids in cell movement.

Microtrabeculae- Tiny fibers which interconnect all of the structures within the cell and help to give the cell shape.

Microtubule- Fibers which extend from the center of the cell to the cell membrane. They are involved in cell reproduction and are part of the composition of cilia and flagella.

Mitochondria- The organelle in which cellular respiration occurs.

Mutation- A heritable change in DNA structure. Mutations may modify the coding region of a gene, the regulatory region, or both; or they may be limited to intergenic regions. They may be as small as single nucleotide changes or as large as deletions of blocks of genes.

Neoplasm- New growth of abnormal tissue.

Nicking- Nuclease action to sever the sugar-phosphate backbone in one DNA strand but not the other at one specific site.

Nitrogen base- Type of molecule that forms an important part of nucleic acid, composed of a nitrogen-containing ring structure. Hydrogen bonds between bases in opposing complementary strands link the two strands of a DNA double helix.

Nucleoid- A DNA mass within a chloroplast or mitochondrion and prokaryotic cells.

Nucleotide- Subunit that polymerizes into nucleic acids (DNA or RNA). Each nucleotide consists of a nitrogenous base; a sugar; and one to three phosphate groups.

Nucleolus- A structure within the nucleus at which ribosomes are created.

Nucleus- The organelle in eukaryotes which contains the cells DNA and thus indirectly controls protein production and the rest of the cell.

Nuclear membrane- Membrane surrounding the nucleus that is covered with pores and controls nuclear traffic.

Nucleosome- Nuclear structures composed of histones around which DNA is wrapped.

Okazaki fragment- Segment of newly replicated DNA produced during discontinuous DNA replication.

Oncogene- A gene that contributes to the production of a cancer.

Oocyte- The gamete in females.

Oogenesis- The process of ovum formation in female animals.

Operator- A DNA region at one end of an operon that acts as the binding site for repressor protein.

Operon- A set of adjacent structural genes in bacteria whose mRNA is synthesized in one piece, together with the adjacent regulatory signals that affect transcription of the structural genes.

Organelle- Any of the structures that occur within a cell, such as nuclei, mitochondria, lysosomes, etc.

Passive transport- A form of transport which allows highly polar molecules to move through the cell membrane without the expenditure of energy. This may occur either through protein channels or facilitated diffusion.

Peroxisome- An organelle similar to a vacuole which contains oxidizing enzymes which can help neutralize toxic substances.

Phagocyte- Collective term for cells that engulf other cells or microorganisms.

Phosphodiester bond- A bond between a two sugar groups and a phosphate group; such bonds form the sugar-phosphate-sugar backbone of DNA and RNA. A diester bond (between phosphoric acid and two sugar molecules) linking two nucleotides together to form the nucleotide polymers DNA and RNA.

Plasma membrane- Outer membrane of cells composed of proteins and a phospholipid bilayer that controls cellular traffic.

Plasmid- A DNA particle in some bacteria that has the ability to replicate with the bacterial cell but that is not part of the bacterial genome. Plasmids are important tools in genetic engineering, as they can be isolated, modified structurally, usually by insertion of genes or other DNA of interest, and inserted back into bacterial cells, where they replicate.

Plastid- Vital organelle that aids in the metabolism of unicellular organisms and plant cells (chloroplasts, chromoplasts, leucoplasts are examples).

Polar body- At the completion of meiosis I in an oöcyte, one of the haploid products is extruded from the main body of the cell to form a *polar body*. It consists of little more than the extruded nucleus and a bit of cytoplasm, surrounded by a plasma membrane. If the ovum is fertilized and meiosis II is completed, one of the meiotic products is again extruded to form a second polar body. Polar bodies do not have a future.

Polymer- A large molecule that is produced by connecting subunits to form a chain. In *homopolymers*, the subunits are identical, e.g. -A-A-A-A-. In *heteropolymers*, two or more kinds of subunits are used, often in an alternating pattern, e.g. -A-B-A-B-A-B-. These are examples of *linear* polymers, since assembly occurs only in one dimension. Both DNA and RNA are linear polymers of nucleotides and can be either homo- or heteropolymers.

Polynucleotide- A polymer composed of nucleotides, either ribonucleotides or deoxyribonucleotides, joined by covalent bonds.

Polypeptide- A chain of amino acids joined together by peptide bonds.

Prokaryotes- Primitive cell type that lacks a nuclear membrane and membrane-bound organelles.

Protein- A complex molecule found in numerous cellular structures that is composed of amino acids.

Prophase- Phase of mitosis in which the chromatin duplicates itself and thickens into chromosomes, the spindle fibers form, and the nuclear membrane disintegrates.

Proto-oncogene- The non-activated form of a cellular oncogene in an untransformed cell. A gene that, when mutated or otherwise affected, becomes an oncogene.

Pseudopod- Extensions of the cytoplasm toward which the rest of the cytoplasm tends to flow. Pseudopodia can be used for movement and the capture of prey.

Recombinant DNA- A novel DNA sequence formed by the joining, usually *in vitro*, of two non-homologous DNA molecules.

Replica plating- A technique to rapidly transfer microorganism colonies from a master plate, in an exact spatial pattern, to a number of further plates.

Replication- DNA synthesis. The process of copying.

Replication fork- The point at which the two strands of DNA are separated to allow replication of each strand.

Ribosome- Structures found mainly in the endoplasmic reticulum whose function is to synthesize protein based upon the code of a messenger RNA molecule.

RNA- *Ribonucleic acid*, a molecule that is a necessary component of the protein synthesis process.

Rolling-circle replication- A model of DNA replication that accounts for a circular DNA molecule producing linear daughter double helices.

S (Svedberg unit)- A unit of sedimentation velocity, commonly used to describe molecular units of various sizes (because sedimentation velocity is related to size).

Satellite- A terminal section of a chromosome, separated from the main body of the chromosome by a narrow constriction.

Semiconservative replication- The mode by which DNA replicates. Each strand acts as a template for a new double helix. The established model of DNA replication in which each double-stranded molecule is composed of one parental strand and one newly polymerized strand.

Sexual- Reproduction involving the union of female and male gametes to form a zygote.

Spermatocyte- A male germ cell that has initiated spermatogenesis.

Spermatogenesis- The process of forming haploid germ cells in males.

Spermatozoön- (Plural *spermatozoa*) A mature male germ cell. Commonly shortened to *sperm*.

Spermiogenesis- The process of forming mature sperm from the haploid products of spermatogenesis.

Spindle- The structure formed during nuclear division that participates in separation of chromosomes into daughter cells. The spindle consists of microtubules that radiate from two bodies (centrioles) that form the poles of the spindle.

Supercoiling- Negative (tending to unwind the helix) or positive (tending to wind the helix) coiling of double-stranded DNA that differs from the relaxed state.

Synapsis- [Greek *synapsis*-connection] The pairing of homologous chromosomes during prophase I of meiosis. The pairing is highly specific, point-by-point matching. During this pairing, crossing over between nonsister chromatids occurs, generating new chromosomal combinations of alleles.

Telophase- Phase of mitosis in which the chromosome pairs have separated and reached opposite poles of the cell as the spindle begins to disintegrate, the nuclear membrane reappears, and the cytoplasm begins to divide.

Vacuole- Membrane-bound sacs within a cell used to hold food particles, water, etc.

Vesicle- A tiny vacuole, often used to carry protein molecules packaged at the Golgi bodies.

Virus- A particle that contains some of the genes required for self replication but that cannot function outside a cell.

Z DNA- A left-handed form of DNA found under physiological conditions in short GC segments that are methylated.

zygote- [Greek *zygotos* joined together] A fertilized egg.

Index